Temporal Information Systems in Medicine

Carlo Combi · Elpida Keravnou-Papailiou
Yuval Shahar

Temporal Information Systems in Medicine

Foreword by Jim Hunter

 Springer

Prof. Dr. Carlo Combi
University of Verona
Department of Computer Science
Strada le Grazie 15
37134 Verona
Italy
carlo.combi@univr.it

Prof. Dr. Yuval Shahar
Ben Gurion University
Medical Informatics Research Center
Department of Information Systems Engineering
84105 Beer Sheva
Israel
yshahar@bgu.ac.il

Prof. Dr. Elpida Keravnou-Papailiou
University of Cyprus
Department of Computer Science
Kallipoleos Street 75
1678 Nicosia
Cyprus
elpida@cs.ucy.ac.cy

ISBN 978-1-4899-8812-6   ISBN 978-1-4419-6543-1 (eBook)
DOI 10.1007/978-1-4419-6543-1
Springer New York Dordrecht Heidelberg London

Printed on acid-free paper

Springer is part of Springer Science+Business Media (www.springer.com)

This book is dedicated to our families, who supported us and waited patiently for a decade while this book was written.

Foreword

*Sed **fugit** interea fugit irreparabile **tempus**, singula dum capti circumvectamur amore.*[†]

I was very pleased indeed when Carlo, Elpida and Yuval asked me to contribute this foreword. Most immediately, I was pleased to be given the chance to encourage you to continue to read and explore what follows in this book.

Our bodies are dynamic systems that change over time if this were not so, the practice of medicine would be easy (or at least easier than it actually is). Clinicians use the past history of signs, symptoms etc., as a diagnostic tool to identify the current state of the patient. Their knowledge of what has happened to other similar patients and the models they possess of disease development, allows them to make some attempt to predict the future. If we, as informaticians, attempt to capture and reproduce these skills, we must be able to represent temporal medical knowledge and apply it to real patient temporal data. This book provides an excellent starting point for anyone wishing to understand the current state of the art and has been written by three of the most respected researchers in the field.

But I was also pleased because it gives me a chance to reflect on the evolution of the field over the past 30 years (*tempus fugit*). Clearly there have been many developments, many of which are documented in the following pages. For me, the most significant has been the emergence of the electronic medical patient record as a reality. The first paper I published in this field[1], took a very simplistic view of disease as a progression through a number of stages. In part it was simple because the field was in its infancy, but it could afford to be simple because we had very little data to test it against.

[†] But meanwhile it flies: time flies irretrievably, while we wander around, prisoners of our love of detail. *Virgil, Georgics.*

[1] Hunter JRW and RK Sinnhuber, "Representation of Disease Development", *Expert Systems 83*, pp 174-183 (1983) [http://www.csd.abdn.ac.uk/jhunter/publications/83rdd.pdf]

In contrast, the core of BabyTalk[2], my most recent project, tries to interpret the complex data entered into a database for a baby in a neonatal intensive care unit. These time-stamped data consist of (i) the physiological parameters (heart rate, oxygen saturation, etc) sampled once per second, (ii) the results of laboratory and blood gas analysis entered automatically every few hours, (iii) discrete coded information and free text entered intermittently by the clinical staff. From having little or no data we have gone (almost) to having too much. Temporal data abstraction (see Chapter 5) becomes crucial in reducing the volume of data. The difference between valid and transaction times (3) is vital as clinical staff enter the fact that action X has been taken, usually without saying when; we have chosen to model the resulting temporal uncertainties in a way similar to that used in the Asbru language (Section 7.5). Humans are fallible and make mistakes; the interpretation of the resulting data can suggest temporal inconsistencies which must be resolved. These problems will be solved, but it will take more time!

As the authors say, the book does not attempt to cover everything that might be classified as relating to "temporal information systems in medicine". They point to temporal probabilistic networks and temporal data mining. From my own experience, I might also add the modelling of physiological systems (both qualitative and quantitative) and the time-series analysis of data sampled at high frequency that one finds in intensive care.

But no matter, what is covered here is well covered. According to the authors, the book has taken a decade to write (*tempus fugit* again) perhaps because they have been "prisoners of their love of detail". I, for one, am glad that they have finally managed to escape, and make the results of their efforts available to the rest of us.

Aberdeen, Scotland,   *Jim Hunter*
January, 2010

---

[2] Portet F, E Reiter, A Gatt, JRW Hunter, S Sripada, Y Freer and C Sykes, "Automatic Generation of Textual Summaries from Neonatal Intensive Care Data", *Artificial Intelligence*, Vol 173, pp 789-816 (2009)

Preface

Writing a collective volume on information systems in medicine from the temporal perspective has been a big challenge to us for a number of reasons. One reason was the growing interest in time-oriented aspects coming from different research communities, inspired by different research issues and problems. A second reason that is not unrelated to the first was the diverse and multidisciplinary nature of the various aspects of temporal information systems in medicine. A third reason was to attempt to strike a balance between the theoretical and methodological issues on one hand, and the challenging aspects of developing applications that can deal with real-life, time-oriented medical problems, on the other hand. A final reason was that such a book, as we envisioned it, simply did not exist. Given the growing work in the field and its significance, we considered the presence of such a book a necessity for giving a uniform integration, under a common umbrella, of the areas of research impinging on temporal information systems in medicine; that is we felt that the absence of such a book was a hindrance to someone wishing to get an overall acquaintance or to be updated on the various matters that arise in this interdisciplinary field.

One does not need to make an elaborate argument as to why it is important to model and reason with time, particularly in the context of medical information systems. Putting it simply, there is one fundamental truth in life: the world is not static. Situations change. Events happen, spontaneously or otherwise, causing changes. Most properties vary with time, i.e. only few properties are time-invariant or static. Thus, histories of happenings/situations/states, or evolutions of processes need to be modeled and reasoned with. Such models can not be void of time. As already indicated, the modeling of time has attracted interest from a number of research communities, each addressing aspects from its specific sphere of interests. The database systems community is interested in how to store, maintain, and query time-oriented data. The Artificial Intelligence (AI) community is interested in formal theories of time, temporal logics, temporal constraints and temporal reasoning techniques (a comprehensive handbook of temporal reasoning in AI was recently published [142]). The Artificial Intelligence in Medicine (AIM) community is interested in the broader issues addressed by the AI community, but from the perspective of producing methodologies and techniques that can be viably deployed for handling medical problems. The Intelligent Data Analysis (IDA) community, which includes many members from the AI, AIM, and even the database systems communities also has a strong interest in time, based on the use of intelligent techniques for analyzing and interpreting time-oriented data, an area that also intersects with the field of temporal data mining. The medical informatics community has overlapping interests with the above communities, since a main focus of its activities is the use of informatics for health care and health management.

Temporal information systems in medicine, a term we have coined to delineate the field of those information systems that are able to store, manage, query, and support different tasks on time-oriented clinical data, call for the use of diverse techniques, methods and approaches to time, emanating from the work of the above research communities. Medical information systems invariably deal with clinical data organized as patient records, and with medical knowledge. Temporal information systems in medicine use time-oriented patient data and time-based medical knowledge. As such, the relevant reasoning processes encompass temporal reasoning as a key feature. Often, the distance between the raw clinical data and the concepts used within medical knowledge is such that data abstraction processes become necessary means for bringing the raw data and the medical knowledge to a comparable conceptual level for facilitating the required reasoning. When the data are time-oriented, the data abstraction process is referred to as *temporal data abstraction*. Data abstraction, or temporal data abstraction, could in fact be used as a stand-alone process for converting raw data to a more conceptually meaningful form, independently of any higher reasoning processes. Thus temporal databases, temporal modeling and reasoning, and temporal data abstraction figure centrally in the engineering of temporal information systems in medicine and are addressed extensively in this book.

The initial idea of writing such a book was conceived about a decade ago when more than a sporadic interest in the topic emerged. The idea gradually matured over a number of meetings between the three authors at various venues, in particular, during several Artificial Intelligence in MEdicine (AIME) conferences. During the period of writing the book, the interest in the field of temporal information systems in medicine has been steadily growing and broadening with new research topics, e.g. the visualization of time-oriented data; this growth was a main reason for delaying the completion of the book in an attempt to include the new developments. However, one can never succeed in producing a complete account of all developments if a field that is very active in research. This book is by no means a full survey of the field of temporal information systems in medicine, since naturally the field is continuously growing. By necessity, some areas are more emphasized. This is not to say that these areas are more important than others. For example, the areas of (temporal) probabilistic networks and of temporal data mining are not covered, although these are important fields. Thus, to some extent, as it often happens, the choice of topics was driven by the interests of the authors and by the available space. In particular, while some of the chapters include a rather broad overview of the subject matter, others, which are more research oriented, present a relatively brief introduction of the topic at hand, and delve more deeply into several detailed examples, or case studies.

In spite of its omissions and other limitations, we hope that this book, which is unique in its scope, has succeeded in bringing together, under a common theme, relevant research results from various research communities, in a mutually beneficial way to these communities. We also hope that this book will trigger further research interest in this interdisciplinary field.

This book should be of interest to a broad readership coming from the disciplines of medical informatics, computer science, information systems, and biomedical engineering. It could form a basis for advanced undergraduate courses in these disciplines, or for more specialized graduate courses. The book is also addressed to researchers in the various subdisciplines of the field of temporal information systems in medicine, both newcomers as well as older researchers in the field.

Limassol, Cyprus,   *Carlo Combi*
October 2009   *Elpida Keravnou*
                                                                                       Yuval Shahar

Acknowledgements

Much of the information and insights presented in this book were obtained through the assistance of many of our students and colleagues. We would like to thank all of them. The authors were supported, while preparing and writing this book, by their respective universities. Special thanks are due to the University of Cyprus for hosting the authors' meetings.

Contents

Acronyms

AGI   Active Guideline Interpreter: a module for the management of guidelines in a commercial electronic medical record system [397]

ARAMIS   American Rheumatism Association Medical Information System: a system for managing data related to the long-term clinical course of patients suffering from arthritis or, more generally, from rheumatic pathologies [379]

AT   Availability Time: a temporal data dimension proposed in [93]

ATD   Abstract Temporal Diagnosis: a logic-based framework for diagnosis, proposed in [150]

ATG   Abstract Temporal Graph: a directed graph whose nodes represent temporal entities

ATN   Augmented Transition Network

BBN   Bayesian Belief Network

BPEL   Business Process Execution Language: a XML-based language for representing process executions [1]

BPMN   Business Process Modeling Notation: a model for representing processes [2]

C-T-A   Causal-Temporal-Action Model: a general model for diagnostic knowledge, proposed in [214]

DDL   Data Definition Language: the language for specifying a database schema and the related constraints

DIRC   Dynamic Induction Relations of a Context interval

DML   Data Manipulation Language: the language for updating and querying a database

DNM   Dynamic Networks Model

DOEM   Delta Object Exchange Model: a model for temporal semistructured data proposed in [63]

EBM   Evidence-Based Medicine

EC   Event Calculus: a theory of time and change proposed in [228]

ER   Entity-Relationship: a conceptual data model for database design

ET   Event Time: a temporal data dimension proposed in [93]

ETNET   Extended Temporal Network: an extension of TNET, proposed in [196]

FOL   First Order Logics

GCH-OSQL   Granular Clinical History - Object SQL: an object-oriented query
            language for clinical data given at different temporal granularities [82]

GEL   Guideline Expression Language: a guideline-specification language [304]

GLIF   Guideline Interchange Format: a guideline-specification language [286]

HCI   Human-Computer Interaction

HDP   Heart Disease Program: a software tool for diagnosing disorders of the
          cardiovascular system, proposed in [247]

KBTA   Knowledge-Based Temporal Abstraction: a methodology for abstracting
          clinical data proposed in [347]

KNAVE   Knowledge-based Navigation of Abstractions for Visualization and Ex-
          planation: a system for managing temporal clinical abstractions [352]

KHOSPAD   Knocking at the Hospital for PAtient Data: an object-oriented web-
           based system for the management of temporal clinical data [318]

IDAMAP   Intelligent Data Analysis in Medicine and Pharmacology: a series of
          international workshops

IOC   Interval Of Certainty

IOU   Interval Of Uncertainty

IPBC   Interactive Parallel Bar Charts: a system for visual data mining on clinical
          data [69]

IV   Information Visualization

M-HTP   Monitoring Heart-Transplant Patients: a system proposed in [235]

MLM   Medical Logical Module: an independent unit of medical knowledge
          [181]

MVI   Maximal validity interval: property's validity interval derived by EC

NCEP   National Cholesterol Education Program: a guideline for the manage-
          ment of hypercholesterolemia [280]

ODMG   Object Database Management Group: an international group who pro-
          posed a standard for object-oriented databases [55]

OEM   Object Exchange Model: a model for semistructured data proposed
          in [296]

OMT   Object Modeling Technique: an object-oriented design methodology re-
          lated to UML

PCAPE   Partner's Computerized Algorithm Processor and Editor: a specification
          and execution framework for guidelines [439]

SEDC   SpondyloEpiphyseal Dysplasia Congenital: a skeletal dysplasia

SIA   Simple Interval Algebra: an interval algebra based on a subset of Allen's
          relations

SDM   State Description Model

SQL   Structured Query Language: the widely known database query language
          [260]

SUS   Standard Usability Score: a scale for evaluating usability of system inter-
          faces [41]

TA   Temporal Abstraction

TAC   Temporal Association Chart: a software for the investigation of temporal
           and statistical associations within multiple patient records [222]
TCS   Temporal Control Structure: a system that supports reasoning in time-
           oriented clinical domains [336]
T4SQL   Time For SQL: a query language for multidimensional temporal data-
           bases, proposed in [94]
TGM   Temporal Graphical Model: a graphical model for representing semistruc-
           tured data dynamics proposed in [287]
TLSQL   Time Line SQL: a query language managing temporal clinical data with
           indeterminacy proposed in [111]
TMM   Time Map Manager: a theory of time and change proposed in [117]
TMS   Truth-Maintenance System
TNET   Temporal Network: a temporal data model for clinical tasks, proposed in
           [197]
TOD   Time Oriented Database: a temporal data model for clinical data pro-
           posed in [426]
TOPF   Time Oriented Probabilistic Function
tPCT   temporal Parsimonious Covering Theory: a temporal extension of a well
           known general theory of abductive diagnosis [418]
TQF   Tabular Query Form: a visual system for expressing logical formulae
           [294]
TSQL2   Temporal SQL: a temporally-extended SQL [387]
TT   Transaction Time: the temporal data dimension modeling the data inser-
           tion/deletion time
TUP   Temporal-Utilities Package: a general-purpose package for managing re-
           lations among temporal intervals [224]
UML   Unified Modeling Language: a modeling language for object-oriented
           systems [30]
VISITORS   VISualizatIon of Time-Oriented RecordS: a system that combines
           intelligent temporal analysis and information visualization techniques
           ([218]
VT   Valid Time: the temporal data dimension modeling the real world validity
           of data
XML   eXtensible Markup Language: the common standard for data representa-
           tion on the web [3]

Introduction

The title of the book "Temporal Information Systems in Medicine" interleaves three terms: "time", "information systems" and "medicine". The book is about the engineering of information systems for medically-related problems and applications, in which time is of central importance and is a key component of the engineering process of such information systems. Computer-based information systems have become a critical part of products, services, operations, and management of organizations; they are vital to problem identification, analysis, and decision making. Information systems, as a field of academic study, began in the 1960s, a few years after the first use of computers for transaction processing and reporting by organizations. As organizations extended the use of information processing and communication technology to operational processes, decision support, and competitive strategy, the academic field also grew in scope and depth. Data modeling and processing have always been at the heart of information systems; nowadays, with the technological advances enabling the storage and processing of huge volumes of data, the tasks of data modeling and processing have acquired even higher significance. Health care organizations and services are part of the backbone of modern societies. The effective and efficient use of information systems in health care organizations and services is therefore an important element of the quality of life in today's societies. This book is concerned with a subset of such information systems, albeit a major subset that deals with dynamic situations or processes, for which the modeling of time is of prime importance. Such information systems should be able to store, manage, query, and support different inference tasks regarding time-oriented clinical data. Moreover, the relevant inference tasks are largely knowledge-based; thus, the required (medical) knowledge must be focused on the temporal dimension.

Given the multiple domains underpinning the theme of the book (information systems, medical systems, databases, knowledge bases, time representation and temporal reasoning, inference, decision support, etc.), we are clearly dealing with a multi-disciplinary subject. This made our task of selecting the topics and subtopics for discussion more interesting and challenging, but at the same time more difficult and prone to errors of omissions. Bearing in mind this potential limitation, we proceeded to make the relevant choices; we hope that the final result, which certainly is unique in scope given the currently available literature, will succeed in meeting the objectives outlined in the preface.

In each chapter, with the exception of Chapter 1, the reader is given guidance for additional reading through bibliographic notes, whilst s/he can further consolidate their knowledge by attempting the exercises at the end of the chapter. Below we introduce the main topics of the book's chapters.

The chapters are organized into four parts: Fundamentals (chapters 1, 2, and 3); Temporal reasoning and maintenance in medicine (chapters 4 and 5); Time in clinical tasks (chapters 6 and 7); and the display of time-oriented clinical information (Chapter 8).

The role of time in medicine (Chapter 1) is outlined by listing the technical tasks, the medical tasks, and the clinical areas, under which research relating to temporal information systems in medicine may be categorized. Overall, three distinct research directions may be identified, each of which involves temporal data modeling: temporal data maintenance, temporal data abstraction and reasoning, and design of medical temporal systems.

Temporal modeling and temporal reasoning, as topics of fundamental importance to the overall theme of the book, are elaborated in Chapter 2. In this chapter, the reader is guided through the basic notions of time and temporal information by discussing the main features of time domains, time primitives, and temporal entities, raising issues such as temporal granularities, indeterminacy, and complex semantic relations between occurrences. The temporal reasoning requirements for the medical domain are then laid out followed by an overview of general ontologies and models for temporal reasoning that represent pioneering work by the Artificial Intelligence community. Three of these general theories of time that have been especially influential in the development of the field of temporal reasoning, namely the interval-based temporal logic developed by Allen, the event calculus developed by Kowalski and Sergot, and the time map manager of Dean and McDermott are then examined against the particular temporal reasoning requirements of the medical domain, using a simple medical example. Given that these general theories of time were not developed for the purpose of medical problem solving, it comes as no surprise that none of them, at least in its initial form, adequately supports the identified requirements for medical temporal reasoning. However, through this simple exercise, we aim to illustrate the intricacies and complexities of medical temporal reasoning. The last section of this chapter focuses on temporal constraints, a topic relevant both for temporal reasoning and for temporal-knowledge bases and temporal databases. Temporal constraints are discussed at a general level through an abstract structure, the Abstract Temporal Graph.

Another topic of fundamental importance to the theme of the book is temporal databases, discussed in Chapter 3. The aim of this chapter is to acquaint the reader with the basic concepts of temporal databases, i.e. databases that deal explicitly with time-related concepts and information. The main temporal dimensions of valid and transaction times are presented, and a classification of temporal databases according to the supported dimensions is given and illustrated by explaining how the traditional relational model, its algebra, calculus, and query languages can be extended to deal with the different temporal aspects. The discussion then shifts to the design of temporal database schemas using the Entity-Relationship and Object-Oriented data models; both the standard versions and the temporal extensions of the given data models are presented. Some basic features of object-oriented query languages and their extensions for dealing with time-oriented data are also discussed.

Following the presentation on the fundamental temporal notions given in chapters 2 and 3 of Part I, which is largely independent of the medical domain, the discussion in Part II focuses on issues relevant to the medical domain, aiming to present temporal maintenance and temporal reasoning with respect to that domain (chapters 4 and 5). Chapter 4 addresses clinical temporal databases by discussing modeling and querying issues related to the management of temporal clinical data collected into medical records. In particular, the chapter introduces some main (relational and object-oriented) data models explicitly proposed for managing clinical data with multiple temporal dimensions and with clinical temporal data given at different and mixed granularities/indeterminacies. The chapter ends by discussing some specific aspects of a real temporal clinical database system allowing the management of follow-up patients who underwent cardiac angioplasty.

Chapter 5 aims to give a comprehensive and critical review of current approaches to the common task of abstraction of time-oriented data in medicine, or *temporal abstraction*. Temporal-data abstraction is a central requirement for medical reasoning that presently receives much and justifiable attention. General theories of time do not typically address the process of temporal-data abstraction, although its role is especially crucial in the context of clinical monitoring, therapy planning, and exploration of clinical databases. More specifically, this chapter aims to give the reader an understanding of the basic types of data abstraction, both atemporal and temporal types, and of the desired computational behavior of a process that creates meaningful abstractions. In addition, the chapter includes a survey on early and more recent approaches dealing with time-oriented medical data in the context of rudimentary knowledge discovery, generation of intelligent summaries of patient data, and patient monitoring. The new concept of *temporal mediator* is also presented. A temporal mediator is a system that aims to support both temporal reasoning and temporal maintenance, thus forming a linkage between a database system (storing patient data and their abstractions) and a decision-support system, while decoupling the temporal-data abstraction process from the specifics of the decision-support process. Chapter 5 also discusses in relevant detail a particular generic ontology and associated method for temporal-data abstraction developed by one of the authors.

The next two chapters (Part III) deal with time in clinical tasks, namely clinical diagnosis (Chapter 6), and automated support to clinical guidelines and care plans (Chapter 7). The chapter on time in clinical diagnosis aims to explain how time and temporal reasoning can feature, with critical advantage, in the task of clinical diagnosis. This is largely done through a number of representative clinical diagnostic systems. The discussion focuses on the representation of time with respect to diagnostic knowledge and patient data, and the relevant temporal reasoning. In case the reader is not already familiar with clinical diagnostic systems, relevant fundamental notions are also overviewed. The principal objectives of Chapter 6 include appreciating the significance of time in clinical diagnostic reasoning, gaining an understanding of the distinction between abductive and consistency-based diagnostic reasoning, familiarizing the reader with such aspects as temporal granularity, uncertainty, incompleteness, and repetition, gaining an understanding of how time is represented in patient data and disorder models (particularly causal-based models), and presenting an overview of the types of temporal constraints arising in clinical diagnostic problems.

Clinical guidelines are a powerful method for standardization and uniform improvement of the quality of medical care. Clinical guidelines are often best viewed as a set of schematic plans, at varying levels of abstraction and detail, for screening, diagnosis, or management of patients who have a particular clinical problem or condition. Clinical guidelines typically represent the consensus of an expert panel or a professional clinical organization, and, as much as possible, are based on the best evidence available. In order for clinical guidelines to be easily accessible and applicable at the point of care, automated support is necessary; moreover, reasoning about time-oriented data and actions is essential for guideline-based care. Chapter 7 discusses in detail the tasks involved in providing automated support to guideline-based care, surveys several of the major current approaches, and exemplifies them by presenting one of these approaches in detail. The chapter also discusses the issue of converting text-based guidelines into a formal, executable format, and concludes with several insights into the future of guideline-based care.

The display and exploration of time-oriented clinical information represents a relatively recent development in the field of temporal information systems in medicine. We have chosen to present the current outcome of this research as the final part of the book (Part IV). More specifically, the last chapter of the book, Chapter 8, discusses the display of time-oriented clinical data and knowledge. Visualization is relevant both for display of time-oriented *data* as well as for presentation of *knowledge* about such data (e.g., temporal constraints and patterns). Furthermore, temporal information can be displayed and explored either for an *individual* patient or for a whole patient *population*. Chapter 8 discusses and exemplifies both these aspects.

Chapter 1
The role of Time in Medicine

1.1 Science or Art? From Knowledge-Intensive to Data-Intensive Applications

Medical tasks, such as diagnosis and therapy, are by nature complex and not easily amenable to formal approaches. The philosophical question "Is medicine science or art?" is frequently posed to show that expert clinicians often reach correct decisions on the basis of intuition and hindsight rather than scientific facts [22]. Medical knowledge is inherently uncertain and incomplete. Likewise, patient data are often ridden with uncertainty and imprecision, showing serious gaps. In addition, they could be too voluminous and at a level of detail that would prevent direct reasoning by a human mind. The computer-based performance of medical tasks poses many challenges. As such, it is not surprising that AI researchers were intrigued with the automation of medical problem solving from the early days of AI. The technology of expert systems is largely founded on attempts to automate medical expert diagnostic reasoning. For example, most people are familiar with the Stanford experiments of the Heuristic Programming Project resulting in the MYCIN family of rule-based systems [140].

Care providers, such as physicians and other medical professionals, are required to perform various tasks such as diagnose the cause of a problem, predict its development, prescribe treatment, monitor the progress of a patient and overall manage a patient. A care provider's decisions should be as informed as possible. In the present age of information explosion, that everyone experiences with the advent of information communication technologies in general, and the web in particular, the only viable means for handling large amounts of information are computer-based. The work of all care providers can benefit substantially from computer-based support. In the early days, the biggest challenge was the modeling of knowledge for the purpose of supporting tasks such as diagnosis, therapy, and monitoring. To a certain extent this is still a challenge. But the information explosion has brought a drastic change in focus from *knowledge-intensive* to *data-intensive* applications [178] and from *systems that advise* to *systems that inform* [328]. The major challenge is no

C. Combi et al., *Temporal Information Systems in Medicine*,
DOI 10.1007/978-1-4419-6543-1_1, © Springer Science+Business Media, LLC 2010

longer the deployment of knowledge for diagnostic or other purposes but the intelligent exploitation of data. The exploitation of medical data, whether they refer to clinical or demographic data, is extremely valuable and multifaceted. For starters, it can yield significant new knowledge, e.g., guidelines and protocols for the treatment of acute and chronic disorders, by summarizing all available evidence in the particular relevant field, a process currently known as evidence-based medicine. It also can provide accurate predictors for critical risk groups based on "low-cost" information, etc. Secondly, it aims to provide means for the intelligent comprehension of individual patients' data, whether such data are riddled with gaps, or are voluminous and heterogeneous in nature. Such data comprehension functions to close the gap (or conceptual distance) between the raw patient data and the medical knowledge to be applied for reaching the appropriate decisions for the patient in question.

In spite of the shift in focus from knowledge-intensive to data-intensive approaches, the ultimate objective is still the same, namely to aid care providers reach the best possible decisions for any patient, to help them see through the consequences of their decisions/actions and if necessary to take rectifying actions as timely as possibly.

The change in focus has given a new dimension of significance to clinical databases and in particular to the intelligent management and comprehension of such data. The truth of the matter is that some researchers had recognized such issues as important from the early days, but widespread recognition of the necessity to exploit medical data is a relatively recent development. Methods for abstraction, query and display of time-oriented data, lie at the heart of this research. Such methods are of relevance to all medical tasks mentioned above. This is why these three tasks feature very prominently in this book. The various medical tasks are also discussed to a greater or lesser extent. Time considerations arise in all cases. Only in very simple applications, it would be justifiable to abstract time away. For example, in diagnostic tasks, abstracting time away would mean that dynamic situations are converted to static (snap-shot) situations, where neither the evolution of disorders, nor patient states can be modeled.

1.2 Temporal Information Systems in Medicine

Time plays a major role in medical information systems: indeed, it is an important and pervasive concept of the real world and needs to be managed in several different ways: events occur at some time points, certain facts hold during a time period, and temporal relationships exist between facts and/or events [289]. Time has to be considered when representing information within computer-based systems [77], when querying information about temporal features of the represented real world [387], and when reasoning about time-oriented data [270].

In the medical-informatics field, temporal data modeling, temporal maintenance, and temporal reasoning have been investigated, to support both electronic medical records and medical decision-support systems [223, 212, 429, 197, 196, 235, 307, 359, 86, 179, 81, 301, 281, 129]

Before going ahead, it is important to provide a kind of definition for the meaning we assign to the expression "temporal information systems": *temporal information systems in medicine are information systems able to store, manage, query, and support different inference tasks on time-oriented clinical data.* Temporal information systems in medicine can be observed from different perspectives. One perspective concerns the kind of *technical tasks* that need to deal with temporal aspects of clinical data. From this perspective, we can distinguish the following applications:

- management of time-oriented data stored in medical records of ambulatory or hospitalized patients [426, 248, 307, 309, 84, 86, 125, 126, 341, 120, 111, 112];
- prediction of future values of clinical data, given past trends [109, 108, 21];
- abstraction of time-oriented clinical data [197, 200, 357, 359, 234, 393];
- time-oriented knowledge-based decision support, such as for systems supporting diagnosis, monitoring, or therapy planning [335, 29, 212, 406, 140, 331, 247, 214, 200, 278].

A second, important, and more "medical" perspective, concerns the *medical tasks* supported by the systems proposed in the literature:

- diagnosis [212, 103, 29],
- therapy administration and monitoring [200, 112, 167],
- protocol- and guideline-based therapy [406, 111, 359, 278],
- patient management [31, 124, 123, 120, 307, 86].

Finally, we can identify a third perspective, the "clinical" perspective, that concerns the *clinical areas* for which time-oriented applications, possibly performing different medical tasks through different technical tasks, have been designed and implemented, such as:

- cardiology [194, 431, 309, 86, 403, 247, 170, 167],
- oncology [140, 406, 197, 357, 111],
- psychiatry [29],
- internal medicine [105, 335, 120, 359, 331],
- intensive care [74, 140, 109, 70, 179],
- cardiac surgery [235],
- orthopaedics [212],
- urology [120],
- infectious diseases [111],
- anaesthesiology [108, 80, 70],
- paediatrics [170, 179],
- endocrinology [7].

1.3 Research on Time, Medicine, and Information Systems

Representing, maintaining, querying, and reasoning about time-oriented clinical data is a major theoretical and practical research area. Temporal reasoning is important to medical decision making (e.g., in clinical diagnosis and therapy planning) and in medical data modelling and managing (e.g., for representing a patient's medical record). In the literature, three research directions, with respect to their focus and the research communities pursuing them, can be easily identified:

- *Temporal data maintenance.* It deals with the storage and the retrieval of data that have heterogeneous temporal dimensions, and typically is associated with the (temporal) database community.
- *Temporal data abstraction and reasoning.* It supports various inference tasks involving time-oriented data, such as planning and execution, and traditionally it has been linked with the artificial-intelligence community. Tasks such as abstraction of higher-level concepts from time-stamped data and management of temporal granularity create an intersection with the previous research direction.
- *Design of temporal medical systems.* A third scientific community, the medical-informatics one, has been greatly stimulated by the results obtained in the areas of temporal databases and temporal reasoning. Since the members of this community often belong also to the computer science/computer engineering area, researchers in medical informatics have often investigated the issues of temporal data maintenance, and temporal reasoning, especially in the context of the different tasks relevant to medical information systems.

Furthermore, all these research directions necessarily involve *temporal data modelling*, since otherwise data can be neither maintained nor reasoned with.

One indication of the significance of research on time-oriented systems in medicine is that the level and the amount of scientific works in this area motivated four different special issues: two for the journal *Artificial Intelligence in Medicine*, one for the journal *Computers in Biology and Medicine*, and one for the *Journal of Intelligent Information Systems* [203, 206, 365, 364]. Another indication is that research focusing on time in clinical applications received attention also from the general computer science field [168, 169, 34, 415, 71, 206, 347, 359, 29, 88, 402].

Temporal information systems can be effectively studied, defined, and implemented only through a multidisciplinary combination of research in the areas of database systems, artificial intelligence, and medical informatics, and their associated communities. Indeed, besides being an important application area for advanced research on intelligent information systems, these systems provide an excellent motivation for research on multiple theoretical and methodological aspects of dealing with the management of and reasoning about complex time-oriented data. We hope that this book will succeed in demonstrating the depth and multidisciplinary nature of the research and application issues related to temporal information systems in medicine. It is our profound belief that an added value of this book is the merging of database theories/methodologies and reasoning methods for solving complex application problems. Medicine provides database researchers and artificial intelligence researchers with fertile ground and vast amounts of available data and knowledge, as well as excellent reasons for integrating these two areas. The case of handling and reasoning about large numbers of time-oriented data is a good example.

In the book, without trying to be exhaustive in any way, we have attempted to put together and describe the main features of temporal information systems in medicine, primarily from a methodological point of view, and on the basis of the relevant scientific literature.

The rest of the book is organized as follows. The material is split into four Parts: Fundamentals, Temporal Reasoning and Maintenance in Medicine, Time in Clinical Tasks, and The Display of Time-Oriented Clinical Information. Under Part I we have tried to present the fundamental concepts underlying the bulk of research in general temporal information systems. Part I includes two chapters, Chapter 2 that presents the fundamental concepts in temporal modeling and temporal reasoning and Chapter 3 that covers central notions in the area of temporal databases. Part II of the book discusses the specifics of clinical temporal databases (Chapter 4) and the abstraction of time-oriented clinical data (Chapter 5) with particular emphasis on time-oriented monitoring and a general approach to knowledge-based temporal abstraction. In addition, in Chapter 5, the new notion of a temporal mediator is proposed as the bridge between temporal reasoning and temporal maintenance. Part III of the book elaborates on the use of time in clinical tasks, in particular for the task of clinical diagnosis (Chapter 6), and the task of automatically supporting clinical guidelines and care plans (Chapter 7). Finally Part IV discusses the rapidly growing new field on the displaying of time-oriented clinical data and knowledge. For each of the chapters, and for the book as a whole, we have tried to give as much as possible a self-contained account on the discussed topic. This has not been easy or even adequately attainable due to the ever growing interest in the field and the field's inherent diversity. New concepts or solutions are being proposed at the fundamental level of temporal databases, temporal reasoning, or temporal maintenance. New interesting applications in clinically-related areas appear all the time. Trying to include everything would result in a never ending task and a book that is forever in a draft, unpublished state. Thus, we took the decision to opt for a "sound", albeit "incomplete" account, necessarily leaving out work that could potentially be included (or would be worth to include) in the book. Needless to say that the discussion on some of the temporal approaches is by necessity of a cursory nature. However, we have tried to be as comprehensive as possible with the literature references and thus the interested reader could dwell deeper into some topic of his/her particular interest by referring to the original sources.

Part I
Fundamentals

Chapter 2
Temporal Modeling and Temporal Reasoning

Overview

In this chapter, the reader is guided through the basic notions of time and temporal information and is presented with some important, general approaches to represent and reason about temporal information. Simple medical examples are used to help the reader to understand the advantages and limitations of the different approaches.

Structure of the Chapter

After a brief introduction, the basic concepts related to the representation of temporal information are presented. The main features of time domains, time primitives, and temporal entities are introduced in Section 2.2. Section 2.3 continues the discussion by presenting some widely acknowledged, general approaches to temporal reasoning and outlining the requirements for temporal reasoning in medicine, pointing out the limitations of some of the considered general approaches. Section 2.4 introduces temporal constraints, a relevant topic both for temporal reasoning and for temporal knowledge-bases and databases.

Keywords

Time domains, instants, intervals, time metrics, linear time, branching time, circular time, granularity, indeterminacy, interval algebra, event calculus, situation calculus, temporal reasoning, temporal representation, temporal entities, temporal constraints, probabilistic reasoning, deterministic reasoning.

C. Combi et al., *Temporal Information Systems in Medicine*,
DOI 10.1007/978-1-4419-6543-1_2, © Springer Science+Business Media, LLC 2010

2.1 Introduction

A common focus of temporal reasoning, temporal abstraction of clinical data, and modeling and managing clinical data, is the definition or the adoption of a set of basic concepts that enable a description of a time-oriented clinical world in a sound and unambiguous way. Several suggestions have emerged from generic fields of computer science, such as artificial intelligence, or the knowledge and data management areas [10, 8, 381, 399, 387, 289]. Within medical informatics, this effort has progressed from an ad-hoc definition of concepts supporting a particular application to the adoption and the proposal of more generic definitions, supporting different clinical applications [140, 212, 197, 196, 235, 111, 86, 359, 214]. For example, the emphasis in the pioneering work of Fagan on the interpretation of real-time quantitative data in the intensive-care domain is on the application-dependent problems, related to the support of a module that suggests the optimal ventilator therapy at a given time [140], while the work described in [214] uses a generic temporal ontology and a general, comprehensive, model of diagnostic reasoning.

2.2 Modeling Temporal Concepts

Time-related representation requirements for medical applications are many and varied because time manifests in different ways in expressions of medical knowledge and patient information. There are two issues here: how to model time per se and how to model time-varying situations or occurrences. In other words, we have here to consider both how to model the concept of time and how to model entities having a temporal dimension [270, 289].

2.2.1 Modeling Time

In general we could say that modeling time as a dense or discrete number line may not provide the appropriate abstraction for medical applications. The modeling of time for the management of, or the reasoning about time-oriented clinical data, requires several basic choices to be made, depending on the needs of the domain.

2.2.1.1 Time Domain

The time domain consists of the set of primitive time entities used to represent the concept of time. It allows one to define and interpret all the other time-related concepts. According to [184], a *time domain* is a pair $(T; \leq)$, where T is a non-empty set of time instants and $\leq$ is a total order on T. It has been extensively debated whether the real time is both either bounded or unbounded and either dense (e.g., isomorphic to real or rational numbers) or discrete (e.g., isomorphic to natural numbers). A time domain is bounded if it contains upper and/or lower bounds with respect to its order relationship. A time domain is dense if it is an infinite set and for all $t', t'' \in T$ with $t' < t''$, there exists $t''' \in T$ such that $t' < t''' < t''$. On the opposite, a time (unbounded) domain is discrete if every element has both an immediate successor and an immediate predecessor. For example, a widely used approach in temporal databases is the use of a basic *timeline*, i.e. a point structure with precedence relation, which is a total order without right and left endpoints. The basic timeline is (partially) partitioned in non decomposable consecutive time intervals, called *chronons* [387, 78]. On the other hand, Allen's intervals are defined on a continuous timeline [10].

2.2.1.2 Instants and Intervals

Usually both the (primitive) concepts of *time point* (or *instant*) and *time interval* have been used to represent time [235, 359, 86, 214, 387]. These concepts are usually related to instantaneous events (e.g. myocardial infarction), or to situations lasting for a span of time (e.g. drug therapy). In defining basic time entities, time points (i.e., instants) are often adopted. Intervals are then represented by their upper and lower temporal bounds (start and end time points). In practice, most systems employed in medical informatics applications have used a time point based approach, similar to McDermott's *points* [259], rather than use time intervals as the basic time primitives, as proposed by Allen [8]. Several variations exist. *Nonconvex intervals* are intervals formed from a union of convex intervals, and might contain gaps (see Figure 2.1). Such intervals are first-class objects that seem natural for representing processes or tasks that occur repeatedly over time. Ladkin defined a taxonomy of relations between nonconvex intervals [232] and a set of operators over such intervals [231], as well as a set of standard and extended time units that can exploit the nonconvex representation in an elegant manner to denote intervals such as "Mondays." [231]. Additional work on models and languages for nonconvex intervals has been done by Morris and Al Khatib [273], who call such intervals *N-intervals*. In the temporal database community non-convex intervals are usually named *temporal elements* [184]. It is interesting to point out that in the database community, complex time structures are sometimes introduced to model in a compact way all the time dimensions related to a fact. For example, in [184] a temporal element is defined as a finite union of n-dimensional time intervals, assuming that a model is able to represent n different (and orthogonal) time dimensions (the main temporal dimensions for temporal databases are introduced and discussed in the next section).

2.2.1.3 Linear, Branching, and Circular Times

Different properties can be associated with a time axis composed by instants. Usually, both in general and clinically-oriented databases, time is *linear*, since the set

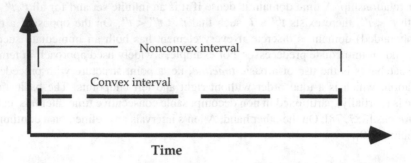

Fig. 2.1 A nonconvex interval. The nonconvex interval comprises several convex intervals.

of time points is completely ordered [111, 86, 214]. However, for the tasks of diagnosis, projection, or forecasting (such as prediction of a clinical evolution over time), a *branching* time might be necessary. In this case, only a partial ordering is defined for times. Such a representation has been found to be useful for pharmacoeconomics, and has been implemented using an object-oriented temporal model, as demonstrated in [158]. *Circular* (or periodic) time is needed when we have to describe times related to recurrent events, such as "administration of regular insulin every morning". In this case no ordering relations are defined for times.

2.2.1.4 Relative and Absolute Times

The position on the time axis of an interval or of an instant can be given as an absolute position, such as the calendric time, when mapped to the time axis used (e.g.: "on November, 3 1996") [426, 197, 105, 85, 111, 359]. This is a common approach adopted by data models underlying temporal databases. However, it is also common to reason with relative time references: "the day after" or "sometimes before that moment".

Relevant to the topic of relative times are several proposals that employ implicit [212, 214, 170] or explicit [348] temporal contexts, which support the representation of relative or context-sensitive temporal clinical information or knowledge. Another concept relevant to this topic is that of *time metrics*: absolute times are generally associated to a metric, being its position given as a *distance* from a given time origin. When a metric is defined for the time domain, relative times can be given quantitatively: "three days after birth".

2.2.1.5 Modeling Temporal Relationships

In modeling temporal relationships, Allen's interval algebra [8] has been widely used in medical informatics [235, 111, 348, 347]. Section 2.3.2.4 describes in some detail Allen's interval algebra. Extensions to Allen's basic thirteen interval relationships have also been proposed [86]. Temporal relationships include two main types: qualitative (interval I1 before interval I2) and quantitative (interval I1 two hours before interval I2). Several general formalisms and approaches [8, 119, 253, 270] have been effectively adopted for satisfying the various needs encountered while modeling temporal relationships. Moreover, temporal relationships can also be classified according to the entities involved: interval/interval, interval/point, point/interval and point/point [270].

2.2.1.6 Modeling Granularities

The *granularity* of a given temporal information is the level of abstraction at which information is expressed. Different units of measure allow one to represent different granularities. One of the first proposals formally dealing with time granularity was made by Clifford [78], in which he provides a particularly clean view of a structure for temporal domains. Using a set-theoretic construction, Clifford defines a simple but powerful structure of time units. He assumes for every temporal domain a *chronon*. By the repeated operation of *constructed intervallic partitioning* - intuitively equivalent to segmentation of the time line into mutually exclusive and exhaustive intervals (say, constructing 12 months from 365 days) - Clifford defines a temporal universe, which is a hierarchy of time levels and units. He also defines clearly the semantics of the operations possible on time domains in the *temporal universe*. It is interesting to note that, unlike Ladkin's construction of discrete time units [231], Clifford's construction does not leave room for the concept of weeks as a time unit, since *weeks* can overlap months and years, violating the constructed intervallic partition properties. A widely accepted definition of temporal granularity, proposed by Bettini et al [28], has been used both for knowledge representation and for temporal data modelling [28, 184, 240, 282, 241]. According to their framework, a granularity is a mapping G from an index set (e.g., integers) to the powerset of the time domain such that:

1. if $i < j$ and $G(i)$ and $G(j)$ are non-empty, then all elements of $G(i)$ are less than all elements of $G(j)$, and
2. if $i < k < j$ and $G(i)$ and $G(j)$ are non-empty, then $G(k)$ is non-empty.

Any $G(i)$ is called a *granule*. The first condition states that granules in a granularity do not overlap and that their index order is the same as their time domain order. The second condition states that the subset of the index set that maps to non-empty subsets of the time domain is contiguous. It is worth noting that the set of granules is always discrete, no matter whether the time domain is discrete or dense. Besides an index, there may be a textual representation of a granule, as in the case, for example, of "June 1999", which refers to the granule composed by the time points contained in that month. Several relationships have been introduced for granularities [27, 160]; for example, according to Bettini et al [27] a granularity G_1 is finer than another granularity G_2 if for each i, there exists j such that $G_1(i) \subseteq G_2(j)$.

2.2.1.7 Modeling indeterminacy

Indeterminacy is often present in temporal information and is related to incomplete knowledge of when the considered fact happened. A frequent need, especially in clinical domains, is the explicit expression of uncertainty regarding how long a proposition was true. In particular, we might not know precisely when the proposition became true and when it ceased to be true, although we might know that it was true during a particular time interval. Sometimes, the problem arises because the time units involved have different *granularities*: the Hb level may sometimes be dated with an accuracy level of hours (e.g., "Wednesday at 5 P.M., October 23, 2002"), but may sometimes be given for only a certain day ("Wednesday, October 23, 2002"). Sometimes, the problem arises due to the naturally occurring incomplete information in clinical settings: The patient complains of a backache starting "sometime during 2001". There is often a need to represent such vagueness. As an example, Console and Torasso [102, 103, 104] present a model of time intervals that represents such partial knowledge explicitly. The model was proposed in order to represent causal models for diagnostic reasoning. The authors define a *variable interval* as a time interval I composed of three consecutive convex intervals. The first interval is *begin(I)*, the second is called *body(I)*, and the third is called *end(I)*. Operations on convex intervals can be extended to variable intervals. We can now model uncertainty about the time of the start or end of the actual interval, when these times are defined vaguely, since the begin and end intervals of a variable interval represent uncertainty about the start and stop times of the real interval; the body is the only interval during which the proposition represented by the variable interval was certainly true.

2.2.2 Modeling Temporal Entities

Let us now consider the more abstract task of modeling temporal entities, i.e., those concepts/things of the real world which must be represented also for their temporal aspects. A rich model providing a number of interrelated basic temporal entities, given at different abstraction levels and with multiple granularities, is often required when dealing with medical temporal information. Many representation issues arise with respect to temporal entities, as detailed below.

2.2.2.1 Defining Temporal Entities

A question that has been investigated in some depth in the literature is: What are the basic (medical) concepts that have temporal dimension? How are they interrelated? In general, we distinguish two different approaches in modeling temporal entities in medical applications: addition of a temporal dimension to existing objects, or creation of model-specific, time-oriented entities. The first approach, originating from research into databases, uses simple, "atomic" temporal entities [111, 86]. This approach is similar to the one underlying the temporal extensions proposed for relational and object-oriented data models: a temporal dimension is added at the tuple/object level or at the attribute/method level [111, 83], as we will see in detail in Chapter 3. The second approach, originating mostly from the area of artificial intelligence, focuses on modeling different temporal features of complex, task-specific entities.

For example, Allen introduces *events*, *properties* and *processes*, to represent different kinds of proposition holding on some intervals (as discussed in Section 2.3.2.4), while McDermott distinguishes between *facts* and *events*, as discussed in Section 2.3.2.3.

Let us now consider some proposals coming from medical applications. Here, several types of compound (abstract) entities are introduced, based on temporal entities that are stored at the database level. For example, in the HyperLipid system [335], patient visits were modeled as instant-based objects called *events*, while administration of drugs was modeled as *therapy* objects whose attributes included a time interval. *Phases* of therapy (inspired by the clinical algorithm modeled by the system) were then introduced to model groups of heterogeneous data, related to both visits and therapies. Events, therapies and phases were connected through a network.

Kahn and colleagues in [197] introduced formally the concept of a Temporal Network (TNET) and later extended it by the Extended TNET, or ETNET model [196]. In both models, a T-node (or an ET-node) models task-specific temporal data, such as a chemotherapy cycle, at different levels of abstraction. Each T-node is associated with a time interval during which the information represented by the T-node's data is true for a given patient.

In the M-HTP system for monitoring heart-transplant patients [235], clinical facts related to a patient are structured in a temporal network (TN) inspired by Kahn's TNET model [196]. Through this network, a physician can obtain different temporal views of the patient's clinical history. Each node of the TN represents an *event* (a visit) or a *significant episode* in the patient's clinical record. An event is time-point based; its temporal location can be specified by an absolute date or by the temporal distance relative to the transplantation event. An episode holds during an interval, during which a predefined property (evaluated by reasoning about several events) holds.

Keravnou and Washbrook introduce *findings*, *features*, and events to distinguish various types of instantaneous and interval-based information (patient-specific or general) [212].

2.2.2.2 Associating Entities to Instants and Intervals

We observe two main approaches in defining *occurrences* of temporal entities, i.e., in associating time with temporal entities. The first approach deals both with instant-related entities and with interval-related entities [235]. The second approach associates clinical entities only with a certain type of time concept, usually an interval, dealing in a homogeneous way also with intervals degenerating to be a single instant [359, 86, 111]. A further distinction exists between the basic time primitives, usually instants (time points), and the time entities that can be associated with clinical concepts [359, 86, 111].

Shoham's approach, for example, is based on the adoption of a set of time points as primitives; predicates, however, such as values of clinical parameters, can only be interpreted over *time intervals*, which are defined as ordered pairs of time stamps (including instants, which are zero-length intervals).

Depending on the underlying properties for time, the actual occurrences of temporal entities can be specified in several different ways:

- *Absolute and relative temporal occurrences*: the existence of some occurrence can be expressed in absolute terms, relative to some fixed time point, by specifying its initiation and termination (e.g.: "Tachycardia on November 3, 1996 from 6:30 to 6:45 p.m."). This is a common approach adopted by data models underlying temporal databases. Similarly, it can be expressed relative to other occurrences, either by qualitative relationships ("angina after a long walk" or "several episodes of headache during puberty") or by quantitative relationships (angina two hours before headache). Incorporation of purely relative time-oriented, interval-based information (especially disjunctions, such as "the patient had vomited before or during the diarrhea episode") within a standard temporal database is still a difficult task.
- *Absolute and relative vagueness, duration, and incompleteness*: an occurrence is associated with absolute vagueness if its initiation and/or termination cannot be precisely specified in a given temporal context; precision is relative to the particular temporal context. Absolute vagueness (called also *indeterminacy*) may be expressed in terms of quantitative constraints on the initiation, termination, or extent of the occurrence, e.g. the earliest possible and latest possible time for its initiation or termination, or the minimum and maximum for its duration: "an atrial fibrillation episode occurred on December 14th, 1995 between 14:30 and 14:45 and lasted for three-four minutes". An occurrence is associated with relative vagueness if its temporal relation with other occurrences is not precisely known but can only be expressed as a disjunction of primitive relations. Incompleteness in the specification of occurrences is thus a common phenomenon.
- *Point and interval occurrences*: An occurrence may be considered a point occurrence in some temporal context if its duration is less than the time unit, if any, associated with the particular temporal context. A point occurrence may be treated as an instantaneous and hence as a non-decomposable occurrence in the given temporal context. Thus an occurrence may be considered an interval occurrence

in some temporal context if its duration is at least equal to the time unit associated with the particular temporal context. Care needs to be taken in associating these concepts to clinical entities, such as symptoms, therapies, and pathologies: a myocardial infarction, for example, could be considered an instantaneous event, within the overall clinical history of the patient, or an interval-based occurrence, if observed during an ICU staying.

2.2.2.3 Semantic Relations between Temporal Entities

Other, more complex, features of temporal entities and of their occurrences need to be suitably considered.

- *Compound occurrences*: repeated instantiations of some type of occurrence, usually, but not necessarily, in a regular fashion, may need to be collectively represented as a periodic occurrence. An abstract periodic occurrence consists of the basic temporal entity and of the "algorithm" governing the repetition. A specific periodic occurrence is the collation of the relevant, individual, occurrences. A temporal trend, or simply trend, is an important kind of interval occurrence. A trend describes a change, the direction of change, and the rate of change that takes place in the given interval of time. An example of a trend could be "increasing blood pressure". A trend is usually derived from a collection of occurrences at a lower level. A temporal pattern, or simply pattern, is a compound occurrence, consisting of a number of simpler occurrences (and their relations). There are different kinds of patterns. A sequence of meeting trends is a commonly used kind of pattern. A periodic occurrence is another example of pattern. A set of relative occurrences, or a set of causally related occurrences, could form patterns. A compound occurrence can in fact be expressed at multiple levels of abstraction. Abstraction and refinement are therefore important structural relations between occurrences. Through refinement an occurrence can be decomposed into component occurrences and through abstraction component occurrences can be contained into a compound occurrence.
- *Contexts, causality and other temporal constraints*: a context represents a state of affairs that, when interpreted (logically) over a time interval, can change the meaning of one or more facts which hold within the context time interval. Causality is a central relation between occurrences. Changes are explained through causal relations. Time is intrinsically related to causality. The temporal principle underlying causality is that an effect cannot precede its cause. Causally unrelated occurrences can also be temporally constrained, as already mentioned. For example, a periodic occurrence could be governed by the constraint that the distance between successive occurrences should be 4 hours.

2.3 Temporal Reasoning

The ability to reason about time and temporal relations is fundamental to almost any intelligent entity that needs to make decisions. The real world includes not only static descriptions, but also dynamic processes. It is difficult to represent the concept of taking an action, let alone a series of actions, and the concept of the consequences of taking a series of actions, without explicitly or implicitly introducing the notion of *time*. This inherent requirement also applies to computer programs that attempt to reason about the world. In the area of natural-language processing, it is impossible to understand stories without the concept of time and its various nuances (e.g., "by the time you get home, I would be gone for 3 hours"). Planning actions for robots requires reasoning about the *temporal order* of the actions and about the *length of time* it will take to perform the actions. Determining the cause of a certain state of affairs implies considering temporal precedence, or, at least, temporal equivalence. Scheduling tasks in a production line, that aim to minimize total production time, require reasoning about *serial* and *concurrent* actions and about time *intervals*. Describing typical patterns in a baby's psychomotor development requires using notions of absolute and relative time, such as "walking typically *starts* when the baby is about *12 months old*, and is *preceded by* standing." Thus, clinical domains pose no exception to the fundamental necessity of reasoning about time.

Temporal reasoning has been used in medical domains as part of a wide variety of generic tasks [60], such as diagnosis (or, in general, abstraction and interpretation), monitoring, projection, forecasting, and planning (as discussed in chapters 6 and 7). These tasks are often interdependent. Projection is the task of computing the likely consequences of a set of conditions or actions, usually given as a set of cause-effect relations. *Projection* is particularly relevant to the *planning task* (e.g., when we need to decide how the patient's state will be after we administer to the patient a certain drug with known side effects). *Forecasting* involves predicting particular future values for various parameters given a vector of time-stamped past and present measured values, such as anticipating changes in future hemoglobin-level values, given the values up to and including the present. *Planning* consists of producing a sequence of actions for a care provider, given an initial state of the patient and a goal state, or set of states, such that that sequence achieves one of the goal patient states. Possible actions are usually operators with predefined certain or probabilistic effects on the environment. The actions might require a set of enabling *preconditions* to be possible or effective. Achieving the goal state, as well as achieving some of the preconditions, might depend on the correct *projection* of the actions up to a point, to determine whether preconditions hold when required. *Interpretation* involves abstraction of a set of time-oriented patient data, either to an intermediate level of meaningful temporal patterns, as is common in the *temporal-abstraction* task or in the *monitoring* task, or to the level of a definite diagnosis or set of diagnoses that explain a set of findings and symptoms, as is common in the *diagnosis* task. Interpretation, unlike forecasting and projection, involves reasoning about only past and present data and not about the future.

From the methodological point of view, one general criterion that can be used when classifying temporal-reasoning research is whether it uses a deterministic or a probabilistic approach [206].

2.3.1 Temporal Reasoning Requirements

Before introducing some relevant generic models for temporal reasoning, let us consider some important (generic) functionalities, a medical temporal reasoning system should include:

- *Mapping the existence of occurrences across temporal contexts*, if multiple temporal contexts are supported and more than one such context is meaningful to some occurrence.
- *Determining bounds for entity occurrences.* The initiation and termination points of absolute existences are usually expressed in (qualitative) terms which need to be translated into upper and lower bounds for the actual points within the relevant temporal context.
- *Consistency detection and clipping of uncertainty.* If the inferences drawn from a collection of occurrences are to be valid the occurrences must be mutually consistent. Inconsistency arises when there are overlapping occurrences that assert mutually exclusive propositions. The inconsistency can be resolved if the boundaries of the implicated occurrences can be moved so that the overlapping is eliminated. In fact the identification of such clashes usually results in narrowing the bounds for the initiation/termination of the relevant occurrences. More generally, inconsistency arises when the (disjunctive) temporal constraints relating a given set of occurrences cannot be mutually satisfied. A conflict is detected when all the possible temporal relationships between a pair of temporal entities are refuted. Temporal constraint propagation, minimization of disjunctive constraints (i.e. reducing the uncertainty), detection and resolution of conflicts are necessary functionalities, as in many other non-medical applications.
- *Deriving new occurrences from other occurrences.* There are different types of derivation. A predominant type is *temporal-data abstraction*, which is described separately in Chapter 5. Other types include decomposition derivations (the potential components of compound occurrences are inferred), causal derivations (potential antecedent occurrences, consequent occurrences, or causal links between occurrences are derived), etc.
- *Deriving temporal relations between occurrences.* Often the temporal relations that hold between occurrences are significant for the given problem solving. Thus if the temporal relation between a pair of occurrences is not explicitly given, it would need to be inferred.
- *Deriving the truth status of queried occurrences.* This functionality brings together many of the other functionalities. A (hypothesized) occurrence, of any degree of complexity, e.g. periodic, trend, compound, etc, is queried against a set

of occurrences (and temporal contexts) that are assumed to be true. The queried
occurrence is derived as true (it can be logically deduced from the assumed occurrences), false (it is counter-indicated by the assumed occurrences), or unknown
(possibly true or possibly false).

- *Deriving the state of the world at a particular time.* The previous functionality starts with a specific set of assumed occurrences and a specific queried
  occurrence. It is considered a necessary functionality because often problem
  solvers seek to establish specific information. Alternatively though, in an inves-
  tigative/explorative mode, the problem solver may need to be informed about
  what is considered to be true at some specific time. The query may be expressed
  relative to another specific point in time which defaults to now, e.g. at time point
  t, what was/is/will be believed to be true during some specified period *p*? This
  functionality may be used to compose the set of assumed occurrences for queries
  of the previous type.

2.3.2 Ontologies and Models for Temporal Reasoning

In this section, we present briefly major approaches to temporal reasoning in philosophy and in computer science (in particular, in the AI area). We have organized
these approaches roughly chronologically.

2.3.2.1 Tense Logics

It is useful to look at the basis for some of the early work in temporal reasoning.
We know that Aristotle was interested in the meaning of the truth value of future
propositions [330]. The stoic logician Diodorus Chronus, who lived circa 300 B.C.,
extended Aristotle's inquiries by constructing what is known as *the master argument*. It can be reconstructed in modern terms as follows [330]:

1. Everything that is past and true is necessary (i.e., what is past and true is necessarily true thereafter).
2. The impossible does not follow the possible (i.e., what was once possible does
   not become impossible).

From these two assumptions, Diodorus concluded that nothing is possible that
neither is true nor will be true, and that therefore every (present) possibility must be
realized at a present or future time. The master argument leads to *logical determin-
ism*, the central tenet of which is that what is necessary at any time must be necessary
at all earlier times. This conclusion fits well indeed within the stoic paradigm.

The representation of the master argument in temporal terms inspired modern
work in temporal reasoning. In particular, in a landmark paper [321] and in subsequent work [322, 323], Prior attempted to reconstruct the master argument using a

modern approach. This attempt led to what is known as *tense logic* - a logic of past and future. In Prior's terms,

Fp: it will be the case that *p*.

Pp: it was the case that *p*.

Gp: it will always be the case that *p* (i.e., ¬F¬p).

Hp: it was always the case that *p* (i.e., ¬P¬p).

Prior's tense logic is thus in essence a *modal-logic* approach (an extension of the first-order logic (FOL) with special operators on logical formulae [136]) to reasoning about time. This modal-logic approach has been called a *tenser* approach [149], as opposed to a *detenser*, or an FOL, approach. As an example, in the *tenser* view, the sentence $F(\exists x)f(x)$ is *not* equivalent to the sentence $(\exists x)Ff(x)$; in other words, if in the future there will be some x that will have a property f, it does not follow that there is such an x now that will have that property in the future. In the *detenser* view, this distinction does not make sense, since both expressions are equivalent when translated into FOL formulae. This difference occurs because, in FOL, objects exist timelessly, time being just another dimension; in tenser approaches, NOW is a point of time in a separate class. However, FOL can serve as a *model theory* for the modal approach [149]. Thus, we can assign precise meanings to sentences such as Fp by a FOL formalism.

An interesting point in the use of time and tenses in natural language was brought out by Anscombe's investigation into the meanings of *before* and *after* [14]. An example is the following: from "The infection was present *after* the fever ended," it does not follow that the fever ended *before* the infection was present. Thus, *before* and *after* are not strict converses. Note that, however, from "The infection started *after* the fever started," we can indeed conclude that the fever started *before* the infection started. Thus, *before* and *after* are converses when they link instantaneous *events*.

2.3.2.2 Kahn and Gorry's Time Specialist

Kahn and Gorry [195] built a general temporal-utilities system, the *time specialist*, which was intended not for temporal *reasoning*, but rather for temporal *maintenance* of relations between time-stamped propositions. However, the various methods they used to represent relations between temporal entities are instructive, and the approach is useful for understanding some of the work in medical domains. The time specialist is a domain-independent module that is knowledgeable specifically about maintaining temporal relations. This module isolates the temporal-reasoning element of a computer system in any domain, but is not a temporal *logic*. Its specialty lies in organizing time-stamped bits of knowledge. A novel aspect of Kahn and Gorry's approach was the use of three different organization schemes; the decision of which one to use was controlled by the user:

1. Organizing by *dates* on a date line (e.g., "January 17 1972")

2. Organizing by special *reference events*, such as *birth* and *now* (e.g., "2 years after birth")
3. Organizing by *before* and *after* chains, for an event sequence (e.g., "the fever appeared after the rash").

By using a *fetcher* module, the time specialist was able to answer questions about the data that it maintained. The time specialist also maintained the *consistency* of the database as data were entered, asking the user for additional input if it detected an inconsistency. Kahn and Gorry made no claims about *understanding* temporally oriented sentences; the input was translated by the user to a Lisp expression. Neither did they claim any particular semantic classification of the type of propositions maintained by the time specialist. Rather, the time specialist presents an example of an early attempt to extract the time element from natural-language propositions, and to deal with that time element using a special, task-specific module.

2.3.2.3 Approaches Based on States, Events, and Changes

Some of the approaches taken in AI and general computer science involve a roundabout way of representing time: Time is represented implicitly by the fact that there was some change in the world (i.e., a transition from one state to another), or that there was some mediator of that change.

The Situation Calculus and Hayes' Histories

The *situation calculus* was proposed by McCarthy [257, 258] to describe *actions* and their effects on the world. The idea is that the world is a set of *states*, or *situations*. Actions and events are functions that map states to states. Thus, that the result of performing the CARE_PROVIDING action in a situation with a suffering patient is a situation where the patient is treated is represented as

$$\forall s True(s, SUFFERING\_PATIENT) \Longrightarrow$$
$$True(Result(CARE\_PROVIDING, s), TREATED\_PATIENT)$$

Although the situation calculus has been used explicitly or implicitly for many tasks, especially in planning, it is not adequate for many reasons. For instance, concurrent actions are impossible to describe, as are actions with duration (note that CARE_PROVIDING brings about an immediate result) or continuous processes. There are also other problems that are more general, and are not specific to the situation calculus [377].

Hayes, aware of these limitations, introduced the notion of histories in his "Second Naive Physics Manifesto" [174]. A *history* is an ontological entity that incorporates both space and time. An object in a situation, or O@S, is that situation's intersection with that object's history [174]. Permanent places are unbounded temporally but restricted spatially. Situations are unbounded spatially and are bounded in time by the events surrounding them. Most objects are in between these two extremes. Events are instantaneous; episodes usually have duration. Thus, we can describe the history of an object over time. Forbus [143] has extended the notion of histories within his qualitative process theory.

Kowalski and Sergot's Event Calculus

Kowalski and Sergot proposed in [228] the Event Calculus (EC), a theory of time and change. EC is an interesting framework, because it is general, well founded and deeply formally studied, and it has also been applied to temporal reasoning in medical domains [212, 235, 206].

From a description of events that occur in the real world and properties they initiate or terminate, EC derives the validity intervals over which properties hold. The notions of event, property, time point, and time interval are the primitives of the formalism: *events* happen at *time points* and initiate and/or terminate *time intervals* over which *properties* hold. Initiated properties are assumed to persist until the occurrence of an event that interrupts them (*default persistence*). An event occurrence, associating the event to the time point at which it occurred, is represented by the *happens(event, timePoint)* clause. The relations between events and properties are defined by means of *initiates* and *terminates* clauses, such as:

$$initiates(ev1, prop, t) \Longleftarrow happens(ev1, t) \wedge holds(prop1, t) \wedge ... \wedge holds(propN, t)$$

$$terminates(ev2, prop, t) \Longleftarrow happens(ev2, t) \wedge holds(prop1, t) \wedge ... \wedge holds(propN, t)$$

The above *initiates* (*terminates*) clause states that each event of type $ev1$ ($ev2$) initiates (terminates) a period of time during which the property *prop* holds, provided that N (possibly zero) given preconditions hold at instant t. The EC model of time and change is concerned with deriving the maximal validity intervals (MVIs) over which properties hold: a validity interval must not contain any interrupting event for the property; a *maximal* validity interval (MVI) is a validity interval which is not a subset of any other validity interval for the property. The clause $mholds\_for(p, [S, E])$ returns the MVIs for a given property p: each MVI is given by a pair $[S, E]$, where S (Start) and E (End) are the lower and upper endpoints of the interval.

Chittaro and Montanari distinguished two alternative ways of interpreting *initiates* clauses in the derivation of MVIs [73, 72]. In the first one (*weak* interpretation), only terminating events are considered as interrupting events, and an initiating event e for property p initiates an MVI, provided that p has not been already initiated by a previous event in such a way that p already holds at the occurrence time of e [73, 72]. For example, both Fig. 2.2b and 2.2d contain two consecutive weakly initiating events (denoted as wI) for the same property, and thus the derived MVI is initiated by the first of the two events. The alternative interpretation (*strong* interpretation) considers also initiating events as interrupting events: therefore, an initiating event e for property p initiates an MVI, provided that there is no subsequent initiating event for p such that p is not terminated between the two events. For example,

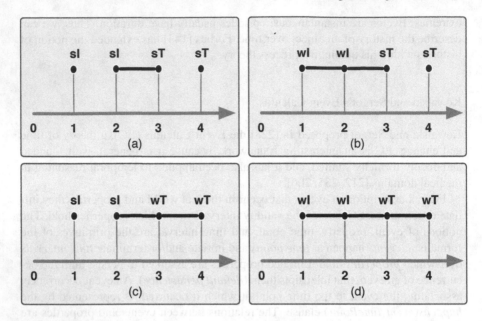

Fig. 2.2 Examples of weakly and strongly initiating and terminating events.

both Fig. 2.2a and 2.2c contain two consecutive strongly initiating events (denoted as *sI*) for the same property, and thus the derived MVI is initiated by the second of the two events. The weak and strong interpretation for *terminating* events give *symmetrical* results: Figures 2.2a and 2.2b show two consecutive strongly terminating events (denoted as *sT*), and the derived MVI terminated by the first of the two events; while Fig. 2.2c and 2.2d show two consecutive weakly terminating events (denoted as *wT*), and the derived MVI terminated by the second of the two events.

As clearly shown by Fig. 2.2, different choices of interpretation for initiating and terminating events may change the derived MVIs. In general, this choice depends on the specific property that needs to be

modeled, as we will also see in the following examples. In particular, weak and strong *initiates* relations can be used to support the so-called temporal *aggregation* and *omission* [73], respectively. For example, consider the problem of monitoring patients who receive a partial mechanical respiratory assistance [235]. A basic requirement of the patient monitoring task is the ability of aggregating similar observed situations. It indeed often happens that data acquired with consecutive samplings do not cause a transition in the classification of the patient ventilatory state. In this case, temporal aggregation requires that the subsequent samples do not clip the validity interval for the patient state. Such a functionality can be easily supported by the EC, provided that a weak interpretation of *initiates* is assumed. Temporal omission is useful when dealing with incomplete sequences of events [73]. As a simple example, consider a patient in a ICU receiving a continuous ECG monitoring, which can be interrupted by patient movements, specific treatments and examinations, and so on; the patient can be connected or disconnected to the device. The situation can be described by means of the property *ECGmonitor(Connection)*, where the value of *Connection* can be *connected* or *disconnected*, and two events: *connect* (resp. *disconnect*), that changes the status of the connection from *disconnected* to *connected* (resp. from *connected* to *disconnected*). While two *connect* (resp. *disconnect*) events cannot occur consecutively in the real world without a *disconnect* (resp. *connect*) event in between, it might happen that an incomplete sequence consisting of two consecutive *connect* events $e1$, $e2$, followed by a *disconnect* event $e3$, is recorded in the database. In such a case, a strong interpretation of *initiates* allows the EC to recognize that a missing *disconnect* event must have occurred between $e1$ and $e2$. However, since it is not possible to temporally locate such an event, the validity of the property *ECGmonitor(connected)* is derived

only between $e2$ and $e3$, and $e1$ is considered as a *pending* initiating event.

2.3.2.4 Allen's Interval-Based Temporal Logic and Related Extensions

Allen [8] has proposed a framework for temporal reasoning, the *interval-based temporal logic*. The only ontological temporal primitives in Allen's logic are *intervals*. Intervals are also the temporal unit over which we can interpret *propositions*. There are no instantaneous events-events are degenerate intervals. Allen's motivation was to express natural-language sentences and to represent plans. Allen has defined 13 basic binary relations between time intervals, six of which are inverses of the other six: BEFORE, AFTER, OVERLAPS, OVERLAPPED, STARTS, STARTED BY, FINISHES, FINISHED BY, DURING, CONTAINS, MEETS, MET BY, EQUAL TO (see Figure 2.3).

It turns out that all of the thirteen relations can be expressed using only a single one, MEETS; for example, *A* BEFORE *B* can be expressed as
$\exists C(A$ MEETS $C \wedge C$ MEETS $B)$ [9].

Incomplete temporal information common in natural-language is captured intuitively enough by a disjunction of several of these relations (e.g., T_1 <starts, finishes, during> T_2 denotes the fact that interval T_1 is contained somewhere in interval T_2, but is not equal to it). In this respect, Allen's logic resembles the event calculus.

Allen defined three types of propositions that might hold over an interval:

1. Properties hold over every subinterval of an interval. Thus, the meaning of Holds(p, T) is that property p holds over interval T. For instance, "John had fever during last night."
2. Events hold only over a whole interval and not over any subinterval of it. Thus, Occur(e, T) denotes that event e occurred at time T. For instance, "John broke his leg on Saturday at 6 P.M."
3. Processes hold over some subintervals of the interval in which they occur. Thus, Occurring(p, T) denotes that the process p is occurring during time T. For instance, "John had atrial fibrillation during the last month."

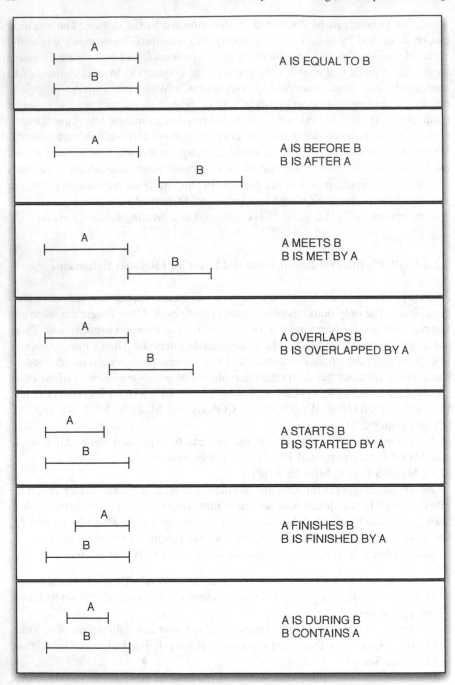

Fig. 2.3 The 13 possible relations, defined by Allen [8], between temporal intervals. Note that six of the relations have inverses, and that the EQUAL relation is its own inverse.

Allen's logic does not allow branching time into the past or the future (unlike, for instance, McDermott's logic).

Allen also constructed a transitivity table that defines the conjunction of any two relations, and proposed a *sound* (i.e., produces only correct conclusions) but *incomplete* (i.e., does not produce all correct conclusions) algorithm that propagates efficiently ($O(n^3)$) and correctly the results of applying the transitivity relations [10].

Unfortunately, the complexity of answering either the question of *completeness* for a set of Allen's relations (finding *all* feasible relations between *all* given pairs of events), or the question of *consistency* (determining whether a given set of relations is consistent) is NP-complete [416, 417]. Thus, in our current state of knowledge, for practical purposes, settling such issues is intractable. However, more work [413] has suggested that limited versions of Allen's relations - in particular, *simple interval algebra (SIA)* networks - can capture most of the required representations in medical and other areas, while maintaining computational tractability. SIA networks are based on a subset of Allen's relations that can be defined by conjunctions of equalities and inequalities between endpoints of the two intervals participating in the relation, but disallowing the $\neq$ (NOT EQUAL TO) relation [413].

Additional extensions to Allen's interval-based logic include Ladkin's inclusion of *nonconvex intervals* [231, 232].

2.3.2.5 McDermott's Point-Based Temporal Logic

McDermott [259] suggested a point-based temporal logic. The main goal of McDermott's logic was to model causality and continuous change, and to support planning.

McDermott's temporal primitives are *points*, unlike Allen's intervals. Time is continuous: The time line is the set of real numbers. Instantaneous snapshots of the universe are called states. States have an order-preserving *date* function to time instants. Propositions can be interpreted either over states or over intervals (ordered pairs of states), depending on their type. There are two types of propositions, facts and events. Facts are interpreted over points, and their semantics borrow from the situation calculus. The proposition (On Patient1 Bed2) represents the set of states where Patient1 is on Bed2. Facts are of the form (T s p), in McDermott's Lisp-like notation, meaning that p is true in s, where s is a state and p is a proposition, and $s \in p$. An event e is the set of intervals over which the event exactly happens: (Occ s_1 s_2 e) means that event e occurred between the states s_1 and s_2 - that is, over the interval [s_1 s_2] - where [s_1 s_2]$\in e$. McDermott's *external* characterization of events by actually *identifying* events as sets of intervals has been criticized (e.g., [149]). Such a characterization seems to define events in a rather superficial way (i.e., by temporal spans) that might even be computationally intractable for certain types of events, instead of relying on their *internal* characterization.

McDermott's states are partially ordered and branching into the future, but are totally ordered for the past (unlike Allen's intervals, which are not allowed to branch into either the past or the future). This branching intends to capture the notion of a known past, but an indeterminate future. Each maximal linear path in such a branching tree of states is a chronicle. A chronicle is thus a complete possible history of the universe, extending to the indefinite past and future; it is a totally ordered set of states extending infinitely in time [259].

2.3.2.6 Shoham's Temporal Logic

Shoham [376], in an influential paper, attempted to clean up the semantics of both Allen's and McDermott's temporal logics by presenting a third temporal logic. Shoham pointed out that the predicate-calculus semantics of McDermott's logic, like those of Allen's, are not clear. Furthermore, both Allen's "properties, seem at times either too restrictive or too general. Finally, Allen's avoidance of time points as primitives leads to unnecessary complications [376].

Shoham therefore presented a temporal logic in which the time primitives are *points*, and propositions are interpreted over time *intervals*. Time points are represented as zero-length intervals, $< t, t >$. Shoham used *reified* first-order-logic propositions, namely propositions that are represented as individual concepts that can have, for instance, a temporal duration. Thus, TRUE(t_1, t_2, p) denotes that proposition p was true during the interval $< t_1, t_2 >$. Therefore, the temporal and propositional elements are explicit. Shoham notes that the simple first-order-logic approach of using time as just another argument (e.g., ON(Patient1, Bed2, t1, t2)), does not grant time any special status. He notes also that the modal-logic approach of not mentioning time at all, but of, rather, changing the interpretation of the world's model at different times (rather like the tense logics discussed in Section 2.3.2.1), is subsumed by reified first-order logic [376, 377, 172]. Shoham provided clear semantics for both the propositional and the first-order-logic cases, using his reified first-order temporal logic. Furthermore, he pointed out that there is no need to distinguish among particular types of propositions, such as by distinguishing *facts* from *events*: Instead, he defined several relations that can exist between the truth value of a proposition over an interval and the truth value of the proposition over other intervals. For instance, a proposition type is downward-hereditary if, whenever it holds over an interval, it holds over all that interval's subintervals, possibly excluding its end points [376] (e.g., "Sam stayed in the hospital for less than 1 week"). A proposition is upward-hereditary if, whenever it holds for all proper subintervals of some nonpoint interval, except possibly at that interval's end points, it holds over the nonpoint interval itself (e.g., "John received an infusion of insulin at the rate of 2 units per hour"). A proposition type is gestalt if it never holds over two intervals, one of which properly contains the other (e.g., the interval over which the proposition "the patient was in a coma for exactly 2 weeks" is true cannot contain any subinterval over which that proposition is also true). A proposition type is concatenable if, whenever it holds over two consecutive intervals, it holds also over their union (e.g., when the proposition "the patient had high blood pressure" is true over some interval as well as over another interval that the first interval meets, then that proposition is true over the interval representing the union of the two intervals). A proposition is solid if it never holds over two properly overlapping intervals (e.g., "the patient received a *full* course of the current chemotherapy protocol, *from start to end*," cannot hold over two different, but overlapping intervals). Other proposition types exist, and can be refined to the level of interval-point relations.

Shoham observed that Allen's and McDermott's *events* correspond to *gestalt* propositions, to solid ones, or to both, whereas Allen's *properties* are both *upward-hereditary* and *downward-hereditary* [376]. This observation immediately explains various theories that can be proved about Allen's properties, and suggests a more expressive, flexible categorization of proposition types for particular needs.

2.3.2.7 Projection, Forecasting, and Modeling the Persistence Uncertainty

The probabilistic approach is typically associated with the tasks of interpretation or forecasting of time-stamped clinical data whose values are affected by different sources of uncertainty [331, 247].

Dean and Kanazawa [118] proposed a model of probabilistic temporal reasoning about propositions that *decay* over time. The main idea in their theory is to model explicitly the probability of a proposition P being true at time t, $P(< P, t >)$, given the probability of $< P, t-\Delta >$. The assumption is that there are events of type E_p that can cause proposition p to be true, and events of type $E_{\neg p}$ that can cause it to be false. Thus, we can define a *survivor function* for $P(< P, t >)$ given $< P, t-\Delta >$, such as an exponential decay function.

Dean and Kanazawa's main intention was to solve the *projection problem*, in particular in the context of the *planning* task. They therefore provide a method for computing a belief function (denoting a belief in the consequences) for the projection problem, given a set of causal rules, a set of survivor functions, enabling events, and disabling events [118]. In a later work, Kanazawa [201] presented a logic of time and probability, $\mathcal{L}_{cp}$. The logic allows three types of entities: domain objects, time, and probability. Kanazawa stored the propositions asserted in this logic over intervals in what he called a *time network*, which maintained probabilistic dependencies among various facts, such as the time of arrival of a person at a place, or the range of time over which it is true that the person stayed in one place [201]. The time network was used to answer queries about probabilities of facts and events over time.

Dagum, Galper, and Horvitz [110, 109] present a method intended specifically for the *forecasting* task. They combine the methodology of static *belief networks* [299] with that of classical probabilistic time-series analysis [424]. Thus, they create a *dynamic network model* (DNM) that represents not only probabilistic dependencies between parameter x and parameter y at the same time t, but also $P(x_t|y_{t-k})$, namely the probability distribution for the values of x given the value of y at an *earlier* time. Given a series of time-stamped values, the conditional probabilities in the DNM are modified continuously to fit the data. The DNM model was tried successfully on a test database of sleep-apnea cases to predict several patient parameters, such as heart rate and blood pressure [107].

2.4 Three Well-Known General Theories of Time and the Medical Domain

As already discussed, three well-known general theories of time, that are justifiably credited for the sparking of widespread interest in time representation and temporal reasoning in the AI community are Allen's interval-based temporal logic [8], Kowalski and Sergot's Event Calculus (EC) [228] and Dean and McDermott's Time Map Manager (TMM) [117]. None of these general theories of time was developed with the purpose of supporting knowledge-based problem solving, let alone medical problem solving. Hence it comes as no surprise that in their basic form, none of these adequately supports the identified requirements for medical temporal reasoning discussed above (see Table 2.1). As a matter of fact various extensions of Allen's logic and the event calculus have been applied to medical problems with lesser or greater success; some of these approaches are mentioned in the sequel. Such attempts resulted in revealing the rather limited expressivity of these theories with respect to medical problems. Their widespread adoption is in fact attributed to their relative simplicity. However, their lack of structuredness both with respect to a model of time as well as a model of occurrences, but more importantly their very limited support for the critical process of temporal data abstraction, renders their applicability in the context of medical problems at large, non viable. Below we quote some of the criticisms of the EC that was expressed by Chittaro and Dojat [70] in their attempt to apply this general theory of time to patient monitoring. In the EC a change in a property is the effect of an event. In real-life a symptom may be self-limiting where no event is required to terminate its existence. The designers went around this problem by introducing so-called "ghost" events. Another limitation encountered was that only instantaneous causality could be expressed. So delayed effects or effects of a limited persistence could not be expressed. The limited support for temporal data abstraction, the lack of multiple granularities as well as the lack of any vagueness in the expression of event occurrences, are also pointed out as issues of concern regarding the expressivity of the EC with respect to the realities of medical problems.

To illustrate further the points of criticism raised, we try to represent some medical knowledge in terms of these general theories. The medical knowledge in question describes (in a simplified form) the skeletal dysplasia (SpondyloEpiphyseal Dysplasia Congenital: SEDC), where a skeletal dysplasia is a generalized abnormality of the skeleton. This knowledge is given below:

"SEDC *presents from birth* and can be lethal. It *persists throughout the lifetime* of the patient. People suffering from SEDC exhibit the following: short stature, due to short limbs, *from birth*; mild platyspondyly *from birth*; absence of the ossification of knee epiphyses *at birth*; bilateral severe coxa-vara *from birth*, *worsening with age*; scoliosis, *worsening with age*; wide triradiate cartilage *up to about the age of 11 years*; pear-shaped vertebral-bodies *under the age of 15 years*; variable-size vertebral-bodies *up to the age of 1 year*; and *retarded ossification* of the cervical spine, epiphyses, and pubic bones."

The text given in italic font refers to time, directly or indirectly.

Table 2.1 Evaluation of General Theories of Time Against Medical Temporal Requirements. key: X - does not support; (V) - supports partly; V - supports

	Allen's Time-Interval Algebra	Kowalski & Sergot's Event Calculus	Dean & McDermott's Time-Token Manager
multiple conceptual temporal contexts	X	X	X
multiple granularities	X	X	X
absolute existences	X	V	V
relative existences	V	X	(V)
absolute vagueness	X	X	V
relative vagueness	V	X	X
duration	X	V	V
point existences	X	V	V
interval existences	V	V	V
periodic occurrences	X	X	X
temporal trends	X	X	X
temporal patterns	(V)	X	(V)
structural relations (temporal composition)	X	X	X
temporal causality	(V)	(V)	(V)

The temporal primitive of Allen's interval-based logic is the time interval, and eight basic relations (plus the inverses for seven of these) are defined between time intervals. The other primitives of the logic are properties (static entities), processes and events (dynamic entities), which are respectively associated with predicates holds, occurring and occur as already discussed:

$$holds(p,t) \Longleftrightarrow (\forall t' in(t',t) \Longrightarrow holds(p,t'))$$
$$occurring(p,t) \Longrightarrow \exists t' in(t',t) \wedge occurring(p,t')$$
$$occur(e,t) \wedge in(t',t) \Longrightarrow \neg occur(e,t')$$

The logic covers two forms of causality, event and agentive causality. Allen's logic is a relative theory of time, where time is structured as a dense time line. In order to represent the SEDC knowledge in terms of Allen's logic we need to decide which of the entities correspond to events, which to properties, and which to processes. The relevant generic events are easily identifiable. These are: *birth(P)*, *age1yr(P)*, *age11yrs(P)*, *age15yrs(P)* and *death(P)* which mark the birth, the becoming of 1 year of age, etc of some patient *P*. Deciding whether to model SEDC and its manifestations as properties or processes is not immediately apparent. In the following representation the distinction into processes and properties is decided on a rather ad hoc basis:

$$occurring(SEDC(P),I) \Longrightarrow occur(birth(P),B) \wedge occur(age1yr(P),O) \wedge$$
$$occur(age11yrs(P),E) \wedge occur(age15yrs(P),F) \wedge occur(death(P),D) \wedge$$
$$started-by(I,B) \wedge finished-by(I,D) \wedge holds(stature(P,short),I) \wedge$$
$$holds(ossification(P,knee-epiphyses,absent),B) \wedge$$
$$occurring(coxa-vara(P,bilateral-severe,worsening),I) \wedge$$
$$occurring(scoliosis(P,worsening),I) \wedge$$
$$holds(triradiate-cartilage(P,wide),W) \wedge started-by(W,B) \wedge finished-by(W,E) \wedge$$
$$holds(vertebral-bodies(P,pear-shaped),F') \wedge started-by(F',B) \wedge before(F',F) \wedge$$
$$holds(vertebral-bodies(P,variable-size),V) \wedge started-by(V,B) \wedge finished-by(V,O) \wedge$$

$occurring(ossification(P,cervical\text{-}spine,poor),I) \wedge$
$occurring(ossification(P,epiphyses,retarded),I) \wedge$
$occurring(ossification(P,pubic\text{-}bones,retarded),I)$

In this formalization, a relative representation has been "forced" on absolute occurrences. The specified events are not consequences of the occurrence of SEDC; their role is to demarcate the relevant intervals. For this (disorder) representation to be viable, the implication should either be temporally screened against the particular patient in order to remove future or non-applicable consequences, or simply such happenings should be assumed to be true by default. A particular limitation of any relative theory of time is inability to adequately model the derivation of temporal trends, or the derivation of delays or prematurity with respect to the unfolding of some process, since the notion of temporal distance which is inherently relevant to both types of derivation is foreign to such theories of time. A statement about a trend, delay, prematurity, etc is a kind of summary statement for a collection of happenings over a period of time. Another limitation of relative theories of time is inability to model absolute vagueness. In the above representation the widening of the triradiate cartilage is expected to hold exactly up to the occurrence of the event "becoming 11 years of age" and also it is not possible to delineate a margin for the termination of the property "pear-shaped vertebral bodies"; instead its termination is expressed in a relative way by saying that this happens before the event "becoming 15 years of age" happens, which does not capture the intuitive meaning of the given manifestation.

The temporal primitive of Kowalski and Sergot's EC is the event. Events are instantaneous happenings which initiate and terminate periods over which properties hold. A property does not hold at the time of the event that initiates it, but does hold at the time of the event that terminates it. Default persistence of properties is modeled through negation-as-failure. Causality is not directly modeled, although a rather restricted notion of causality is implied, e.g. an event happening at time t causes the initiation of some property at time $(t+1)$ and/or causes the termination of some (other) property at time t. The calculus can be applied both under a dense or a discrete model of time. The EC representation of the SEDC knowledge consists of a number of clauses like the following:

$initiates(birth(P),ossification(P,knee\text{-}epiphyses,absent),t) \Longleftarrow$
$\qquad\qquad\qquad happens(birth(P),t) \wedge holds(SEDC(P),t)$

$initiates(birth(P),stature(P,short),t) \Longleftarrow$
$\qquad\qquad\qquad happens(birth(P),t) \wedge holds(SEDC(P),t)$

$terminates(death(P),stature(P,short),t) \Longleftarrow$
$\qquad\qquad\qquad happens(death(P),t) \wedge holds(SEDC(P),t)$

$initiates(birth(P),coxa\text{-}vara(P,bilateral\text{-}severe,worsening),t) \Longleftarrow$
$\qquad\qquad\qquad happens(birth(P)) \wedge holds(SEDC(P),t)$

$terminates(age15yrs(P),vertebral\text{-}bodies(P,pear\text{-}shaped),t) \Longleftarrow$
$\qquad\qquad\qquad happens(age15yrs(P),t) \wedge holds(SEDC(P),t)$

Many of the criticisms discussed above with respect to Allen's logic apply to the EC as well. Properties in the EC are analogous to Allen's properties. They are essentially 'static' entities. Evolving situations such as temporal trends, or retardation in the execution of some process, or more generally continuous change, cannot be adequately modeled within pure EC. For example, the above clause concerning coxa-vara talks about some worsening being initiated, and also, based on the various axioms of the EC, it can be inferred that the worsening holds at every instant of time. What is initiated is "bilateral severe coxa-vara" while the worsening of this condition is a kind of meta-level inference on the continuous progression of this condition. Furthermore, absolute vagueness is not addressed, and as with Allen's logic, the SEDC knowledge is not represented as an integral entity but as a sparse collection of 'independent' happenings.

The temporal primitive of Dean and McDermott's TMM is the point (instant). The other temporal entity is the time-token that is defined to be an interval together with a (fact or event) type. A time-token is a static entity. It cannot be structurally analyzed and it cannot be involved in causal interactions. A collection of time-tokens forms a time map. This is a graph in which nodes denote instants of time associated with the beginning and ending of events and arcs describe relations between pairs of instants. This ontology can be applied both under a dense or a discrete model of time. Below we represent part of the SEDC knowledge as a time map. The granularity used is years and the reference point (denoted as *ref*) is birth. The first argument of the time-token predicate is the (fact or event) type and the second is the interval. Predicate elt expresses margins (bounds) for the beginnings and endings of intervals, with respect to *ref*.

$(time\text{-}token(SEDC, present)I)$
$(time\text{-}token(coxa\text{-}vara, bilateral\text{-}severe)C)$
$(time\text{-}token(coxa\text{-}vara, worsening)C')$
$(time\text{-}token(ossification, epiphyses, retarded)E)$
$(time\text{-}token(triradiate\text{-}cartilage, wide)W)$
$(time\text{-}token(vertebral\text{-}bodies, pear\text{-}shaped)V)$
....
$(elt(distance(begin\ C)\ *ref*)0, 0)$
$(elt(distance(end\ C)\ *ref*), *pos\text{-}inf* *pos\text{-}inf*)$
$(elt(distance(begin\ C')\ *ref*)?, ?)$
$(elt(distance(end\ C')\ *ref*)?, ?)$
$(elt(distance(begin\ W)\ *ref*)0, 0)$
$(elt(distance(end\ W)\ *ref*)10, 11)$
$(elt(distance(begin\ V)\ *ref*)0, 0)$
$(elt(distance(end\ V)\ *ref*)?, 14)$
$(elt(distance(begin\ E)\ *ref*)?, ?)$
$(elt(distance(end\ E)\ *ref*)?, ?)$
....

Again the SEDC process per se and its manifestations are represented as independent occurrences. The expression of absolute temporal vagueness is supported (see instances of predicate elt above), but no mechanism for translating qualitative expressions of vagueness into the relevant bounds based on temporal semantics of properties is provided. In the above representation "up to about the age of 11 years" is translated, in an ad hoc way, to the margin (10 11) while for "under the age of 15 years" it is not easy to see what the earliest termination ought to be. The points raised above regarding the representation of trends, process retardations, etc., apply here as well. Again this is because the types associated with the tokens capture either instantaneous events, or static, downward hereditary, properties.

Thus, the important reasoning process of temporal data abstraction is not supported by any of the three general theories of time considered.

2.5 Temporal Constraints

In the AI community there has been substantial interest in networks of constraints, in particular in arc and path consistency algorithms, over the past 30 years and many authors have contributed to the development of the relevant ideas (see for example [119, 251, 269, 271]). Generally, the work on consistency algorithms focuses on computational matters and not so much on the constraints themselves. As such, the constraints used are often of a relatively simple form, such as ranges for the temporal distances concerned. In medical tasks such as clinical diagnosis, one needs to address more complex forms of temporal constraints. For example, simple ranges capture uncertainty but in a rather categorical or discrete way. A simple range cannot model the fuzziness that often arise in clinical domains, such as when different ranges with varying degrees of typicality are required. Furthermore, clinical temporal constraints could be of mixed types and more importantly they could involve different granularities.

In this section we define an abstract structure for the representation of temporal constraints. In Chapter 6 we discuss particular instantiations of this structure of relevance to clinical diagnosis. The abstract structure, referred to as an *Abstract Temporal Graph*, or ATG for short, on one hand places the different types of constraints within the same, and thus unifying, framework and on the other hand enables the analysis and differentiation of the various types of constraints. By viewing temporal constraints in an abstract and more holistic way, it is possible to adopt, and appropriately adapt, well known constraint consistency algorithms from the general literature. Such algorithms can be further refined for particular instantiations of the abstract structure.

The problems we wish to address in this section are the following:

1. Checking the consistency of a set of constraints.
2. Deciding the satisfiability of some constraint with respect to a set of constraints that are assumed to be mutually consistent.

The first problem concerns the validation of the temporal consistency of a body of knowledge (e.g. the knowledge comprising the model of some disorder), or a set of data (e.g. the data on some patient). The second problem concerns the evaluation of the temporal consistency of some hypothesis against the evidence.

Abstract Temporal Graph

An Abstract Temporal Graph (ATG) is a directed graph whose nodes represent temporal entities (events, occurrences, etc.) and its arcs are labeled with the possible temporal constraints between the given pairs of nodes.

Let C be the domain of binary temporal constraints. The elements of C are mutually exclusive; only one of these can give the relationship that actually holds between the existence of one temporal entity and that of another temporal entity. C is either a finite or an infinite set. At the general level of discussion we consider the elements of C to be abstract entities processed by the following access functions:

1. $id : C \times C \rightarrow \{true, false\}$
   Function id returns $true$ only if its arguments are identical.
2. $inverse : C \rightarrow C$
   Every element in C has an inverse that is also an element of C; hence the domain of constraints is considered symmetrical. For example the inverse of $before$ is $after$ and of 1 day (meaning 1 day before) is -1 day (meaning 1 day after). Function $inverse$ returns the inverse of its argument. Furthermore, $inverse(inverse(c)) = c$.
3. $transit : C \times C \rightarrow 2^C$
   The arguments of function $transit$ refer to three (normally distinct) temporal entities, say n_i, n_j and n_k; the first argument, c_{ik}, represents the constraint from n_i to n_k and the second, c_{kj}, the constraint from n_k to n_j. The function returns the disjunctive constraint from n_i to n_j, i.e. the transitivity of the given (atomic) constraints.

In addition, special constant $self\_ref$ denotes the element of C that gives the constraint of any temporal entity with itself, e.g. $equal$ or 0 $days$, etc. Any self-referencing arc in an ATG would have as its label the set $\{self\_ref\}$. Furthermore, $inverse(self\_ref) = self\_ref$.

A more formal definition of an ATG is now given.

Definition 1 — An *Abstract Temporal Graph* (ATG) is a graph consisting of a finite set of nodes, $n_1, n_2, ...n_m$, denoting temporal entities (of the same type), and a finite set of directed arcs. A directed arc from n_i to n_j is labeled with a set of temporal constraints, $tc_{ij} \subseteq C$, denoting a disjunctive constraint from n_i to n_j. An ATG has access functions *match* and *propagate* for processing disjunctive constraints.

1. $match : 2^C \times 2^C \rightarrow 2^C$
   Function *match* returns the common elements between the two disjunctive constraints, in other words their intersection. If the function returns the empty set, denoting a complete mismatch between the argument constraints (none of the arguments is the empty set), a conflict is signalled.

> **match**(C_i, C_j)
> $\quad R \leftarrow \{\}$
> $\quad$ for $c_i \in C_i$ do
> $\quad\quad$ for $c_j \in C_j$ do

$\qquad$ if $id(c_i, c_j)$ then $R \leftarrow R \cup \{c_j\}$

$\quad$ return R

2. $propagate : 2^C \times 2^C \rightarrow 2^C$

Function $propagate$ returns the transitivity of its argument constraints that respectively represent the pairwise (disjunctive) constraints between three (normally distinct) temporal entities.

$\quad$ **propagate**(C_i, C_j)

$\qquad$ $R \leftarrow \{\}$

$\qquad$ for $c_i \in C_i$ do

$\qquad\qquad$ for $c_j \in C_j$ do

$\qquad\qquad\qquad$ $R \leftarrow R \cup transit(c_i, c_j)$

$\quad$ return R

If in an ATG, label tc_{ij} has only one element, there is no uncertainty (from the perspective of C) as to the relationship from n_i to n_j. If however $tc_{ij} = C$, there is complete ignorance of this relationship. Unconnected nodes in an ATG can always be connected via arcs with labels set to C.

There are two extreme cases. One is when for every pair of nodes, n_i and n_j, tc_{ij} is a singleton. This is the case where there is complete temporal knowledge and no uncertainty whatsoever (always with respect to C). This is what we strive to reach, as in reality there is only one possible scenario. The other extreme case is when for every pair of nodes, n_i and n_j, $tc_{ij} = C$. This is the case of complete ignorance regarding temporal information, i.e. everything is unknown or not given and hence everything is completely temporally unconstrained. In this case there is no point in connecting the nodes and hence the ATG degenerates into a set of unconnected nodes.

Definition 2 — A *fully connected ATG* is an ATG for which every pair of nodes n_i and n_j such that $i \neq j$ is connected in both directions and each connection is labeled (possibly with the

$\quad$ entire set of constraints, C).

Any ATG can be easily converted to a fully connected ATG, simply by adding the relevant arcs and labeling them with C. However, inverse arcs are redundant since their labels can be obtained directly from their counterparts. Hence inverse arcs may be deleted; which arc is actually deleted from each pair of counter arcs could be decided on the basis of some ordering of the nodes.

Definition 3 — An *ordered ATG* is an ATG whose nodes $n_1, n_2, ..., n_m$ form a topological ordering and for every pair of nodes n_i and n_j such that $i < j$ (i.e. n_i precedes n_j in the topological ordering), there is a labeled connection from n_i to n_j. Pairs of nodes, n_i, n_j, such that $i \geq j$ are not connected. The $(m-1)$ arcs connecting nodes, that are consecutive under the topological ordering, i.e. the arcs from n_i to n_{i+1}, for $i = 1, ..., m-1$, are referred to as *basic arcs* because each of them represents the sole path between the given pairs of nodes.

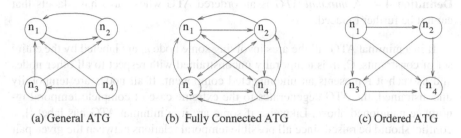

(a) General ATG   (b) Fully Connected ATG   (c) Ordered ATG

Fig. 2.4 Abstract Temporal Graphs

If one of the nodes in an ATG denotes *now*, presumably this node will either be the first or the last in the chosen topological ordering of the nodes, so that either it points to all other nodes or it is pointed by all other nodes.

An ATG can be converted to an ordered ATG, by adding any missing connections in the specified direction and deleting any connections in the opposite direction. Figure 2.4 shows a general ATG and its fully connected and ordered versions, where the ordering of the nodes is n_1, n_2, n_3, and n_4. The arc labels are omitted in the figure. When an arc from n_i to n_j is added, its label is set to tc_{ji}^{-1} (see below), provided there is an arc from n_j to n_i, otherwise the label is set to C. No self-referencing arcs are included.

A fully connected ATG can be reduced to its ordered version by deleting the inverse arcs according to the chosen ordering of its nodes. If m is the number of nodes, the fully connected ATG has $m(m-1)$ arcs (since self-referencing arcs are excluded) while the ordered ATG has half this number of nodes. Alternatively we can say that in the ordered ATG, $(m-1)$ arcs emanate from n_1, $(m-2)$ from n_2, ..., and 1 from n_{m-1}, giving a total of $m(m-1)/2$ arcs. Hence in the ordered ATG, the first node, n_1, points to $n_2, ..., n_m$, and no node points to it, an intermediate node, n_i, is pointed at by $n_1, ..., n_{i-1}$ and points to $n_{i+1}, ..., n_m$ for $i = 2, ..., m-1$ and the last node, n_m, is pointed at by every other node and points to none.

If in some ATG, a pair of nodes, n_i, n_j, is connected in both directions, the labels of the opposite arcs, tc_{ij} and tc_{ji}, must be consistent. Consistency can be interpreted in a strict sense as $match(tc_{ij}, tc_{ji}^{-1}) = tc_{ij}$, or in a more liberal sense as $match(tc_{ij}, tc_{ji}^{-1}) \neq \{\}$, where $tc^{-1} = \{inverse(c) \mid c \in tc\}$ gives the inverse of tc in C; in particular $C^{-1} = C$. The strict interpretation requires complete match, while the liberal one is satisfied with a match on just one constraint and thus a conflict is raised only if there is no match. In the following algorithms we use the liberal interpretation[1].

---

[1] An alternative interpretation is to use the union of the labels, i.e., $tc_{ij} \leftarrow tc_{ij} \cup tc_{ji}^{-1}$. This interpretation extends the possibilities.

Definition 4 — A *minimal ATG* is an ordered ATG whose arcs have labels that cannot be further reduced.

If in a minimal ATG, all the arcs involving some node n_i are labeled by the entire set of constraints, C, n_i is temporally unconstrained with respect to all other nodes and as such it represents an unconnected component. If all nodes are temporally unconstrained, the ATG degenerates to the extreme case of complete temporal ignorance mentioned above. Likewise if some arc in a minimal ATG has label {}, a conflict should be raised since all possible temporal relations between the given pair of temporal entities have been refuted.

Starting from a general ATG, the goal is to turn it into its minimal form by propagation and matching of constraints. Below we examine some algorithms for this task.

The first algorithm, *minimize_full*, extends the general ATG to its fully connected version (including self-referencing arcs as well, the presence of which simplifies the expression of the algorithm), then performs the propagation and matching of constraints and finally deletes the inverse and self-referencing arcs to obtain the ordered ATG, with minimal constraints. The algorithm is an adaptation of the well known Floyd-Warshall algorithm [119].

> **minimize_full**
>   (* add connections *)
>   for $i = 1$ to m do
>   for $j = 1$ to m do
>   if $i = j$
>   then add a self-referencing arc at node n_i
>   with label set to {*self_ref*}
>   else if there is an arc from node n_i to node n_j
>   then do nothing
>   else if there is an arc from node n_j to node n_i
>   then add an arc from n_i to n_j with label set to tc_{ji}^{-1}
>   else add an arc from node n_i to node n_j with label set to C
>   (* propagate constraints *)
>   repeat
>   for $k = 1$ to m do
>   for $i = 1$ to m do
>   for $j = 1$ to m do
>   (* critical step *)
> $tc_{ij} \leftarrow match(tc_{ij}, propagate(tc_{ik}, tc_{kj}))$
>   if $tc_{ij} = \{\}$ a conflict is raised
>   until no arc label is reduced (* i.e. nothing changes *)
>   (* remove self-referencing and inverse arcs *)
>   for $i = 1$ to m do
>   for $j = i$ to m do
>   if $i = j$ remove self-referencing arc at node n_i
>   else do

$$tc_{ij} \leftarrow match(tc_{ij}, tc_{ji}^{-1})$$

if $tc_{ij} = \{\}$ a conflict is raised

else delete arc from n_j to n_i

The critical step of the algorithm is taken to be the combined application of functions *propagate* and *match* with respect to triples of nodes, namely $match(tc_{ij}, propagate(tc_{ik}, tc_{kj}))$, which is executed m^3 times, m being the number of nodes, at each round of the repeat cycle. The number of executions of the critical step can be reduced if the minimization of constraints is done, not with respect to the fully connected ATG, but with respect to the ordered ATG. Performing the minimization on the fully connected ATG, in some sense defeats the purpose of having the ordering which is not just to reduce the space complexity by halving the number of arcs, but also to reduce the speed of execution of the minimization process.

The second algorithm, *minimize_ordered*, first converts the general ATG to its ordered version and then does the minimization by propagating and matching constraints with respect to triples of distinct nodes. In this algorithm the critical step is executed $(m^3 - 3m^2 + 2m)/6$ times, at every round of the repeat cycle. Thus although both algorithms have the same complexity, namely $O(rm^3)$, where r is the number of times the repeat cycle is executed, in real terms the critical step will be executed substantially fewer times in the second algorithm, assuming that m is of the order of tens rather than hundreds.[2] However, there is a point of caution. In the fully connected ATG, there are at least two distinct paths between every pair of nodes and the propagation is bidirectional. In the ordered ATG the propagation is unidirectional and thus the labels of the basic arcs (recall that a basic arc represents the sole path between the given pair of nodes — see Definition 3) stay invariant under this propagation. Basic arcs influence (except the last in sequence), but are not influenced by the propagation. Thus if a basic arc has label C prior to the execution of the minimization process, it will continue to have this label at the end of it. In other words nothing more is learned about that arc. Thus the results of the two algorithms are not equivalent,[3] since the first does a complete minimization but not necessarily the second, unless the labels of the basic arcs are given in minimal form to start with.

minimize_ordered

    let $n_1, n_2, ..., n_m$ be the specified ordering of the nodes

    (* covert to ordered form *)

    for $i = 1$ to $(m - 1)$ do

        for $j = i + 1$ to m do

            if nodes n_i and n_j are unconnected

                then add an arc from n_i to n_j with label set to C

            else if there is only an arc from n_i to n_j

---

[2] This assumption is not unrealistic, especially when disorders are modeled separately in terms of their own temporal graphs.

[3] In [119] two constraint networks are equivalent if they give the same solution set, where a solution set is the set of all feasible scenarios. If the arc labels are minimal, every temporal relationship included should participate in at least one feasible scenario.

then do nothing
else if there is only an arc from n_j to n_i
    then add an arc from n_i to n_j with label set to tc_{ji}^{-1}
    and delete the arc from n_j to n_i
else do (* there is a bidirectional connection *)
    $tc_{ij} \leftarrow match(tc_{ij}, tc_{ji}^{-1})$
    if $tc_{ij} = \{\}$ a conflict is raised
    else delete arc from n_j to n_i
(* propagate constraints *)
repeat
    (* repeat for every intermediate node in the ordering *)
    for $k = 2$ to $(m-1)$ do
        (* repeat for every incoming arc to node n_k *)
        for $i = 1$ to $(k-1)$ do
            (* repeat for every outgoing arc from node n_k *)
            for $j = (k+1)$ to m do
                (* critical step *)
                $tc_{ij} \leftarrow match(tc_{ij}, propagate(tc_{ik}, tc_{kj}))$
until no arc label is reduced (* i.e. nothing changes *)

Let us now return to the problems given at the beginning of the section.

Checking the Consistency of a Set of Constraints

The solution of the first problem, checking the consistency of a set of constraints is given by algorithm minimize_full or minimize_ordered. If during the execution of these algorithms, function *match* returns an empty set denoting a complete mismatch, a conflict is raised. Complete mismatch means that the disjunctive constraint relating two temporal entities, obtained via some route in the ATG, is in complete disagreement with the constraint, for the same pair of entities, obtained via another route in the ATG. In other words all possible temporal relations between the two entities have been refuted. The minimization algorithms detect the presence of some inconsistency but do not say which of the (original) constraints are responsible for it.[4]

We can sketch an algorithm for determining the possible causes of the inconsistency, as follows. Its aim is to identify minimal subsets of arc labels, each of which when omitted results in the resolution of the conflict. Omission of a label means that the particular arc label is replaced with C, which says that "If this is an erroneous label causing a conflict, it should be replaced with the label of complete ignorance.". First the algorithm assumes that a single label is the cause. Each label that is not equal to C is omitted in turn, and every time the minimization algorithm is ran to see if the omission has erased the conflict. If no single label is responsible for the

---

[4] There could be multiple possibilities as to the cause of the inconsistency; the identification of the actual cause amongst them may require external means.

conflict, the next step is to omit pairs of labels (not equal to C) in turn, and so forth. This is a complex algorithm. If k is the number of labels, originally not equal to C, in the worst case all these labels will be mutually inconsistent, meaning that the minimization algorithm will be ran $2^k - 1$ times. Obviously the use of *minimize_ordered* instead of *minimize_full* reduces the speed of execution, not only because this algorithm executes the critical step fewer times, but also because k would be expected to be half that for the fully connected ATG.

Deciding the Satisfiability of Some Constraint

The solution of the second problem, deciding the satisfiability of some constraint with respect to a set of constraints that are assumed to be mutually consistent, is as follows. It is based on the assumption that the queried constraint and the ATG representing the set of constraints are of the same form, i.e. the temporal entities are of the same type and the domain of constraints, C, is the same. In most applications of this problem, the queried constraint would describe a datum about the patient and the ATG would represent the model of some disorder.

Let n_i and n_j be the temporal entities implicated in the queried constraint, and let qc be the (disjunctive) constraint itself (from n_i to n_j). The temporal entities, n_i and n_j, could respectively denote the start and end of some symptom, or the starts of two distinct symptoms, etc. We distinguish the following cases:

1. Both n_i and n_j appear as nodes in the ATG.
2. Only one or none of these temporal entities appears as a node in the ATG.

In the first case, the solution is given as follows:

> convert the ATG to minimal form
> if there is an arc from n_i to n_j in the ATG
> then if $match(tc_{ij}, qc) \neq \{\}$
>   then the queried constraint, qc, is satisfied
>   else it is not satisfied
> else if $match(tc_{ji}, qc^{-1}) \neq \{\}$
>   then the queried constraint, qc, is satisfied
>   else it is not satisfied

In the second case, the liberal approach is to say that the queried constraint is satisfied by default (especially if the sentence denoted by the queried constraint expresses normality), and the strict approach is to say that it is not satisfied (except again when the queried constraint expresses normality). Which approach is taken would depend on whether satisfiability is interpreted as absence of explicit conflict, or as explicit consistency.

Summary

In this chapter we have overviewed the fundamental notions of temporal modeling and temporal reasoning. First, we discussed the modeling of basic temporal concepts, starting with the modeling of time and moving to the modeling of temporal entities, i.e., objects/facts having some temporal dimension. A model of time should take into consideration the following aspects: the time domain, the representation of instants and intervals, the structure of time (linear, branching, or circular), the representation of absolute and relative times and the representation of relations. Temporal granularity and indeterminacy are also important aspects when modeling time. Analogous considerations apply to the modeling of temporal entities and of their occurrences; different kinds of temporal entities were discussed as well as their association with the time domain and their semantics. Then, we discussed temporal reasoning starting with listing the required functionalities for medical temporal reasoning and moving to the presentation of the most influential ontologies and models for temporal reasoning (tense logics, time specialist, situation calculus, event calculus, interval-based and point-based temporal logics). In order to demonstrate the multiple aspects of medical temporal reasoning, a simple medical example was analyzed against three well-known general theories of time. We ended the chapter by discussing temporal constraints through an abstract representation, i.e. the abstract temporal graph. The main purpose of this chapter was to acquaint the reader with the necessary basic background with respect to temporal modeling and temporal reasoning, to facilitate the coverage of the remaining chapters of this book.

Bibliographic Notes

Apart from the specific references that are mentioned in this chapter, additional discussion on temporal logic is given in Rescher and Urquhart's excellent early work in temporal logic [330]. The AI perspective has been summarized well by Shoham [376, 377]. An overview of temporal logics in the various areas of computer science, and of their applications, was compiled by Galton [149]. Van Benthem's comprehensive book [414] presents an excellent view of different ontologies of time and their logical implications.

Problems

2.1. Using Prior's notation, we can write the two following predicates:
$$P(\exists \ patient \ diagnosis(patient, \text{tuberculosis}))$$
$$\exists \ patient \ P(diagnosis(patient, \text{tuberculosis}))$$

1. Explain in your own words exactly what each expression means, and why is their meaning different in Prior's logic. Create two convincing examples to demonstrate that there might be two mutually exclusive interpretations.
2. Explain why the distinction between the two expressions is meaningless in standard First Order Logic (predicate calculus).
3. Write, using Tense Logic notation, an expression that means "Patient Jones will have had the operation." Use expressions such as "procedure(Jones, Operation)".

2.2. Explain how, using only the MEETS relation between temporal intervals, and one or more interval variables and existential quantifiers, we can define the following relations:

1. The relation A DURING B
2. The relation A OVERLAPS B
3. What is the computational advantage of using 13 different temporal relations as opposed to only one relation? Think about applications such as theorem proving, planning, temporal queries, storage and retrieval of data.

2.3. Create a set of situation calculus axioms to express the following facts:

1. The effect of the action of entering the hospital room by a patient, when the patient is at the door, is that the patient is inside the hospital room. Use expressions such as At(Door, Patient), Enter(Patient, HospitalRoom), Within(HospitalRoom, Patient).
2. Explain the semantics of the predicates and/or functions you are using in terms of sets of states.
3. Can we express in the situation calculus the fact that the patient is looking around while moving from one room to another one? How, or Why?

2.4. What is the representational advantage of a reified logic, in which temporal arguments are explicitly separated from the rest of the predicate?

2.5. Consider the following 4 propositions:
i) "Mark had a complete removal of the appendix on January 15 1988 between 6 to 9 PM."
ii) "Joe has earned 3000 Euros during February 1999."
iii) "Mary had mild anemia during March to May 1997."
iv) "Peter was occasionally using insulin shots during July and August 1995."

1. Indicate in a small table what are the temporal-proposition properties of each proposition with respect to the properties *downward-hereditary* (dh), *upward-hereditary* (uh), *gestalt* (g), *concatenable* (c), and *solid* (s) as defined by Shoham.
2. What is the ontological type of each proposition according to Allen?
3. What is the ontological type of each proposition according to McDermott?

2.6. Study in more detail the classical approaches to temporal reasoning (Allen's time-interval logic, Event Calculus, etc.) together with Shoham's criticism. Analyze similarities and differences between these approaches and examine their appropriateness with respect to some medical domain and task you are familiar with.

2.7. Implement the notion of an ATG as an abstract data type, together with the discussed minimization algorithms. Apply your code to some example temporal graphs (see also Chapter 6).

Chapter 3
Temporal Databases

Overview

In this chapter, the reader is acquainted with the basic concepts of temporal databases, i.e. databases that deal explicitly with time-related concepts and information. Examples of clinical data with *valid* and *transaction times* are given. A classification of temporal databases according to the supported temporal dimensions is presented. The reader, then, learns how the relational model and the related algebra, calculus, and query languages have been extended to deal with the different time aspects. The presentation then focuses on the main data models employed in the conceptual design, the Entity-Relationship data model and the Object-Oriented data models. The reader learns how to design a temporal database schema by both the standard (atemporal) data models and their corresponding temporally-oriented extensions. Finally, the reader is guided through some basic features of object-oriented query languages and their extensions for temporal information. A section on further topics related to temporal databases sketches some research directions on multimedia and semistructured temporal data.

Structure of the Chapter

In this chapter we introduce the main concepts of temporal databases. After the introduction to the chapter, which motivates the rest of the chapter and introduces the reader to the main research topics considered in the area of temporal databases, we describe the two main time dimensions considered in this area: valid and transaction times. A classification of temporal databases according to the time dimensions they support is then presented. Then, we consider in some detail how the relational model, the most widely studied data model, can be extended to consider time aspects: a discussion on the basic features of related temporal algebras and calculi is introduced. We then show the main issues that need to be considered in

C. Combi et al., *Temporal Information Systems in Medicine*,
DOI 10.1007/978-1-4419-6543-1_3, © Springer Science+Business Media, LLC 2010

using/extending the widely used relational query language SQL, both for querying and manipulating data.

We then move to the Entity-Relationship data model, which is employed in the conceptual design of databases: several time aspects can be highlighted when conceptually designing the database schema.

Basic features of object-oriented data models and query languages are introduced and some insights are provided about the management of complex temporal information. Main emerging standard efforts in the area of object-oriented database systems, such as UML (Unified Modeling Language) and ODMG (Object Database Management Group), are considered, focusing on their use and extension for managing/modeling temporal information.

The chapter ends with a brief description of some more advanced topics, namely the management of multimedia temporal data and of semistructured temporal data.

Keywords

Temporal databases, valid time, transaction time, temporal relational model, temporal algebra, temporal calculi, Entity-Relationship data model, object-oriented data models, UML, ODMG, object-oriented query languages.

3.1 Introduction

The storage and management of information that has several temporal aspects have been extensively investigated within the research area of *temporal databases* (see, for example, [398, 139, 163, 187, 340, 384, 115]). The study of temporal dimensions of data started with much work on the definition of temporally-oriented extensions of the relational model; besides the modeling aspects, several studies considered the issues related to the extension of relational algebra and calculus to explicitly manage temporal dimensions of data. Even though most research in the temporal database field focuses on temporal relational models and query languages, several topics of the database area have been dealt with from the perspective of modeling and management of time-related aspects. As an example, time and databases have been explored both (i) at the conceptual level, considering and extending the Entity-Relationship data model and the object-oriented data models and the related query languages, and (ii) at the logical level, studying temporal relational and object-oriented data models and languages, and other theoretical topics, such as temporal functional dependencies, and (iii) at the physical level, with the design of access methods to temporal data.

To confirm the growth of research efforts in the area of temporal databases, we cite here two efforts of some groups in the temporal database community. The first one is related to the definition and update of a "consensus glossary", where basic concepts and related terms adopted in the temporal databases community are described in some detail [184] . This effort started in 1993 with a workshop on temporal databases promoted by Rick Snodgrass. This enabled the research community to build its work on a more sound ground, where concepts and terminology are shared among people and a deep discussion is stimulated on the different topics.

The second effort worth mentioning deals with the definition of a temporal query language based on the widely used relational query language SQL (Structured Query Language) [387]. TSQL2 was proposed in 1995 [387] and is based on a joint effort of several people from the temporal database community. The goal behind the design of TSQL2 is the development of a specification for a widely acknowledged extension to SQL-92, the standard of the SQL query language [260].

The cited efforts, together with the fast increase of works dealing with time-related topics in the area of databases, confirm the feeling that time is a basic topic that has to be faced both from a theoretical and a methodological point of view as well as from the application-oriented perspective. In this chapter, we restrict ourselves to the basic concepts and the main research directions in temporal databases, which have an important role for the modeling and implementation of temporal information systems for medicine. From a wide perspective we can say that all the work in the temporal databases area can be important for medical information systems: nevertheless, in this chapter, after an introduction to the basic features of temporal databases, we focus on concepts, issues, and solutions proposed for the conceptual and logical modeling and querying of temporal information, basically considering widely known models, as the relational model, the Entity-Relationship model, the object-oriented ones, and the related languages. Further important topics as that of modeling and managing multimedia temporal data and semistructured temporal data are briefly discussed at the end of the chapter. We decided to not consider other specific advanced topics as access methods to temporal data, temporal deductive databases, active temporal databases and so on. The reason for this choice is twofold: first, we prefer to enter in some detail on the basic and more important concepts in this field, thus trying to avoid the risk of saying very few and superficial words on several things; second, we want to discuss to some extent those arguments, which we perceive to be closer to people working on temporal information systems in healthcare, usually involved in the database design at the conceptual and logical level.

3.2 Valid and Transaction Times

Two basic, orthogonal, temporal dimensions have been recognized in the temporal database community: the first one, called *transaction time*, refers to the moments at which a fact is either entered into or deleted from the database; the second one, called *valid time*, is related to the time at which the fact is/was true in the modeled reality. Valid and transaction times are widely recognized as the two basic temporal dimensions of temporal databases. Furthermore, there exists a consolidated terminology about temporal databases dealing with valid and/or transaction times, as witnessed by the Consensus Glossary of Temporal Database Concepts [184]. Valid and transaction times are defined as follows.

Definition 3.1. Valid time. The *valid time* (VT) of a fact is the time when the fact is true in the modeled reality.

Definition 3.2. Transaction time. The *transaction time* (TT) of a fact is the time when the fact is current in the database and may be retrieved.

Valid time is usually provided by the database users, while transaction time is system-generated and supplied.

To explain the meaning of these two basic temporal dimensions, we consider a simple example, taken from a general clinical scenario.

Example 3.1. On August 10, 1997, 3:00 p.m., 3:30 p.m., and 4:00 p.m., respectively, three patients convey to the physician their symptom history and the physician enters immediately this data into the database: Mr. Rossi had palpitation symptoms from July 19, 8:30 p.m. to July 20, 1997, 7 a.m., headache from July 1, 8:45 p.m. to July 10, 1997, 1 p.m. and nausea on August 10, 1997, from 10 a.m. to 2 p.m.; Mr. Smith had nausea on June 16, 1997 from 5:30 to 7:45 p.m., and headache on July 5, 1997, from 9:00 a.m. to 5:30 p.m.; Mr. Hubbard had lower extremity edema from May 19, 1997, 5 p.m. to May 21, 9 a.m. By mistake, the physician enters only "edema" instead of "lower extremity edema" for the symptom of Mr. Hubbard. On October 23, 1997, at 4 p.m., Mr. Hubbard goes back to the physician and tells him that he has to declare other symptoms he previously didn't considered as important: chest pain on June 12, 1997 from 8:30 a.m. to 9:15 a.m., and chest and arm discomfort from June 27, 1997 8 a.m. to June 29, 1997 10 p.m. Moreover, the physician corrects the data about the previous symptoms the patient talked about, i.e. the lower extremity edema.

Table 3.1 Database of patients' symptoms.

Patient	Symptom	VT	TT
Rossi	palpitation	[97Jul19;8:30, 97Jul20;7:00]	[97Aug10;15:00, ∞)
Rossi	headache	[97Jul1;8:45, 97Jul10;13:00]	[97Aug10;15:00, ∞)
Rossi	nausea	[97Aug10;10:00, 97Aug10;14:00]	[97Aug10;15:00, ∞)
Smith	nausea	[97Jun16;17:30, 97Jun16;19:45]	[97Aug10;15:30, ∞)
Smith	headache	[97Jul5;9:00, 97Jul5;17:30]	[97Aug10;15:30, ∞)
Hubbard	edema	[97May19;17:00, 97May21;9:00]	[97Aug10;16:00, 97Oct23;16:00)
Hubbard	chest pain	[97Jun12;8:30, 97Jun12;9:15]	[97Oct23;16:00, ∞)
Hubbard	chest and arm discomfort	[97Jun27;8:00, 97Jun29;22:00]	[97Oct23;16:00, ∞)
Hubbard	lower extremity edema	[97May19;17:00, 97May21;9:00]	[97Oct23;16:00, ∞)

In this example, we are able to clearly distinguish two different kinds of time: the first one is related to the moments at which the physician enters data into the database, just after the talks with the patients. The *transaction time* models this temporal dimension. The second kind of time is related to when the patients suffered from the reported symptoms. The *valid time* models this second temporal dimension. It is worth noting that these two kinds of time are orthogonal: the physician can enter data about symptoms when they are still ongoing, just after the end of the symptoms as well as at a time point some days (or months) after the end of symptoms. Furthermore, the chronological order of the insertion of symptom data into the database is independent of the chronological order of the appearance of the symptoms themselves.

Without assuming any data model, we can say in general that, in the physician's database, the piece of information related to a patient symptom (i.e., the considered fact) must be timestamped by both the valid and the transaction times [187]. In the following sections we discuss how to apply timestamping in the different data models (e.g., in the relational model it is possible to timestamp either attributes or tuples).

Let us assume that, in representing this data, we exploit the relational data model: without loss of generality, we will timestamp tuples. With these assumptions, the example can be modeled as in Table 3.1. Both valid and transaction times are represented by intervals, in columns VT and TT, respectively. The lower bound of the transaction time is the time at which the data item is entered into the database, while the upper bound is the time (if any) at which the data item is logically deleted from the database. The occurrence of the special symbol ∞ as the ending point of an interval means that the interval is still ongoing (the ending point of such an interval is also denoted by NOW or u.c., for until changed, in the literature).

Besides transaction and valid times, the concept of *user-defined time* is widely used in the literature to mean an attribute having time as domain, whose semantics is completely unknown to the system: for example, the birth date of a patient or the date of the first hospital admission of a patient could be two user-defined times. User-defined time is not supported by the (temporal) database system.

3.3 A Taxonomy for Temporal Databases

On the basis of the distinction between valid and transaction times, a widely accepted taxonomy has been proposed in the literature in the past years [187, 184]. According to this taxonomy we can distinguish different categories of databases (and of database systems) with respect to their support of the temporal dimensions of data:

- *Snapshot databases.* These databases represent only the current state of the modeled world: the only way the user can represent time-oriented data is to use user-defined times to model explicitly the temporal dimensions of data. The system does not support any time semantics, but some time domain. Figure 3.1 provides the representation of the snapshot database related to Example 3.1, without considering any temporal dimension.

- *Valid-time databases.* These databases, also known as *historical databases*, support only the valid time, i.e., the history of database updates is not managed (i.e., chronology of data insertions and deletions); only the temporal dimensions of the modeled facts is considered. By this kind of databases, for example, we can represent only the intervals during which the patients suffered from the given symptoms (see Example 3.1), while it is not possible to keep a trace of when the physician entered into and/or deleted data from the database. Figure 3.2 depicts the (logical) valid-time database corresponding to Example 3.1: a flat table cannot represent a valid-time database anymore; a three-dimensional "cube" suitably represents the valid-time database. The third dimension allows us to define the validity of each row of the table. In the figure, the parts in gray on the lateral face of the cube highlight the periods during which the facts represented by the corresponding rows are valid; the parts in gray on the upper face of the cube highlight the periods when (some) facts represented into the database are true in the modeled world.

- *Transaction-time databases.* These databases are also known as *rollback databases* in the literature. In this kind of databases, only the transaction time is supported: we can keep trace of the update operations performed on the database, but we are not able to represent the temporal dimensions of the represented facts. Considering Example 3.1, transaction-time databases allow us to represent only the times at which the physician enters and/or deletes patients' data: the system is not able to support the representation of the intervals during which the symptoms were present. Figure 3.3 depicts the logical representation of the transaction-time database related to Example 3.1. All the states of the database are (logically) stored: the current and the previous ones.

- *Bitemporal databases.* These databases support both transaction and valid time: in our example (see Example 3.1), the system is able to support both the time during which the symptoms were present and the time at which symptom data was entered and/or deleted into the database. Figure 3.3 provides the logical representation of the bitemporal database related to Example 3.1: with respect to the previous transaction-time database, each state of the database is now a "cube", as for valid-time databases. Another graphical representation of bitemporal (conceptual) databases has been proposed by Jensen and Snodgrass [186]: facts are represented as boxes on a cartesian plane where x- and y- axes correspond to valid and transaction times. Figure 3.5 shows the first row related to the patient Hubbard in Table 3.1 and its update.

In the currently used terminology, the general expression *temporal database* stands for a database supporting some kind of temporal dimension [184]. It is worth noting that the graphical representation of the database content in a table, like that provided in Table 3.1, is not related to a specific temporal database system: in other words, we can establish whether the considered database is snapshot, transaction-time, valid-time, or bitemporal, only when we explicitly identify which temporal dimensions the system is able to support: for example, if the attributes VT and TT are managed explicitly by the user, we have a snapshot database, while if both VT and TT are system supported, we have a bitemporal database.

Patient	Symptom
Rossi	palpitation
Rossi	headache
Rossi	nausea
Smith	nausea
Smith	headache
Hubbard	chest pain
Hubbard	chest and arm discomfort
Hubbard	lower extremity edema

Fig. 3.1 Logical representation of a snapshot database of patients' symptoms.

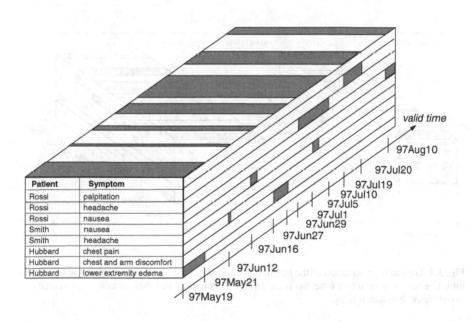

Fig. 3.2 Logical representation of the valid-time database of patients' symptoms.

Depending on the modeled application domain, valid and transaction times can be related to each other in different ways. In [185], Jensen and Snodgrass propose several classifications (*specializations*, in the authors' terminology) of bitemporal

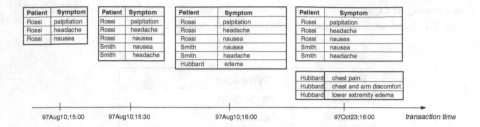

Fig. 3.3 Logical representation of the transaction-time database of patients' symptoms.

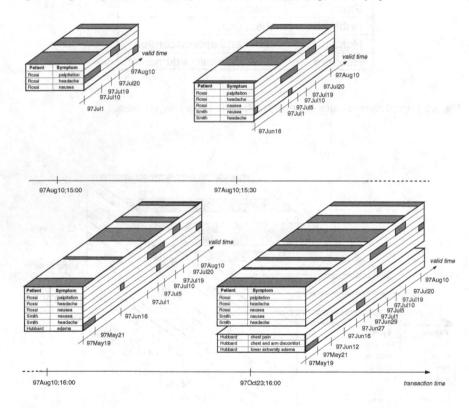

Fig. 3.4 Logical representation of the bitemporal database of patients' symptoms. In the figure the time line for the transaction time has been divided in two parts and then suitably represented on two different horizontal lines.

relations (even though the relational model is used, these classifications are independent from the adopted data model), based on the relationships between valid and transaction times of timestamped facts. They take into account both the relationships that exist between valid and transaction times of each single tuple of a relation and the relationships that exist between the valid and/or transaction times of

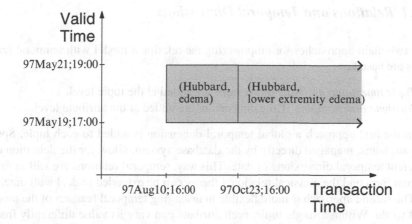

Fig. 3.5 Graphical representation of tuples according to the bitemporal conceptual model [186].

different tuples of the relation. As an example, on the basis of the relationships that exist between the starting and/or ending points of the valid and transaction times of each single tuple of a relation, Jensen and Snodgrass characterize classes of relations, as retroactive, delayed retroactive, predictive, early predictive, retroactively bounded, strongly retroactively bounded, delayed strongly retroactively bounded, general, degenerate, and so on [185]. For instance, given the starting point VT_s of the valid time and the starting point TT_s of the transaction time, a relation is *retroactive* if, for each tuple of the relation, the relationship $VT_s \leq TT_s$ holds, while a relation is *delayed retroactive with bound $\Delta t \geq 0$* if, for each tuple of the relation, the relationship $VT_s \leq TT_s - \Delta t$ holds. According to such a criterion, the relation of Table 3.1 is *retroactive*, because $VT_s < TT_s$ for every tuple, i.e., the physician enters symptom data after the appearance of the symptoms themselves.

3.4 Extending the Relational Model with Time

Different proposals deal with adding valid and/or transaction times to relational databases. In this section we focus first on the two main approaches for empowering the relational data model with temporal dimensions, and then we describe the main features of relational algebrae and calculi proposed for temporal relational databases. The last part of this section provides the main features of query languages proposed for temporal relational databases.

3.4.1 Relations and Temporal Dimensions

The two main approaches for empowering the relational model with temporal features are based on:

- *Tuple timestamping*. Time dimensions are added at the tuple level.
- *Attribute timestamping*. Time dimensions are added at the attribute level.

In the first approach a global temporal dimension is added to each tuple. Special attributes, managed directly by the database system, allow for the definition of different temporal dimensions of data. This way, temporal relations are still in first normal form and the classical relational theory can be extended to deal with time.

The second approach is more flexible in managing temporal features of the modeled world. Within a single tuple, each attribute can vary its value differently from the other attributes. While it is more natural to model time-varying real-world entities, the resulting tuples are not in first normal form. Attribute domains are complex and are composed of couples of values and timestamps. This way, the classical (flat) relational theory cannot be directly used to deal with such kind of data.

3.4.1.1 Tuple Timestamping

Tuple timestamping means that the relation schema has some special attributes, which add (several) temporal dimensions to each tuple of any relation instance of the relation schema. Different choices can be made about the domain of temporal dimensions. The temporal domain can be composed of:

- time points (chronons);
- time intervals;
- temporal elements.

Let us consider the following example related to a hospitalization database; we will show how to timestamp tuples of this database according to the above different temporal domains for valid and transaction times (hereinafter, for simplicity we will use simple integers as time units instead of dates).

Example 3.2. To manage the quality of care at the hospital level, the head of the hospital considers the history of patients' hospitalizations, as maintained in a suitable bitemporal database. At time 4, the administrative operator enters into the database the fact the Mr. Hubbard is hospitalized in the Intensive Care Unit from time 4 to time 9; at time 5, he enters the fact that Mr. Rossi was in the Intensive Care Unit from time 2 to time 3. At time 8, new information is added into the database: Mr. Rossi was in the Cardiology ward from time 1 to time 1 and now he is in the Cardiology ward since time 4; Mr. Smith is in the ward of Internal Medicine from time 5 to time 9 and then he will move to the Neurology ward, where he will stay from time 9 up to a not yet defined time. At time 10, the operator enters the (wrong) fact that Mr. Hubbard has been hospitalized in the Cardiology ward since time 10; at time 12 the operator corrects the previous error and enters the correct fact that Mr. Hubbard has been hospitalized in the Pneumology ward since time 10; at time 14, the operator is informed that Mr. Hubbard left the Pneumology ward at time 12, and consequently he corrects data into the database.

Point-based timestamping. The relation depicted in Table 3.2 represents the history of patient hospitalizations given in Example 3.2. Tuples are timestamped (both for transaction and valid times) with time points. When time points are used for tuple timestamping, several assumptions have to be made on the database content and on the modeled world. In our case, for example, we have to assume that a patient is only in one ward at a given (valid) time point. At time point 10, Mr. Hubbard is hospitalized in the Pneumology ward and the erroneous information that Mr. Hubbard was in the Cardiology ward has been implicitly (logically) deleted. Moreover, we have to assume that the hospitalization in a ward stops when the patient is hospitalized in another ward. This way, when we have to represent that a patient finished any hospitalization we need to add a tuple containing the name of the patient, the value null (or similar) for the ward name, and the suitable times, as depicted in Table 3.2 for patient Hubbard. More generally, the following assumptions have to be made:

- a tuple with a new (more recent) valid time ends the validity of the tuple (if present) with the same key (in our case, the patient name) and the closest preceding valid time;
- the insertion of a new tuple makes no more current the possibly existing tuple with the same key and the same valid time.

Representing valid and transaction times with time points allows the description of simple real world situations. More particularly, it is suitable for representing the history of attribute values (e.g., the hospital ward) of a key (e.g., the patient name). This kind of representation can be useful for clinical time series as, for example, the temperature of a patient during a hospitalization or the monitoring of vital signs in intensive care units, where the assumption is that a measurement is significant for a parameter (diastolic, systolic blood pressures, heart rate, and so on) until a new measurement is performed.

Interval-based timestamping. On the other hand, considering the timestamping by intervals, it is possible to model more complex situations, where we need to explicitly represent deletions, multiple values, and isolated valid intervals. Table 3.3 represents the already considered relation for hospitalizations using intervals instead of points for timestamping tuples.

The relation depicted in Table 3.3 highlights how the interval-based timestamping improves data expressiveness: current tuples (identified by symbol ∞ in the value of transaction time) can be directly observed, while in the previous situation the identification of the current tuples required the examination of the whole relation. Tuples representing still valid facts (identified by the symbol ∞ in the value of valid time) can be directly observed, while with the point-based timestamping the whole relation had to be scanned. Null values for ending the validity of a tuple

Table 3.2 Database of patients' hospitalizations; timestamping with time points.

Patient	Ward	VT	TT
Rossi	Cardiology	1	8
Rossi	Cardiology	4	8
Rossi	Intensive Care Unit	2	5
Smith	Internal Medicine	5	8
Smith	Neurology	10	8
Hubbard	Intensive Care Unit	4	4
Hubbard	Cardiology	10	10
Hubbard	Pneumology	10	12
Hubbard	null	12	14

Table 3.3 Database of patients' hospitalizations; timestamping with time intervals.

Patient	Ward	VT	TT
Rossi	Cardiology	[1, 1]	[8, +∞]
Rossi	Cardiology	[4, +∞]	[8, +∞]
Rossi	Intensive Care Unit	[2, 3]	[5, +∞]
Smith	Internal Medicine	[5, 9]	[8, +∞]
Smith	Neurology	[10, +∞]	[8, +∞]
Hubbard	Intensive Care Unit	[4, 9]	[4, +∞]
Hubbard	Cardiology	[10, +∞]	[10, 11]
Hubbard	Pneumology	[10, +∞]	[12, 13]
Hubbard	Pneumology	[10, 12]	[14, +∞]

are avoided. The improvement in data expressiveness is counterparted by the more
complex structure of the relation schema: time-related attributes are now intervals,
which can contain the special symbol ∞ for still current intervals.

Table 3.4 Database of patients' hospitalizations; timestamping with time elements.

Patient	Ward	VT	TT
Rossi	Cardiology	[1, 1] ∪ [4, +∞]	[8, +∞]
Rossi	Intensive Care Unit	[2, 3]	[5, +∞]
Smith	Internal Medicine	[5, 9]	[8, +∞]
Smith	Neurology	[10, +∞]	[8, +∞]
Hubbard	Intensive Care Unit	[4, 9]	[4, +∞]
Hubbard	Cardiology	[10, +∞]	[10, 11]
Hubbard	Pneumology	[10, +∞]	[12, 13]
Hubbard	Pneumology	[10, 12]	[14, +∞]

Timestamping with temporal elements. The adoption of temporal elements is
the last step in improving data expressiveness in tuple timestamping. In this case,
we are allowed to use finite sets of disjoint intervals for adding temporal dimensions

to tuples. Table 3.4, for example, represents the hospitalizations of Example 3.2, where the valid time is modeled by temporal elements. In this case, the history of hospitalizations of a given patient in a given ward (e.g., Mr. Rossi in the Cardiology ward) can be collected in a single tuple.

3.4.1.2 Attribute Timestamping

This approach allows one to separately represent the valid time of each single attribute within a tuple. The value of any attribute is thus complex and is composed of an atemporal part (e.g., the value of a lab exam) and a time dimension, which can be represented in a similar way to what we described above for tuple timestamping, i.e., by single time points, time intervals, or time elements. Depending on the adopted approach, some temporal data models distinguish two different categories of attributes, time-varying and constant. In other proposals, all the attributes have a temporal dimension. In this case, even though all the attributes have their proper temporal dimension, usually each tuple is *homogeneous*, i.e. the time span over which an attribute has (different) value(s) within a tuple is the same for each attribute of the considered tuple. Table 3.5 depicts how the information of Example 3.2 is represented with attribute timestamping on valid and transaction times. In this case, we are using bitemporal elements for timestamping attributes. Each timestamp has the format $\langle valid\_time, transaction\_time \rangle$, where $valid\_time$ is a temporal element (expressed as union of intervals) and $transaction\_time$ is a time interval.

Table 3.5 Database instance of patients' hospitalizations; attribute timestamping with (bi-dimensional) temporal elements.

Patient	Ward
Rossi ⟨[1, +∞], [5, +∞] ⟩	Cardiology ⟨[1, 1] ∪ [4, +∞], [8, +∞] ⟩
	Intensive Care Unit ⟨[2, 3], [5, +∞]⟩
Smith ⟨[5, +∞], [8, +∞]⟩	Internal Medicine ⟨[5, 9], [8, +∞]⟩
	Neurology ⟨[10, +∞], [8, +∞]⟩
Hubbard ⟨[4, 12], [4, +∞]⟩	Intensive Care Unit ⟨[4, 9], [4, +∞]⟩
	Cardiology ⟨[10, +∞], [10, 11]⟩
	Pneumology ⟨[10, +∞], [12, 13]⟩
	Pneumology ⟨[10, 12], [14, +∞]⟩

3.4.2 Temporal Algebras and Calculi

According to the temporal relational data models, which have been proposed in the literature, several extensions to relational algebra and to relational calculus have been advocated. Without giving a detailed description of any temporal algebra/calculus, we briefly describe here the main features of the proposed temporal algebras and calculi.

In extending the relational algebra to deal with temporal data, we can identify two different steps: (i) the first one deals with extending the standard (atemporal) relational operators to manage tuples having temporal dimensions; (ii) the second one is related to the introduction of new specific operators for managing explicitly the temporality of tuples.

In the first step, a given temporal algebra has to define the semantics of the usual relational operators (union, intersection, difference, projection, selection, join, cartesian product). For example, let us assume that we have to define the union operation in a temporal algebra based on tuple-timestamping. Informally, we could say that the union of two temporal relations is composed of the tuples of the first relation and those of the second relation. What happens for those tuples which have the same values for the atemporal attributes, but different timestamps in the two relations? The answer depends, obviously, on the kind of timestamping used by the temporal database system: if the timestamping is provided through temporal elements, the natural choice is to compose a single tuple with the same values of atemporal attributes and with the union of the timestamps of the two tuples as its timestamp. If timestamping is through intervals, we have two distinct tuples in the result in case the two timestamps are not intersecting; when the two timestamps of the operands intersect, the usual choice in temporal database systems is to perform the *coalescing* operation, which consists in timestamping a single tuple by an interval, which contains all the time (sub)intervals which are contained at least in one of the two given timestamps.

Let us consider now the join operation (on atemporal attributes). In this case, there are several problems to solve. First, we should decide whether timestamps have to be considered in joining tuples. A solution could be that we only join tuples with intersecting timestamps. Another solution could allow joining of tuples with disjoint timestamps. Second, we have to decide the result of a join on two temporal relations. The implicit requirement (due to the fact that we are defining an algebra) is that the result is a temporal relation. The schema of the result should be composed of a timestamp and of the set of (atemporal) attributes which are in the two relations. The tuples of the result relation are composed of the (atemporal) attribute values of those tuples which satisfy the join condition. A further decision is on the timestamp value of such tuples. The timestamp could be defined either by default (e.g., the timestamp of the tuple of the first/second operand; the intersection of timestamps of the two tuples) or by the user, through some suitable parameters specified in the join operation.

Regarding the second step, several new operators have been proposed in the literature, to deal explicitly with some temporal aspects of data. Among them, we focus here on the operators *timeslice* and *coalesce*. The *timeslice* operator allows one to consider only a part of the database content, on the basis of the temporal aspects. More precisely, with the *timeslice* operator we can perform a temporal selection, by considering only those tuples which have the temporal dimension (usually, the valid time) contained in the time interval specified with the operator.

Let us consider the relation *Pat_Hosp*, depicted in Table 3.3 and related to Example 3.2. We could be interested in considering only those patients, who were hospitalized in the period [4, 10]. According to the standard relational algebra, we can define the following selection, returning the tuples depicted in Table 3.6:

$$Res\_Pat\_Table \leftarrow \sigma_{(4 \leq begin(VT) \leq 10) \lor (4 \leq end(VT) \leq 10)} Pat\_Hosp$$

Although the adoption of operators of standard relational algebra is always possible, the explicit definition of complex selection conditions can be awkward and not convenient for defining complex temporal conditions. The *timeslice* operation, expressed by the symbol τ, can be suitably defined as:

$$\tau_t(R) \equiv_{def} \sigma_{(begin(t) \leq begin(VT) \leq end(t)) \lor (begin(t) \leq end(VT) \leq end(t))}(R)$$

Table 3.6 Database after a valid-time timeslice.

Patient	Ward	VT	TT
Rossi	Cardiology	[4, +∞]	[8, +∞]
Smith	Internal Medicine	[5, 9]	[8, +∞]
Smith	Neurology	[10, +∞]	[8, +∞]
Hubbard	Intensive Care Unit	[4, 9]	[4, +∞]
Hubbard	Cardiology	[10, +∞]	[10, 11]
Hubbard	Pneumology	[10, +∞]	[12, 13]
Hubbard	Pneumology	[10, 12]	[14, +∞]

If needed, transaction time can also be considered in timeslice operation. In this case, the operation is often referred to as *rollback*. The *rollback* operation is usually performed with respect to a time point rather than a time interval. It allows one to represent the state of the database at a specified time point.

For the *rollback* operation, usually the symbol ρ is adopted:

$$\rho_t(R) \equiv_{def} \sigma_{(begin(TT) \leq t \leq end(TT))}(R)$$

Let us consider, for example, the need of knowing which data was stored into the database at time 9: the result of the operation $Res\_Pat\_Table \leftarrow \rho_9 Pat\_Hosp$ is depicted in Table 3.7.

Table 3.7 Database after a transaction-time timeslice.

Patient	Ward	VT	TT
Rossi	Cardiology	[1, 1]	[8, +∞]
Rossi	Cardiology	[4, +∞]	[8, +∞]
Rossi	Intensive Care Unit	[2, 3]	[5, +∞]
Smith	Internal Medicine	[5, 9]	[8, +∞]
Smith	Neurology	[10, +∞]	[8, +∞]
Hubbard	Intensive Care Unit	[4, 9]	[4, +∞]

Temporal relational calculi extend the standard relational calculus to deal with temporal aspects of tuples. The main difference between an algebra and a calculus is that, while the algebra is a *procedural* language, i.e. it allows the user to suitably perform operations in order to reach the desired result, the calculus is a *declarative* language, i.e. it allows the user to describe the features of the desired result, without specifying how to reach it. Similarly to what has been described for temporal algebras, a generic temporal calculus extends the classic (atemporal) relational calculus both by introducing specific operators and functions for the temporal domain, and by considering the temporal dimension of the result.

Let us consider, for example, the equivalent expression to the valid-time timeslice operation $\tau_{[4,10]} Pat\_Hosp$:

$$\{r \mid (Pat\_Hosp(r) \wedge \exists t(t \in r[VT] \wedge 4 \le t \le 10)\}$$

In this simple expression, the temporal dimension of the result is simply the temporal dimension of each tuple in relation *Pat_Hosp* that satisfies the logical condition. The temporal condition on valid time is expressed by considering the valid time interval as a set of time points. Note that in the standard relational model it is not possible to consider non-atomic attributes such as those representing the valid and transaction times.

3.5 Temporal Relational Query Languages

The management of temporal dimensions has an impact on many aspects of query languages, ranging from the definition of statements of the Data Definition Language (DDL), which allow one to express constraints among the different temporal dimensions, to the definition of statements of the Data Manipulation Language (DML), supporting both the specification of temporal search conditions and the definition of the temporal dimensions of the query result.

Without loss of generality, we start with the standard SELECT command of SQL and introduce, step by step, the specific extensions/modifications needed to deal with temporal dimensions. Let a temporal database be defined as a set of *temporal* relations, i.e. relations with both valid and transaction times.

In order to allow comparisons between temporal dimensions, two choices are possible. The first choice is to deal with temporal and non-temporal data in the WHERE clause [387]. The second choice is to constrain purely temporal conditions to appear in a specific clause WHEN [279][1]. In case we opt for the second alternative (this allows one to exploit specific tools to manage temporal data at both the logical and physical level), another design issue is the specification of the temporal dimensions we can refer to in the WHEN clause. The most general option is to allow both the temporal dimensions to be used in the WHEN clause.

By default, a query is evaluated with respect to the current state of the database, i.e., on the tuples whose transaction time is a special interval having the (not yet

---

[1] Obviously, to make it possible to check conditions mixing temporal and non-temporal data, we must allow temporal dimensions to occur, whenever necessary, in the WHERE clause.

defined) right bound containing the current time (usually identified by the value $+\infty$ or *uc*, for *until changed*). Obviously, there must be the possibility of evaluating the query over both the current and the past states of the database and/or of the information system. This requirement can be managed by adding suitable qualifiers to the FROM clause. If we are interested in querying the database about its past states, we can use the well-known AS OF clause, which has been added to many temporal query languages proposed in the literature (cf. [387]).

Example 3.3. Let us consider a scenario where the physician has to pair data about patient symptoms, given in Table 3.8, with data about prescribed therapies, reported in Table 3.9. The attribute *P_id* in both tables contains a code identifying each single patient.

Table 3.8 Database of patient symptoms: the relation *pat_sympt*.

P_id	symptom	VT	TT
1	headache	[97Oct1, ∞]	[97Oct10, 97Oct14]
2	vertigo	[97Aug8, 97Aug15]	[97Oct15, 97Oct20]
2	vertigo	[97Aug10, 97Aug15]	[97Oct21, ∞]
1	headache	[97Oct1, 97Oct14]	[97Oct15, 97Oct20]
1	headache	[97Oct1, 97Oct14]	[97Oct21, ∞]

Table 3.9 Patient therapy database: the relation *pat_ther*.

P_id	therapy	VT	TT
1	aspirin	[96Oct1, 96Oct20]	[97Oct10, ∞]
2	paracetamol	[97Aug11, 97Aug12]	[97Oct15, ∞]

Let us consider, for example, the following query on the relation depicted in Table 3.8

```
SELECT *
FROM pat_sympt S
AS OF 97Oct20
```

which would return the contents of the relation *pat_sympt* as of October 20, 1997. The result of the query is reported in Table 3.10.

Further issues arise when more than one relation comes into play in a query. As an example, let us assign a proper meaning to the cartesian product of the relations that appear in the FROM clause. A possible choice is to impose that temporal tuples of different relations can be joined only when both their transaction times and their valid times overlap (*sequenced* semantics, as detailed in the following). As for the

Table 3.10 Tuples of the relation *pat_sympt* (Table 3.8) as of October 20, 1997.

P_id	symptom	VT	TT
2	vertigo	[97Aug8, 97Aug15]	[97Oct15, 97Oct20]
1	headache	[97Oct1, 97Oct14]	[97Oct15, 97Oct20]

result of the cartesian product, the language should allow the user to choose among different ways of executing the cartesian product between temporal relations (e.g., one can opt for the usual cartesian product of relations, where the temporal attributes are treated as the standard ones). Furthermore, besides the standard (atemporal) join, the language should support some forms of temporal join, where the join condition allows the user to join tuples on the basis of different temporal features.

Finally, when several temporal relations are involved in a query, the language must provide some mechanisms to allow the user to obtain a consistent result, that is, a temporal relation endowed with valid and transaction times. While the transaction time of the resulting relation can be obtained (according to the standard definition of the cartesian product given above) as the intersection of the transaction times of the considered tuples, valid time must be explicitly defined by the user, even though the system can provide some default rules.

Example 3.4. On the basis of Tables 3.8 and 3.9, the physician wants to determine the symptoms of patients for which the valid time intervals of symptoms and therapies overlap.

The query of Example 3.4 can be formulated as follows:

```
SELECT symptom, S.P_id
FROM pat_sympt S, pat_ther T
WHEN NOT(VALID(S) BEFORE VALID(T)) AND
     NOT(VALID(T) BEFORE VALID(S))
WHERE S.P_id = T.P_id
```

where BEFORE(·) is the well-known relation of Allen's Interval Algebra [10].

The WHEN clause can actually be replaced by a suitable temporal join in the FROM clause, as shown in the following equivalent formulation of the query:

```
SELECT symptom, S.P_id
FROM pat_sympt S TJOIN pat_ther T ON
     NOT(VALID(S) BEFORE VALID(T)) AND
     NOT(VALID(T) BEFORE VALID(S))
WHERE S.P_id = T.P_id
```

In order to properly evaluate the query, some default criteria must be specified to determine the values of the valid times of the resulting relation, whenever the user does not provide any explicit rule. As an example, the valid time of each tuple belonging to the resulting relation can be defined as the intersection of the valid times of the corresponding tuples belonging to the relations that occur in the FROM clause.

According to such a criterion, the result of the query of Example 3.4 is the relation of Table 3.11.

Table 3.11 Patient symptoms that overlap therapies.

P_id	symptom	VT	TT
2	vertigo	[97Aug11, 97Aug12]	[97Oct21, ∞]

However, we expect that, in most cases, the user will explicitly define the way in which the valid times of the resulting relation must be computed, taking into account the meaning he/she assigns to the query and the relative data. As an example, in the case of Example 3.4, the user can formulate the following query:

```
SELECT symptom, S.P_id WITH VALID(S) AS VALID
FROM pat_sympt S TJOIN pat_ther T ON
    NOT(VALID(S) BEFORE VALID(T)) AND
    NOT(VALID(T) BEFORE VALID(S))
WHERE S.P_id = T.P_id
```

In such a way, the user assigns to the valid time of the resulting relation the values that this attribute has in the *pat_sympt* relation. Indeed, the valid time of the tuple belonging to Table 3.12 is the time interval during which the patient with *P_id* equal to 2 suffered from vertigo episodes, while the valid time of the tuple belonging to Table 3.11 represents the time interval during which both vertigo episodes and therapeutic actions occurred.

Table 3.12 Patient symptoms that overlap therapies (revisited).

P_id	symptom	VT	TT
2	vertigo	[97Aug10, 97Aug15]	[97Oct21, ∞]

Further specific clauses have already been proposed in the literature to explicitly support valid and transaction times in bitemporal databases [279, 387], such as TIME SLICE and MOVING WINDOW.

Among the several proposals dealing with temporal query languages, TSQL2 is the result of the research efforts of several people from the temporal database community, which were grouped in a language definition committee chaired by Richard Snodgrass, one of the pioneers in temporal database research [387]. TSQL2 is an extension of the widely accepted standard SQL-92 [260].

As regards the above mentioned choices for temporal query languages, TSQL2 does not add any clause similar to WHEN. Selection conditions involving valid times are defined within the WHERE clause. The same thing happens for selections involving transaction times. Valid and transaction times of tuples can be referred to by the VALID(·) and TRANSACTION(·) constructs, respectively.

For example, let us consider the query of Example 3.4. Its expression in accordance with the TSQL2 syntax would be:

```
SELECT symptom S.P_id
VALID VALID(S)
FROM pat_sympt S pat_ther T
WHERE S.P_id = T.P_id AND VALID(S) INTERSECT VALID(T)
```

TSQL2 allows one to specify the kind of table required as a result of the query. For example, while the default result of a query on bitemporal relations is a bitemporal relation, the keyword SNAPSHOT in the SELECT clause allows one to have an atemporal relation as query result.

More generally, TSQL2 allows the specification of several kinds of relations:

- snapshot relation, without temporal dimensions;
- valid-time state relations, i.e. relations having valid times expressed with intervals;
- valid-time event relations, i.e. relations having instantaneous valid times;
- transaction-time relations, i.e. relations having tuples timestamped by the transaction time;
- bitemporal state relations, i.e. relations timestamped by both the transaction time and the (interval) valid time;
- bitemporal event relations, i.e. relations timestamped by both the transaction time and the (instantaneous) valid time.

As an example, the definition of the relation depicted in Table 3.4, according to the TSQL2 syntax, would be:

```
CREATE TABLE Hospitalization (Patient CHAR(30),
Ward CHAR(20))
AS VALID STATE DAY AND TRANSACTION
```

where, besides the part of the standard SQL statement, the construct AS VALID is used to specify the kind of relation the system has to manage. In our example, we defined a bitemporal state relation. The granularity for valid times is that of days, while the granularity for transaction time is system-dependent and therefore is not specified in the statement CREATE TABLE.

Temporal aspects have also been considered in the development of the most recent ISO standard version of SQL. A new component, called SQL/Temporal, was formally approved in July 1995 as part of the SQL3 draft standard [133] and then withdrawn [173]. As a matter of fact, most commercial relational database systems provide some extensions to the relational model to represent times and dates through suitable data types and to the SQL query language to allow manipulation of stored time values, but these specifications are not yet completely standardized across vendors.

The SQL3 SQL/Temporal component temporally extends the basic relational model, e.g., by adding the data type PERIOD and the two temporal dimensions VT and TT, and it supports temporal queries [386]. SQL3 is fully compatible with the previous versions of SQL, so that migration from these atemporal versions is simple and efficient, and existing code can be reused without any additional intervention.

SQL3 has also been designed to manage temporalities, by providing suitable semantics for the query. Indeed, temporal queries can be specified according to two different semantics, namely, sequenced and non-sequenced semantics, that explicitly deal with the above mentioned issues in evaluating a temporal query.

According to the *sequenced semantics*, an SQL statement is executed over a given temporal dimension instant by instant, that is, it takes into consideration one time instant (a state of a relation) at a time. As an example, a primary key constraint with a sequenced semantics checks that, at any time instant, there are no tuples with null values on the key attributes and there are no pairs of tuples with the same values on the key attributes. The *sequenced semantics* allows one to reuse existing SQL code by performing a query at any time instant of the whole temporal axis.

The *non-sequenced semantics* is more complex than the sequenced one. According to it, an SQL statement accesses a temporal relation as a whole. While sequenced semantics provides suitable tools that facilitate the access to data, non-sequenced semantics manipulates all pieces of temporal information in the database in a single step. In the general case, the outcome of a non-sequenced query is an atemporal relation. However, the user can possibly define the temporal dimension(s) of the resulting relation, thus obtaining a temporal relation.

The non-sequenced semantics for the VT and TT temporal dimensions is imposed by the token NONSEQUENCED followed by the specification of the considered temporal dimensions, i.e., VALIDTIME and TRANSACTIONTIME, respectively.

As for compatibility requirements, we must distinguish between upward compatibility and temporal upward compatibility. *Upward compatibility* constrains any new SQL standard to be syntactically and semantically compatible with the existing one [18]. According to this requirement, SQL3 manages atemporal relations, both syntactically and semantically, as SQL does[2]. *Temporal upward compatibility* constrains any code available for an atemporal relation to produce the same result if executed over the temporal extension of that relation. SQL3 enables one to add one or even two temporal dimensions to an atemporal relation. In such a case, any query that originally was running over an atemporal relation, now runs over a temporal relation. In order to reuse previous queries defined over the atemporal relation, their semantics must not be changed and only current information must be taken into account.

---

[2] It must be noticed that upward compatibility can be seen as a particular case for the non-sequenced semantics, which take into consideration only information associated with the current state.

3.6 Advanced Temporal Data Models and Query Languages

After having introduced the main concepts of temporal databases and some approaches extending the relational data model and the relational query languages to deal with temporal information, we now consider in some detail more complex temporal data models, as those based on the data model and on some object-oriented data models. These models allow one to represent complex information, which could not suitably be represented by the relational data model. In this section, we focus mainly on the valid time of the considered temporal information.

3.6.1 Extending the ER Model

The Entity-Relationship (ER) data model is the main data model adopted for conceptual modeling, i.e. for the database design at an abstract level, focusing mainly on the information we need to represent without considering any detail related to the computer-based application [135]. The main concepts of the ER data model are the *entity* and the *relationship*.

An *entity* represents any concept, real thing, or object which must be represented in the database application. An entity is characterized by the values of its *attributes*: for example, the entity representing the patient John Silver has attributes *name*, *surname*, and *birthdate* having values 'John', 'Silver', and 'October 3, 1957', respectively. Attribute values can be atomic (e.g., values for the attribute *surname* are simple strings, which are not divisible), composite (e.g., values for the *birthdate* are structured according to the components *year*, *month*, and *day*), and multivalued (e.g., values for a further attribute *phone numbers* can a be a list of different phone numbers for the patient). Composite and multivalued attributes can be freely nested to build *complex* attributes.

Entities can be associated by *relationships*. A relationship represents any kind of association, we have to represent in the database. A relationship is characterized by the entities it associates and (possibly) by the values of some attributes. For example, the relationship between the entity representing the patient John Silver and the entity representing the Cardiology division is characterized by the two mentioned entities and by the values 'myocardial infarction' and 'emergency' for the attributes *admission diagnosis* and *hospitalization category*, respectively.

Entities sharing the same features, i.e. the same set of attributes, are described by their *entity type*. An entity type is characterized by a name and a set of attributes. Usually, one or more subsets of attributes of a given entity type have a *uniqueness constraint*. For each collection of entities of a given entity type, it is not possible to have two or more entities having the same combination of values for the given subsets of attributes. These attributes are called *key attributes*.

Entity types, called subclasses, can be defined as a refinement of another entity type, which is called its *superclass*. For example, an entity type *patient* can be defined as a *specialization* of the entity type *person*. A patient is a person with some other specific features (as hospitalization ID, accepting physician, and so on). From another point of view, we say that the entity *person* is a *generalization* of the entity *patient*. Two aspects have to be considered in specialization/generalization. From an *extensional* point of view (i.e., considering the set of all entities of a given entity type) entities of a specialized entity type are also members of the set of entities of the corresponding superclass. From an *intensional* point of view (i.e., considering the attributes, which describe properties characterizing any entity), a specialized entity type *inherits* all the features, i.e. the attributes, of its superclass. Several new attributes can be defined in the subclass, to properly specialize it.

Relationships having similar features, i.e. the same attributes and the same entity type for the associated entities, are described by *relationship types*: a relationship type among *n* entity types defines a set of relationships among entities of these types. A relationship type is mainly characterized by a name, a set of attributes, a relationship degree, a cardinality ratio, and a participation constraint. The relationship degree is the number of participating entity types. Relationships with degree 2 (binary relationships) and 3 (ternary relationships) are usually considered. The cardinality ratio specifies the number of relationships an entity of a given entity type can participate in. It is usually expressed by a couple (*min*, *max*), where *min* (resp. *max*) is the minimum (resp. maximum) number of relationships. The participation constraint specifies whether at least one relationship for any entity of a given entity type must exist. A participation is total when a relationship must exist for each entity of a given entity type, otherwise the participation is partial.

The ER data model is provided with a powerful graphical notation, to describe the schema of a database. ER diagrams provide a graphical representation of the conceptual schema of the considered database, based on the previously introduced concepts of entity and relationship. Different graphical notations have been proposed in the database literature. In the following we will adopt one of the most widely accepted notations [16]. Notations corresponding to the above concepts will be introduced during the modeling of the example that follows.

Let us consider now the conceptual design of a database having different temporal aspects. The first step will be the exploitation of the standard (atemporal) ER data model.

Example 3.5. We are required to provide the conceptual design of a database, which allows us to represent information related to follow-up patients. For these patients, usual demographic data is collected (surname, name, birthdate, Address, phone number). The history of patient addresses must be kept by the database, for epidemiological studies. Further data is related to the code, the hospital assigns to a patient, the status of the patient when he was enrolled in the follow-up, and the date when the patient was enrolled. At each visit, which is identified by a specific administrative code, data on physiological measurements (Heart Rate, Systolic and Diastolic Pressure) are acquired, together with some comments from the physician, who is visiting the patient. The date at which the visit happens must be recorded. Patient symptoms are recorded for each patient. We need to store the name and a brief description of each symptom. As for temporal aspects of data, the interval during which the patient had a given symptom must be stored in the database.

    The first step is the conceptual design of the database through the ER data model.
It consists of identifying the main concepts/objects we need to represent, that can
be described in a self-contained manner. In other words, we need to identify those
objects which have an autonomous existence in the part of the real world we are considering. These objects are modeled by suitable entity types. In our case, it seems
appropriate to introduce the entity types *Patient*, *Symptom*, and *Visit*. Entity types
are graphically represented by boxes filled with a suitable label, corresponding to
the name of the entity type. The second step is related to the identification of relationships among entities. Thus, we identify relationship types *presence*, which associates entities of *Patient* and *Symptom*, and *follow-up*, which associates entities of
Patient and *Visit*. Relationship types are graphically represented as diamonds connected to the entity types, the relationship type associates, by lines. Now, we have
to specify for each entity and relationship types, the attributes representing them.
In our case, attributes can be easily identified within the natural-language sentences
of the requirements in Example 3.5. For example, attributes of entity type *Visit* are
Vis_code, which is a key attribute, *HR*, *SBP*, *DBP*, for the physiological measurements, and *Comments*, for some additional information about the visit. All these attributes are atomic and are graphically represented as small circles linked by a line
to the entity/relationship they refer to. Filled circles stand for single key attributes,
while a filled circle with a line linking several attributes stands for a key consisting
of all the linked attributes. A composite attribute is graphically represented by an
oval connected to the attributes composing it and to the entity/relationship it refers
to.
    Let us now consider more closely the temporal aspects of the information in Ex-
ample 3.5. We can distinguish temporal dimensions for different parts of the conceptual schema. Entity types *Patient* and *Visit* have a temporal dimension, represented
by the enrollment and visit dates, respectively. Ad-hoc attributes (*Enroll_date* and
Vis_date, respectively) allow us to represent these temporal aspects. It is worth noting that *Enroll_date* and *Vis_date* are composite attributes (*Year*, *Month*, and *Day*
being their (component) atomic parts). To be able to model histories of addresses
for patients, we need to associate a temporal dimension to single attributes. In this
case, the solution we provide consists of a multivalued composite attribute *Address*.
It can have several composite values, composed of *City* and *Street*, for the atemporal
part, and *Year*, *Month*, and *Day*, for the temporal dimension.
    The last issue we have to face, is related to the temporal dimension we have to
model for the relationship type *presence*. We need to represent the interval during
which a given patient had a given symptom. The addition of two suitable composite
attributes to the relationship type *presence*, to represent the start and the end of the
interval during which the patient had the symptom, seems to be a simple solution.
However, it is incorrect; indeed, a binary relationship type, as *presence*, represents
a set of associations between entities. So, we can represent such a set as a set of
couples (*entity1*, *entity2*), where *entity1* and *entity2* are values of key attributes of
the corresponding entity types. According to this definition, since the multiple presence of the same element in a set is not possible, two entities can be associated
at most one time. In our case, let us assume that the patient having SSN equal to

'SND56' had the symptom 'chest pain' three times, the first occurrence from May 3 to May 4, 1999, the second from September 12 to September 15, 1999 and the third from January 1 to January 3, 2000. As depicted in Figure 3.6, *SSN* and *Name* are the key attributes for entity types *Patient* and *Symptom*, respectively. The association between the patient and the symptoms is, thus, represented by the couple ('SND56', 'chest pain') for any specification of the interval during which the symptom was present. In other words, with this solution, we are able to represent only a single association between a patient and a symptom; the above specified real-world situation cannot be modeled. The right solution, then, is to model the interval of appearance of a symptom as an independent entity; the entity type *Sympt_span* has two attributes, collectively key attributes, which represent the start and the end of an interval. The entity type *Sympt_span* is related to the entity types *Symptom* and *Patient*, by the relationship type *presence*, which becomes a ternary relationship type. This solution allows us to represent all the occurrences of a symptom for the same patient, because each relationship of the set is composed by a triple, which includes also the key attributes of *Sympt_span*.

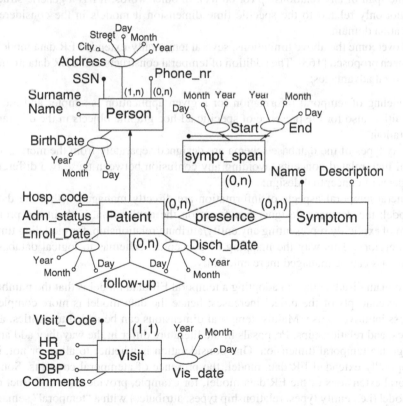

Fig. 3.6 ER schema of a database about follow-up patients, as required in Example 3.5.

The approach we have described in designing a database schema by the ER data model, does not assign any special role to temporal information. The atemporal constructs of the ER data model are employed to represent temporal aspects of data; even though this approach gives a suitable database schema, it is easy to observe that the provided ER diagrams suffer from two main limitations:

- the ER diagrams, as the one shown in Figure 3.6, are difficult to read and to correctly understand, because the temporal dimensions of data require the insertion of additional entities, attributes, and relationships, to be represented;
- application-dependent concepts, represented by entities, relationships and attributes, are merged in the diagram with more general concepts, related to temporal aspects.

The ER diagram depicted in Figure 3.6, indeed, has specific entities and attributes for temporal aspects, as we previously discussed, merged with application-related entities and relationships. For example, the entity type *Sympt_span*, which represents the time span over which a patient had a given symptom, could be used to represent the time span of any relationship (or entity). In other words, it has a generic structure, not only related to the specific time dimension it models in the considered application domain.

To overcome the above limitations, several temporally-extended ER data models have been proposed [163]. The addition of temporal constructs to the ER data model has several advantages:

- modeling of temporal information for a given application is simpler and more intuitive, also for the presence of specific ad-hoc graphic objects in the diagram notation;
- entity types of the database schema are designed separately from the more general time-related concepts, avoiding any confusion between these two different aspects of conceptual design;
- general temporal aspects of information are directly managed within the data model; the database designer can focus on the application, without the problem of explicitly representing any entity/attribute/relationship, for modeling time dimensions. This way the mapping of conceptual schemas to logical database schemas can be managed more efficiently.

The main disadvantage in adopting a temporal ER data model is that the number of basic concepts of the model increases: hence the data model is more complex and less intuitive to use. Mainly, temporal dimensions can be added to entities, attributes, and relationships. Proposals in the literature differ in the way they add and manage the temporal dimension. One consideration is whether to allow or not, in a temporally-extended ER data model, the presence of atemporal concepts. Some temporal extensions of the ER data model, for example, provide each construct of the model (i.e., entity types, relationship types, attributes) with a "temporal" semantics [164, 90]. In this case, any construct is time-dependent (e.g., an entity has a timespan; an entity type can be related to several entities with the same attribute values, but with different, disjoint timespans).

If not every model construct is temporal, another consideration is where to add temporal dimensions. Attention has to be paid to the various temporal constraints existing among different entities and/or relationships, as we show below. If the data model allows both temporal and atemporal concepts, the first step in designing the conceptual database schema is the identification of the entity types, relationship types, and attributes which have a temporal dimension.

Let us consider a generic temporal ER model. Both temporal and atemporal constructs are allowed; entity types, relationships, and attributes can be either time-varying or not. Let us assume that the valid time is the temporal dimension considered by our data model. Figure 3.7 gives the database schema for Example 3.5. A small watch near a graphic symbol denotes the temporal construct corresponding to the given graphic symbol. For example, in the diagram, the temporal attribute *Address* of entity type *Person* allows the management of the history of addresses for a person; the temporal entity type *Patient* inherits from *Person* (which is atemporal) and is related to the entity type *Symptom* through the temporal relationship type *presence*. Intuitively, a constraint must hold for each entity of type *Patient* and the corresponding relationships of type *presence*; the valid time of the relationship must be during the valid time of the entity, because it is not meaningful that a patient had a symptom before starting to be a patient. It is worth noting that the definition of the valid time of entities of type *Patient* is application-dependent: the valid time start could be set either by the patient's birthdate or by the enrollment date of the patient.

3.6.2 Temporal Object-Oriented Data Models and Query Languages

Object-oriented technology applied to the database field has some useful features - abstract data type definition, inheritance, complex object management - in modeling and managing complex information, as that related to clinical medicine [54, 217, 30]. Object-oriented design methodologies and standards have received much attention in the last decade: we mention here the Object Modeling Technique (OMT) methodology and the related Unified Modeling Language (UML) [30], and the database-oriented Object Database Management Group (ODMG) standard [54, 55].

Even though different approaches and proposals in the literature do not completely agree on the meaning of different terms used by the "object-oriented" community, there is a kind of agreement on the basic concepts of any object-oriented data model. Below, we introduce the basic concepts of the object-oriented approaches, using mainly the terminology and the graphic notation adopted in the previously cited, widely adopted approaches.

An *object* can model any entity of the real world, e.g., a patient, a therapy, a time-interval. The main feature of an object is its *identity*, which is immutable, persists during the entire existence of the object, and is independent of the *properties* or the *behavior* of the object. For example, objects representing two patients with the

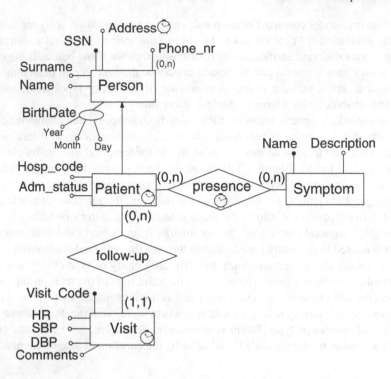

Fig. 3.7 Temporal ER schema of a database about follow-up patients.

same name will have the same value for the property representing the patient name; however, they are distinguished by their identity.

The identity is usually provided by the database system through an identifier, called OID (Object IDentifier). An object is characterized by a state, described by properties (*attributes* and *relationships* with other objects) not accessible from the outside, and by a behavior, defined by *methods*, describing modalities, by which it is possible to interact with the object itself [217].

Objects are created as instances of a *class*. A class (named also *type* in some proposals) describes (i) the structure of properties through attributes and relationships, and (ii) the behavior of its instances through methods applicable to objects, i.e. instances of the class. The *extent* of a class is the set of all objects of that class stored into the database at a given moment.

An object attribute is a named property of a class and describes the values, which can be a part of the object state.

Methods describe operations which can be applied to objects of the considered class. Each method has a declaration (also called *signature*), consisting of a name, a set of parameters, identified by name and their class, and a result, identified by its class. Like attributes, code associated to the execution of a method is not accessible from outside. This feature is called, in the object-oriented literature, *encapsulation*.

Further features of object oriented databases are inheritance, polymorphism, management of complex objects, and persistence [217]. Let us now illustrate the basic concepts by modeling a database satisfying the requirements expressed in Example 3.5. UML provides several graphic notations, for specifying different aspects of system design at the conceptual level. Below we use the class diagram, which allows us to design a database schema. Figure 3.8 depicts the schema of the relevant database. A class is denoted by a box with the class name (centered, in boldface) in the top part of the box. The second part of the box (optional, represented in Figure 3.8) contains the attribute names; an optional third part of the box contains the names of methods. Associations, which represent a group of links (with common semantics) between objects, are represented as lines between classes, the connected objects belong to. The name of an association, which is optional, is given in italics in the class diagram. The cardinality ratio (which is called "multiplicity" in the context of the UML model) is denoted by a symbol at the end of the considered association line; it specifies the number of objects of a class that can be associated to a single object of the related class. In the class diagram, a solid ball ("many") means zero or more. A hollow ball denotes "zero or one" cardinality. The lack of a symbol means "exactly one". More specific cardinality ratios can be annotated by explicit quantities (as "1..*", or "2..6").

Among the approaches adopted in dealing with the temporal dimension of data in object oriented data models and query languages [383], we focus here on i) the direct use of the object oriented data model, and on ii) the modeling of the temporal dimensions of data through ad-hoc data models. The first approach is based on the claim that the rich (and extensible) type system usually provided by object-oriented database systems allows one to represent temporal dimensions of data as required by the different possible application domains. The second approach, instead, tries to provide the user with a data model where temporal dimensions are first-class citizens, avoiding the user the effort of modeling from scratch temporal features of data for each considered application domain.

Among the proposed object-oriented systems able to deal with temporal information, OODAPLEX [430] and TIGUKAT [292] adopt the direct use of an object-oriented data model. In these systems suitable data types (i.e., classes) allow the database designer to model temporal information. For example, in modeling different concepts of time, OODAPLEX allows the usage of the supertype *point*, from which each user-defined or system-supported data type inherits, to represent different temporal dimensions. TIGUKAT models the valid time at the level of object and of collections of objects. However, for each single application it is possible to use the rich set of system-supported classes, to define the real semantics of valid (or transaction) time. The query language defined in TIGUKAT is able to follow the more recent standards existing for SQL; the proposed query language TQL extends SQL statements to support the object-oriented data model of TIGUKAT; in a similar way to that proposed in OODAPLEX, TQL doesn't define specific constructs to explicitly manage temporal aspects of a query [292]. Among the proposed object-oriented query languages, OQL, proposed by the ODMG [55], gained some interest

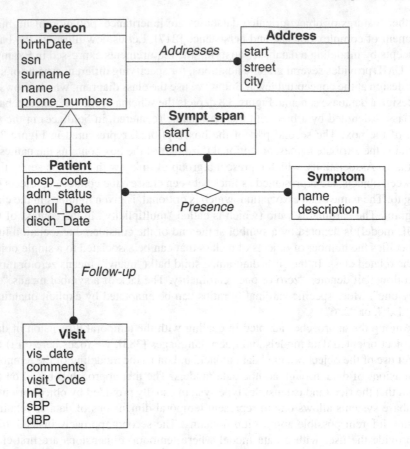

Fig. 3.8 Object-oriented schema of the database about follow-up patients.

both in the research and the commercial areas: it has a syntax similar to that of SQL, even though it is based on a fully fledged object data model.

Example 3.6. Considering the object-oriented schema depicted in Figure 3.8 and related to Example 3.5, the physician wants to determine the names and the enrollment dates of those patients who had heart rate values higher than 100 bpm during visits occurred at least 10 days after the enrollment of patients.

The query of Example 3.6 can be expressed as follows, according to the OQL syntax:

```
SELECT P.surname, P.enroll_Date
FROM Patient P, P.Follow-up V
WHERE V.hR > 100 AND V.vis_date > (P.enroll_Date + 10 days)
```

It is worth noting that the similarity with SQL is only at the syntactic level, since OQL is based on the object identity for accessing data, while SQL performs a value-based access to data. Another interesting feature of OQL is the use of paths (expressed via the dot notation) also in the FROM clause: this way, because of the implementation of relationships between objects through the object identity, it is possible to limit some object variables (V in this case) to range only on some subset of objects of a given class (in this case, only the objects of class *Visit* belonging to the set P.follow-up for a given object of class class *Patient* assigned to the variable P).

Besides the direct use of the object oriented data model, as we have shown, another approach which has been widely adopted in dealing with the temporal dimension of data by object oriented data models and query languages consists of temporally-oriented extensions, which allow the user to explicitly model and consider temporal dimensions of data [383]. As an example of this approach, Figure 3.9 depicts the schema related to the requirements of Example 3.5 according to a slightly revised UML notation. The temporal data model underlying this notation allows the distinction between *atemporal* classes (such as *Symptom*), which do not have a temporal dimension, and *temporal* classes (such as *Patient*), which have a valid time, expressed by the attribute *validTime*, which is managed directly by the data model. In other words, the data model provides a sound semantics for temporal classes, i.e. classes having the attribute *validTime*, and the user can focus on the application-dependent aspects of the database schema. In general, both classes, associations, and attributes could be extended to manage temporally-oriented features of information.

This approach, based on the definition of suitable temporal data models, is present in many proposals dealing with the definition of temporal query languages, like TOOSQL [333], OQL/T [396]. TOOSQL supports both valid and transaction time; TOOSQL is, then, a bi-temporal query language [184]. In modeling valid time, the TOOSQL data model encodes attribute values as time sequences, composed of (*value, temporal element*) pairs [333]; proposed extensions to SQL are related to all the statements, both for query and for updating. In more detail, the SELECT statement has some new clauses: WHEN, to define constraints on the valid time of retrieved objects; GROUP TO, to link new temporal sequences to other temporal sequences previously stored in the database; MOVING WINDOW, to consider aggregate data (e.g. the mean salary) with respect to an interval having a specified duration and moving on the valid time of an attribute; TIME SLICE, to consider only those objects valid in the specified interval. Finally, the clauses ROLLBACK and WITHOUT CORRECTIONS allow us to query the database about the transaction time [333]. In TOOSQL it is somehow possible to use different time granularities, usually statically predefined during the definition of types. OQL/T is a query language for object-oriented knowledge bases [396]. This language supports associations between objects and constraints on the valid time; many functions and temporal relationships have been introduced to define in a simple way complex temporal conditions.

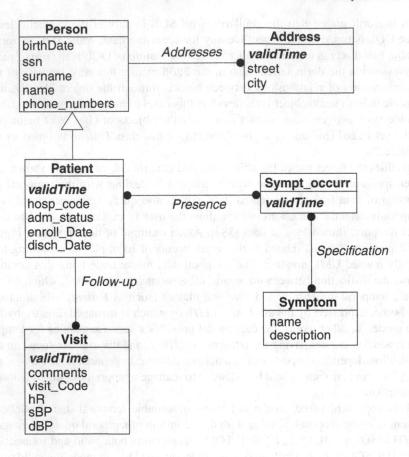

Fig. 3.9 Temporal object-oriented schema of the database about follow-up patients.

3.7 Further Topics: Managing Multimedia and Semistructured Temporal Data

Let us end this chapter with some short description of two further research topics related to the area of temporal databases. We underline and mention them here for their specific importance for the medical and clinical domain. Indeed, in the medical area a huge amount of information comes in the form of images, videos, graphics, and even audio data, i.e., in the form of *multimedia data*. All these kinds of data are then further integrated with other (possibly multimedia) data, processed to derive high level descriptions of their content, and interpreted according to the current state of the art of the medical knowledge. Another basic feature of medical and clinical data is that often they are stored according to different formats, with different structures, or even with missing or implicit structures: it is the case, for example, of the different forms filled by clinicians, where the structure of the form is recorded with data, or of the discharge report, which is often written in natural language style, even though some common basic structures could be retrieved. Moreover, slightly different formats and different record structures could be used in different clinical organizations: this often arises the need of integrating these heterogeneous data to be able to access them in a unified way. To this regard, the database research community focused on methods for representing and querying *semistructured data*, i.e., data that are neither raw data nor strictly typed data with an absolute schema fixed in advance [3]; sometimes these data have some kind of implicit structure. Semistructured data are often represented by using data models based on directed labeled graphs [296, 45, 297, 3]: data are organized in graphs where nodes denote either objects or values, and edges represent relationships between them. From a more practical point of view, semistructured data can be considered as the formal basis for the study of XML (eXtensible Markup Language), which is widely used as the common standard for data representation on the web [3]. In the following, we will consider the main aspects related to the management of temporal features of multimedia and semistructured data, respectively.

3.7.1 Managing Multimedia Temporal Data

In the field of multimedia database research, topics related to the integration of textual and visual information, as well as that of temporal aspects of textual and visual data and of their relationships, have been considered in several studies. Each of them highlighted different aspects of the problem depending on the modeling needs and the focus of the considered applications: some approaches consider only static images [52], while others consider also video [122, 191, 288], and even audio data [122]; some data models allow the association between visual data and simple unstructured notes [334], while some other data models consider the capability of modeling information related to visual data as structured entities [122, 288, 239]. Moreover, some proposals are based on (generic or ad-hoc) temporal data models, allowing one to manage temporal aspects related to multimedia data [52, 122, 191]. As regards temporal aspects in multimedia presentations, where problems arise in synchronizing for playout heterogeneous data such as text, video, and audio, different data models and tools have been proposed [23, 24, 244, 239].

In modeling video and images we can distinguish different levels of abstraction: at the lowest abstraction level video data is stored as an uninterpreted (raw) stream of bytes. At an higher level of abstraction a video stream is perceived as a flow of images (frames), displayed at a suitable frequency (frame-rate). A further level of abstraction arises if we assume that in a video stream we can identify, according to its content, different subparts, which can be combined with other subparts of other video streams to obtain new videos. Similar approaches in defining different levels of abstraction in video databases have been discussed in some widely known proposals [122, 191, 89]. Some multimedia data models, as for example those described in [191, 89], consider clips as subparts of a video, where some objects are present.

Each clip, then, identifies a scene: a partial ordering in the sequence of clips must hold; this way, it is not possible to model scenes containing sub-scenes. Usually, an object is simply associated to (i.e., present in) a sequence of frames, and thus to each subsequence of the sequence. A more sophisticated approach has been proposed to be able to deal with different semantics of the associations of objects, or more generally observations, and frame subsequences: as an example, it is possible to manage complex objects, as the one modeling the observation "cardiac cycle of the patient during a cardiac angiography", which maintains its proper meaning only if associated to the whole identified frame sequence and not to some subsequence [89]. In this direction, it is interesting to note that we could start by considering the approach proposed by Shoham for the semantics of temporal propositions (see also Chapter 2 for details). Several constraints have to be considered for the valid time of these connected data: for example, the valid time of a clip can be derived from the valid time of the overall video, by knowing the frame rate of the video and the position of the first frame of the clip within the frame sequence of the video.

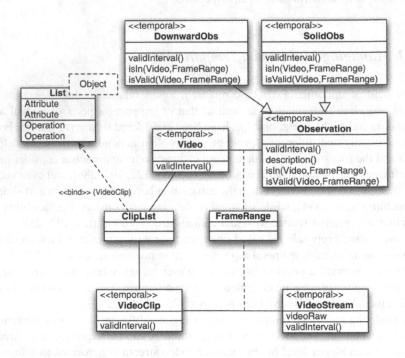

Fig. 3.10 The UML schema for a temporal object-oriented database integrating temporal data and videos.

Figure 3.10 depicts the UML class diagram for types for an object-oriented multimedia database systems managing multimedia temporal data and their relationships. Classes Video, VideoClip, and VideoStream realize three different levels of abstraction, from the higher one, representing the composition of different parts of different streams of frames, to the lowest one, managing the storing of raw data, respectively. Valid times of objects of abstract classes are derived from the valid times of the composing low level raw objects, representing video streams. Classes Observation and their two subclasses SolidObs and DownwardObs represent different temporal semantics of observations related to videos. In particular, an observation is downward(-hereditary) when, if it is associated to a specific range of frames, it holds on the given range of frames and also on any sub-range of the given range and even on single frames. On the other side, if a liquid observation holds on a given range of frames, it cannot hold on any intersecting frame range. Methods isIn and isValid, redefined in each class inheriting from Observation, allows to specify whether an observation is simply associated to a frame range or it even holds on it. The class FrameRange represents the specification of both the association between videos and observations and the association between videoclips and videostreams.

3.7.2 Managing Semistructured Temporal Data

As for the classical database field, also in the *semistructured data* context it has been considered the possibility to extend previously proposed models in order to take into account the evolution of data through time. Again, the development of methods for representing and querying changes in semistructured data and specifically in XML data have been often based on graph-based data models. Even though the previously introduced concepts of valid and transaction times have been considered as a starting point, these proposals focus on different aspects and propose different strategies to deal with temporal dimensions of information. Semistructured temporal data models can provide the suitable infrastructure for an effective management of time-varying documents on the web: indeed, semistructured data models play the same role of the relational model for the usual database management systems. Thus, semistructured temporal data models could be considered as the reference model for the logical design of (data-intensive) web sites, when time-dependent information plays a key-role. For these reasons, it is important to consider also for semistructured temporal data models a complete formalization of the constraints related to the considered temporal dimensions, as it happened in the temporal database area for temporal relational models [187].

Figure 3.11 depicts a portion of a possible semistructured temporal database related to visits, symptoms, and therapies of patients. In the considered representation of semistructured temporal data, only non-leaf nodes (represented through a box) have a temporal dimension, namely the valid time, denoted through an interval under the name of the node. As it is usual for semistructured data, both schema and instance come together in the graph-based representation. Moreover, the tree-structure allows one to represent different co-existing possible structures: in the figure, for example, the nodes Patient have partially different structures, represented by nodes at the lower levels. Furthermore, different constraints could be considered among valid times of different nodes: for example, in the considered scenario, the valid time of the root element `PatientDB` is derived by considering all the valid times of nodes `Patient`, to mean that the root is valid during the minimal time interval containing all the `Patient` valid times.

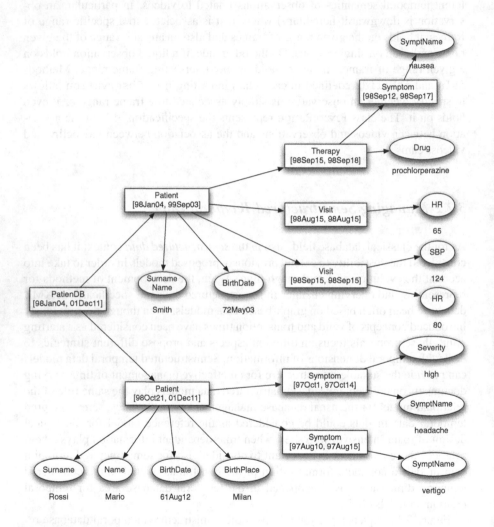

Fig. 3.11 A Temporal semistructured database for patients' symptoms, therapies, and visits.

Research contributions concerned with temporal aspects in semistructured databases consider different temporal dimensions and adopt different data models and strategies to capture the main features of the considered notion of time [12, 62, 63, 130, 162, 261, 287, 332, 421].

The Delta Object Exchange Model (DOEM) proposed in [62, 63] is a temporal extension of the Object Exchange Model (OEM) [296], a simple graph-based data model, with objects as nodes and object-subobject relationships represented as labeled arcs. Change operations (i.e. node insertion, update of node values, addition and removal of labeled arcs) are represented in DOEM by using *annotations* on nodes and arcs of an OEM graph for representing the history. Intuitively, annotations are the representation of the history of nodes and edges as it is recorded in the database. This proposal takes into account the *transaction time* dimension of a graph-based representation of semistructured data. DOEM graphs (and OEM graphs as well) do not consider labeled relationships between two objects (actually, each edge is labeled with the name of the unique pointed node).

The Temporal Graphical Model (TGM) [287] is a graphical model for representing semistructured data dynamics. This model uses *temporal elements*, instead of simple intervals, to keep trace of different time intervals when an object exists in the reality. In [287] the authors consider some issues (e.g., admitted operations and queries) related to the representation of *valid time*.

The Temporal XPath Data Model [12] is an extension of the XPath Data Model capable of representing history changes of XML documents. In particular, this approach introduces the *valid time* label only for edges in the XPath model. In [130], the authors propose to extend Web servers in order to deal with transaction time (i.e., modification time) of resources, such as XML documents. In [162, 421] some proposals devoted to represent and query histories of XML documents are introduced. In [182], Hunter proposes a framework for merging potentially inconsistent (semi) structured text using temporal knowledge. In [46] the authors propose an archiving tool for XML data which provides meaningful change descriptions and is able to support different basic functions concerning the evolution of data. For example, it is possible to retrieve a specific version of data from the archive and to query the temporal history of any element. A proposal for the management of valid time for XML document is described in [261, 332]: the authors consider several aspects of the topic, ranging from data modeling and query languages, to the implementation of indexing structures. In [332], the authors introduce in an informal way some consistency conditions for graphs representing temporal XML documents. Consistency is mainly based on the fact that each *snapshot* graph (i.e., the graph composed by nodes valid at a given time) must be a rooted tree. The authors approach for modeling valid time consistency of XML data is limited to manage a *sequenced semantics* [384] of XML data: a graph representing a temporal XML document is a compact representation of several snapshot graphs, each of them representing the XML document valid at a specific time point.

In [97, 96] some first attempts are faced towards the characterization of constraints between valid times of nodes and edges for semistructured temporal data; the same approach has been then applied to the management of semistructured temporal (and multimedia) clinical data [98, 95].

Some proposals consider explicitly bitemporal semistructured temporal models [11, 131]. In [131] the authors propose a graph-based model which uses labeled graphs to represent semistructured databases and the peculiarity of these graphs is that each edge label is composed by a set of descriptive properties, i.e. meta-data (e.g. name, transaction time, valid time, security properties of relationships). Edges can have different properties: a property can be present in an edge and missing in another one. This proposal is very general and extensible: any property may be used and added to adapt the model to a specific context. In particular, the model allows one to represent temporal aspects and to consider only a temporal dimension or multiple temporal dimensions: to this regard, some examples of constraints which need to be suitably managed to correctly support semantics of the time-related properties are provided, both for querying and for manipulating graphs. We can state that in principle this proposal can be used as a bitemporal model, although some other constraints must be specified in order to guarantee the consistency of the specific concepts included in the set of labels of the model itself. The issue of representing valid and transaction times in XML document is faced in [11]: the Bitemporal XML Data Model is an extension of the XPath Data Model capable of representing the history of both the domain and the database of XML documents. In particular, this approach introduces the pair of *valid time, transaction time* labels only for edges in the XPath model.

Summary

In this chapter we have overviewed the fundamental notions of temporal databases. First, we introduced the basic temporal dimensions, namely that of valid and transaction times, and the taxonomy for temporal databases, we can derive from these dimensions. Adding these temporal dimensions to the existing data model has been the challenge since several years. We, thus, overviewed how to extend the relational model and the related algebra and calculus. Besides these theoretical aspects, several query languages have been proposed to manage temporal data: in particular, we faced in this chapter the issue of extending the widely known SQL. Other advanced data models and languages are needed for dealing with temporal data at design level and for complex applications: in this direction, extensions to the ER data model have been described as well as the main features of object-oriented models and languages extended to consider temporal information. In this part, ER and UML-based diagrams have been used. We ended the chapter by briefly discussing temporal aspects of multimedia and semistructured data that cannot be neglected by systems dealing with real world temporal clinical data. The main purpose of this chapter was to make the reader familiar with the basic concepts of temporal databases, to facilitate the coverage of the chapters of this book centered on the specific management of temporal clinical data.

Bibliographic Notes

Several books and survey papers provide an overall view of research efforts in the temporal database area. Some books are related to the efforts of the community to assess some basic and acknowledged terminology and related concepts: in such a way, further research is sound with respect to the shared basic concepts and comparisons among different proposals are made easier. We would like to mention here; the book on theory, design, and implementation of temporal databases [398] that is one of the first attempts to provide an overall view of different research directions for temporal databases; the book edited by Rick Snodgrass on the TSQL2 query language [387]; the book edited by Etzion, Jajodia and Sripada, which collects the results discussed in the temporal database community during several conferences and workshops [139], the book by Rick Snodgrass on the development of time-oriented databases by SQL [384] and the book by Date, Darwen, and Lorentzos on temporal databases and the relational model [115] deal with the management of temporal data through (atemporal) relational database systems, even though from different perspectives.

As for papers, besides the seminal one by Snodgrass and Ahn [385], we mention here the survey by Özsoyoglu and Snodgrass, within a special section of papers on temporal and real-time databases [290], that tries to consider and analyze the huge amount of scientific work dealing with the management of time and temporal information in temporal and real-time databases [291]; moreover, Jensen and Snodgrass describe the main results and acknowledged concepts of the temporal database community in [189].

Furthermore, temporal database issues are considered in other, partially overlapping, research areas, such as that of spatio-temporal databases [4, 226] and that of temporal data warehouses [157].

Among the several conferences which consider temporal database issues in their topics of interest, two events have emerged as the most important ones: the series of biannual international symposia on spatial and temporal databases (SSTD) that publish the proceedings in the LNCS series by Springer, and the series of the annual symposia on temporal representation and reasoning (TIME), having the proceedings published by IEEE Computer Press, that has emerged as the meeting point for the scientists working on time in computer science, with a specific track on temporal databases.

Finally, it is worth noting that in the Encyclopedia of Database Systems [246], edited by Ling Liu and Tamer Özsu for Springer, there are 81 entries in the section related to Temporal Databases, edited by Christian Jensen and Richard Snodgrass: these entries, written by several scientists active in the temporal database area, confirm the reached soundness and introduce the reader to the well established concepts of this research area [188].

Problems

3.1. Examine the following table from a bitemporal database, for a given patient:

Table 3.13 Bitemporal database for a given patient

Parameter name (event name)	Value/unit (attribute:value)	Valid start time	Valid stop time	Transaction time
WBC count	9600/cc	96/1/17:9:00a.m.	96/1/17:9:00a.m.	96/1/17:5:00p.m.
Hemoglobin	6.9 gr/100cc	96/1/17:9:00a.m.	96/1/17:9:00a.m.	96/1/17:6:00p.m.
Blood transfusion	dose:2 units	96/1/18:10:00p.m.	96/1/18:11:30p.m.	96/1/19:1:00a.m.
Hemoglobin	10.2 gr/100cc	96/1/19:9:00a.m.	96/1/19:9:00a.m.	96/1/19:5:00p.m.
Hemoglobin	12.1 gr/100cc	96/1/19:9:00a.m.	96/1/19:9:00a.m.	96/1/20:8:00p.m.
WBC count	7800/cc	96/1/21:9:00a.m.	96/1/21:9:00a.m.	96/1/21:6:00p.m.
WBC count	8100/cc	96/1/18:9:00a.m.	96/1/18:9:00a.m.	96/1/22:8:00a.m.

Assume that the current time is 96/1/22:6:00p.m. Assume also that, in this particular case, the more recent the update of a datum is, the more it is considered reliable.

1. What would be the result of two rollback (transaction-time) queries that ask, at the current time, what were the latest known hemoglobin values and white blood-cell (WBC) counts from the point of view of a physician looking at the database on 96/1/19:6:00p.m.?
2. What would be the result of these two queries about the valid values of the same parameters on 96/1/19:6:00p.m, from the point of view of the current time?
3. How would the database look like in the case of a rollback database? What would be the result, if any, of the two queries?
4. How would the database look like in the case of a historical database? What would be the result, if any, of the two queries?

3.2. It is often the case that interventions start at a particular valid time, but their stop time is unknown when that time is recorded in the database (e.g., the patient is enrolled, until further notice, in a specific experimental therapy protocol). How would you represent that from the aspect of valid time(s)? What might be the implication for the transaction time(s)?

3.3. It is often the case that database tuples, such as a table raw containing the results of a particular laboratory test, have to be deleted from the patient's record after some time. How would you represent that occurrence using valid and transaction times, and what would be the meaning of the new representation? (Hint: Consider adding another transaction-time column.)

3.4. Read the following clinical scenario.

Dr. Jones has examined Ms. Smith in her clinic on May 7, 1997, 11:00am (Ms. Smith came since she had symptoms of a urinary-tract infection). Dr. Jones recorded a fever of 102° Fahrenheit at that time. (Subsequently, the fever was measured and recorded as 98.6° on May 8, 1:00am.) Half an hour later, she has drawn a blood sample and sent it to the laboratory, and asked Ms. Smith to provide her with a urine sample. Dr. Jones proceeded to immediately test the urine for traces of white blood cells (WBC) (result: "highly positive"). She has also sent the urine sample to the laboratory, asking to culture it. On May 8 at 7:00a.m., Dr. Jones called the laboratory and recorded in her notes that the WBC count in Ms. Smith's blood sample was 11, 500/cc. On May 10, 11:00a.m., Dr. Jones added also that she just learned that the urine culture had been positive for E. Coli bacteria, and that the bacteria were sensitive to Sulfisoxazole. Dr. Jones called Ms. Smith two hours later and told her to start taking that drug (dose: 1gr, frequency: four times a day), which she has previously prescribed for her, but asked that she wait before using it. On May 11, 9:00a.m., Dr. Jones got another call from the laboratory; it seems that there was an error in the first report; in fact, the bacteria were resistant to Sulfisoxazole but were sensitive to Ampicilin. Dr. Jones promptly called Ms. Smith and asked her to stop the Sulfisoxazole and to start from noon an Ampicilin regimen (dose: 500mg, frequency: four times a day). She recorded the details of the administrations of the two drugs in her notebook.

Create a table in a bitemporal database (similar in format to some of those shown in this chapter) replacing Dr. Jones' notebook, which will describes all parameters and events mentioned in the scenario. For simplicity's sake, assume that the start and stop times of valid intervals are recorded simultaneously during the same transaction time.

Part II
Temporal Reasoning and Maintenance in Medicine

Chapter 4
Temporal Clinical Databases

Overview

This chapter introduces the reader to some important issues and topics of temporal clinical databases. More specifically the chapter focuses on modeling and querying issues related to the management of temporal clinical data collected into medical records. The reader, then, learns how some temporal relational and object-oriented models have been suitably defined to deal with the specific temporal aspects of clinical data. The presentation then focuses on some main data models explicitly proposed for managing clinical data with multiple temporal dimensions and with clinical temporal data given at different and mixed granularities/indeterminacies. The reader learns how to design and query a temporal clinical database both by the temporal relational model and by temporal object-oriented data models with the related query languages, respectively. Finally, the reader is guided through some specific aspects of a real temporal clinical database system allowing the management of follow-up patients who underwent cardiac angioplasty.

Structure of the Chapter

This chapter deals with the task of management of time-oriented clinical data. The relevant literature has progressed from the early systems, which were mostly application dependent, to more general approaches, that, even when applied to the solution of real problems in the management of time-oriented clinical data, have a more generalizable value and inherent soundness. An important role in this direction is that of research efforts in the general field of temporal databases, as detailed in the previous chapter. According to this perspective, in this chapter we first give a brief historical overview of the field; we then deal with two different aspects, which have been considered in the design of temporal clinical database systems, namely that of multiple temporal dimensions of clinical data and that of different

C. Combi et al., *Temporal Information Systems in Medicine*,
DOI 10.1007/978-1-4419-6543-1_4, © Springer Science+Business Media, LLC 2010

temporal granularities in clinical data. Relational and object-oriented methodologies and technologies are discussed and suitably extended and adopted to deal with the above mentioned topics: both theoretical, methodological, and technical issues will be faced, underlining the general aspects and discussing the possible related solutions. The presence of several multiple temporal dimensions of clinical data is an important issue which has been considered both in the temporal database community and in the medical informatics one. This chapter introduces the main temporal dimensions proposed in the literature, by highlighting through clinical examples their importance when dealing with clinical information. The handling of variable temporal granularity (i.e., time units) is a recurring task in managing clinical data. The approaches we describe consider temporal data at different granularities and with indeterminacy.

The management of data of some specific clinical domains is also considered; in particular, this chapter considers several examples taken from a generic patient record and discusses the design and implementation of a web-based object-oriented system managing data from follow-up patients, who underwent angioplasty.

Keywords

Temporal clinical information, granularity, indeterminacy, temporal data models, temporal query languages, relational model, object-oriented data model, multiple temporal dimensions, angioplasty, medical records, web-based clinical applications.

4.1 Introduction

Researchers in the medical informatics field investigated temporal data maintenance, mainly to support electronic medical records. Indeed, a wide variety of applications need to deal with temporal aspects of clinical data for the management of time-oriented data stored in medical records of ambulatory or hospitalized patients [429, 197, 196, 426, 248, 307, 309, 84, 86, 125, 126, 341, 120, 111, 112, 389, 405, 44].

Studies of time-oriented data-centric applications have been performed in multiple clinical areas: cardiology [194, 309, 86], oncology [111, 197], psychiatry [29], internal medicine [120, 335], intensive care [74, 70] urology [120], infectious diseases [111], anesthesiology [70, 80]. Various clinical tasks are supported by the systems proposed in the literature: diagnosis [29], therapy administration and monitoring [112, 200, 405], protocol- and guideline-based therapy [111], and patient management [426, 123, 124, 120, 307, 86, 44].

Moving on to the task of management of time-oriented clinical data, we observe that from the early systems, which were mostly application dependent, the proposals progressed to more general approaches, that, even when applied to the solution of real problems for the management of time-oriented clinical data, have a more generalizable value and inherent soundness [426, 31, 197, 309, 51, 227, 111, 123, 86, 83, 124, 82, 158, 81, 113, 389, 405, 44]. Initially, systems that were designed to manage temporal clinical data were based on the flat relational model [426, 31]. These systems, such as Wiederhold's TOD and Blum's Rx, were based on timestamping database tuples: the date of the visit was added to the specific attribute values.

One of the first applications of databases to clinical domains, explicitly addressing the time representation problem, is the Time Oriented Database (TOD) model [426], originally developed at Stanford University during the 1970s. This model has been adopted, for example, by the American Rheumatism Association Medical Information System (ARAMIS), to manage data related to the long-term clinical course of patients suffering from arthritis or, more generally, from rheumatic pathologies [379]. TOD uses a "cubic" vision of clinical data: values of data related to a particular patient visit are indexed by patient identification number, time (visit date), and clinical parameter type. Specialized time-oriented queries enable researchers to extract, for particular patients, data values that follow certain simple temporal patterns (e.g., increase at some rate). Assignment of a temporal dimension at the tuple level is a method common to many applications of clinical databases [37, 125, 309].

Kahn et al. [197, 196] proposed a specific query language, TQuery, for data that is structured by a specific data model, named TNET. Even though TQuery was patient oriented and was not based on a generic data model, it was one of the first proposals for an extension of query languages so as to enable the system to retrieve complex temporal properties of stored data. Most query languages and data models used for clinical data management were application-dependent; thus, developers had to provide ad-hoc facilities for querying and manipulating specific temporal aspects of data [196].

The following proposals on temporal clinical databases present a more general approach. Extensions of common data models, and in particular of the relational model, are based also on the general database-field literature, in which temporal databases have been attracting special attention since several years [399, 79, 387, 289, 139].

Another distinction useful for the characterization of much of the research in maintenance of temporal data is whether the main topic is the definition of temporal data models [197, 307, 86, 88] or the definition and design of temporally-oriented query languages [111, 112, 83, 80, 127, 94]. Both topics are discussed in the medical informatics literature, although with a more focused interest in data modeling. In medical informatics, attention had been paid mostly to historical databases (which emphasize valid time), extending relational or object-oriented models [111, 83, 80, 82, 158, 127, 389].

The management of temporal clinical data has some specific features which, even though shared with other application areas, make important the specific study of temporal databases for medical data. Quite surprisingly, in our experience these specific features are neither directly related to the specific medical content of clinical records to manage, nor do they depend on the specific clinical environment to consider: they are *general* features, which, however, are so important for the medical field that it is mandatory to consider them in designing temporal database systems for medical data [17, 5, 389, 364, 437]. In the following, we briefly introduce some relevant specific issues, we have to face when dealing with temporal clinical information.

A first set of issues is when we need to model temporal clinical data.

- Often temporal dimensions of clinical data are multiple and cannot be modeled only through valid and transaction times. For example, we could need to associate to the fact "the patient suffered from asthma" both the time when the fact happened, e.g., "in Summer, 1998" (i.e., the valid time), the time when the fact has been inserted into the database, e.g., "August, 17, 2002, 16:23:00" (i.e., the transaction time), and the time when the fact has been notified by the patient to his physician e.g., "July, 10, 2002, 10:00". It is worth noting that this last temporal dimension cannot be captured neither by the valid nor the transaction times.
- The temporal dimension is expressed sometimes by using intervals, for facts having a span of time, and sometimes by using instants, for events occurring at a time point. Moreover, this temporal dimension can be expressed in different and heterogeneous ways: the used time axis, for example, has different time units ("in 1998 the patient had a stroke", "at 4:00 p.m. on June 2000, the patient had an episode of amnesia: it lasted 15 minutes"); in other cases the temporal location is expressed with some vagueness ("between 18:45 and 19:13 of May 25, 2007 the patient had for 150 seconds atrial fibrillation"). We say that the temporal dimension of clinical information is given at different *granularities*, i.e. with different units of measure, and with *temporal indeterminacy*, i.e. with some vagueness in defining its temporal location [88]. Furthermore, different granularities and/or indeterminacy may be used in several different ways to define temporal intervals ("on October 23, 2006 at 18:21 for 35 seconds", "for 5 hours until 13:20 of November 23, 2005", "it started between 14:25 and 14:38 and ended between 18:21 and 18:36 of December 23, 2005"). We also have to consider that this heterogeneity may be present even for the same kind of clinical information: an episode of ventricular fibrillation, for example, can be identified, in a context-varying way, both by the definition of its starting and ending instants, and by the day of its occurrence and by its duration, expressed with the time unit of seconds.
- Clinical information may consist both of natural language, "qualitative" sentences ("the patient suffered from asthma in Summer, 1998", "the patient finished on 11/09/99 a therapy with anticoagulants, lasting from two months"), and of quantitative parameter values ("on May 28, 2001, at 17:25 the physician got a heart rate of 78 bpm from the patient", "blood sample of September 19, 2001: cholesterol in serum is 212 mg/dl").
- In dealing with information having different granularities or indeterminacy, it is possible that in some cases temporal relations cannot be asserted for sure. It is not possible, for example, establish whether "cerebral stroke on November 21, 2003" is before or after "an episode of painless vision loss in November 2003".

A second set of issues is related to the needs arising when physicians have to query the clinical database to support, for example, clinical decisions they have to take.

- It happens that several different temporal dimensions could be involved in the definition of a query on temporal clinical data. For example, when it is necessary to evaluate the quality of care provided by a hospital ward, it is important to consider both when patients' therapy, symptoms, visits and so on occurred and when temporal clinical data describing these facts were at disposal for clinical decision making: these temporal dimensions, as well as other possible temporal dimensions like the transaction time, have to be considered together as they provide different, orthogonal, and complementary information.
- The evaluation of clinical information is often performed according to criteria involving indeterminacy and granularities. For example, in the definition of unstable angina [40], different granularities are involved: "angina at rest, for more than 20 minutes", or "new onset, within two months, exertional angina, involving marked limitations of ordinary physical activity", or "increasing angina within two months from the initial presentation", or "post-myocardial infarction angina, at least 24 hours after". Finally, therapy prescriptions involve different time units and/or indeterminacy: e.g., nitroglycerin is recommended "for the first 24 to 48 hours in patients with acute myocardial infarction and CHF (Congestive Heart Failure), large anterior infarction, persistent ischemia, or hypertension", but in intensive medical management of unstable angina, "the heparin infusion should be continued for 2 to 5 days", and morphine prescriptions are given by using minutes as time unit.
- In querying the system about the stored clinical information we usually use granularities which are different from and not related to the ones used when storing data. In the same query different granularities and/or indeterminacy may be used. Queries can consider conditions both explicitly related to absolute time, and about qualitative or qualitative temporal relations between facts. It must be possible, in other words, to define queries on the database that would be expressed in natural language by sentences as: "We want to identify those patients that suffered from chest pain in 2007 and who, in the ten months after that event, had episodes of ventricular fibrillation lasting no more than 3 seconds" or "We want to know the patients that in the years between 1990 and 1995 had an ocular stroke followed by another stroke in the following seven months, and who were treated with blood thinners after the second intervention for at least 45 days".
- The examples provided in the previous item introduce another issue that needs to be faced: that of *absolute and relative time windowing*. Indeed, when querying clinical data it is often necessary to consider only clinical data belonging to a specified temporal window. Temporal windows could be either *absolute* (e.g., ".. in the years between 1990 and 1995 ..") and, thus, holding for all the query results, or *relative* (e.g., ".. had an ocular stroke followed by another stroke in the following seven months ..") and, thus, referring to (multiple) time periods depending on the clinical events considered for the query results (as, in the above example, the first ocular stroke of each considered patient.)

4.2 Multiple Temporal Dimensions in Clinical Databases

Even though valid time (VT) and transaction time (TT) suffice for many database applications, they are often inadequate to cope with the temporal requirements of complex organizations such as hospitals and healthcare institutions. In these contexts, one often needs to model both the time at which someone/the health information system becomes aware of a fact (availability time) and the time at which the fact is stored into the database. The latter is captured by TT, while the *availability time*, AT, has been introduced and dealt with in [93].

Definition 4.1. The *availability time* (AT) of a fact is the time interval during which the fact is known and believed correct by the information system.

As in several other domains, medical decisions are taken on the basis of the available information, no matter whether it is stored in the database or not. ATcaptures this temporal dimension. Since there can be facts which are erroneously considered true by the health information system, ATmust be an interval: the starting point of ATis the time at which the fact becomes available to the information system, while its ending point is the time at which the health information system[1] realizes that the fact is not correct.

In [215, 216], Kim and Chakravarthy introduce a fourth temporal dimension, called *event time*, to distinguish between retroactive and delayed updates. In [93], Combi and Montanari refine it by showing that two event times are needed to suitably model relevant phenomena. The choice of adding the event time as a separate temporal dimension has been extensively debated in the literature [289] and is discussed in detail in [93].

Definition 4.2. The *event time* (ET) of a fact is the occurrence time of a real-world event that either initiates or terminates the validity interval of the fact.

The event times are the occurrence times of events, e.g., decisions and actions, that respectively initiate and terminate relevant facts.

The concepts of availability time and of event time are useful to clarify the relationships between Kim and Chakravarthy's (initiating) event time and the related notion of *decision time*, which has been originally proposed by Etzion and his colleagues in [137, 148] and later refined in subsequent work, e.g., in [138]. The *decision time* of a fact is the time at which the fact is decided in the application domain of discourse. More precisely, it can be defined as the occurrence time of a real-world event (i.e., decision), whose happening induces the decision of inserting a fact into the database. Decision time has been considered in medical informatics in [362, 363]; moreover, it is worth noting that the concept of decision time is considered also in the medical literature with a sligthly different meaning: indeed, decision time has been defined as the time span between the onset of some symptom and the

---

[1] It is worth noting that an information system of an organization is composed of all the information, software and hardware tools, and human resources devoted to the management and processing of data and information needed by the organization for reaching its goals.

beginning of the decided action/therapy [180]. Even with the second meaning, it is straightforward to observe that the introduction of four temporal dimensions allows the representation of this time span, as it could be derived from the start of VTof the considered symptom and from the initiating event time of the considered action/therapy. Some studies have been done in medical informatics to analyse in a deep way the decision time and its more appropriate values, i.e. the most suitable moments for taking an action on a given patient [252]. In this direction, storing both event and availability times has relevant effects when trying to evaluate the quality of the care provided by the health organization, as they allow one to estimate different decision times involved with the health care processes on a patient and to correctly consider the information available when deciding an action for a patient, as this information could not correspond to that stored into the database when the decision was taken.

Focusing on a real-world example, let us consider the following scenario, we have to represent into a clinical temporal database.

Example 4.1. Patient Hubbard, identified by the attribute PatId having value p2, was visited on June 30, 2008 and high systolic and diastolic blood pressures were revealed (an SBP of 170 mm Hg and a DBP of 120 mm Hg were measured several times that day), due to a strong emotional stress when he was driving on June 28. Even though the physician was aware of the patient's situation since June 30, data were inserted into the database only some days after, on July 2. On June 30, the physician prescribed atenolol (50 mg/day) to the patient since that moment. Due to insertion mistakes, data represent a wrong dosage (5 mg/day) and a wrong start time of the therapy (June 30, 2007). These (wrong) data were inserted on July 2 into the database. On July 4, patient Hubbard was visited again and his blood pressure had normal values. The physician stored immediately this information into the database. On July 12, patient Hubbard calls the physician and says him that he stopped the therapy on July 11, due to an increasing chest pain since the day before: the physician prescribed to the patient a therapy with lisinopril, 10 mg/day until July 15, when the patient will need a further therapy redefinition. On July 13, the physician enters data about the new lisinopril-based therapy and about the end of the atenolol-based therapy. On July 17, the physician discovers that data about the atenolol therapy were wrong and corrects them.

The example highlights several and different temporal dimensions. For example, it is quite straightforward to identify both availability and transaction times as different temporal dimensions: indeed, the moment when the physician becomes aware of the patient's situation (e.g., hypertension) is different from the moment the corresponding data are inserted into the database. Moreover, it is possible, for example, to store the moment when the physician decides for a therapy or when the patient decides for the therapy interruption: in the considered case, the physician decides on June 30 for an immediate start of an atenolol-based therapy, while on July 10 the patient had chest pain that motivates a delayed stop of the therapy (the delay is of one day: the therapy ends on July 11). Initiating and terminating event times for the therapy allow one to represent these two different decisions and events, respectively.

Finally, the valid time models the period when the patient has to assume the prescribed drug; valid time also models when the patient had hypertension.

4.2.1 Modeling Temporal Data with Multiple Dimensions

The temporal data model, we will discuss here, considering all the four temporal dimensions is a straightforward extension of the relational model: a similar approach and considerations could be adopted even when we have to consider an object-oriented or an object-relational model.

In this temporal relational model, any relation, besides the user-defined attributes, is equipped with the four temporal dimensions. VT, TT, and AT are represented as intervals, while ET is represented by a pair of attributes, that respectively record the occurrence time of the initiating event and of the terminating one.

The temporal data model encompasses a single type of key constraints, namely, the *snapshot key* constraint (*key* for short). Given a temporal relation R, defined on a set of (atemporal) attributes X and on the special attributes VT, TT, AT, ET_i, and ET_t, and a set $K \subseteq X$ of its atemporal attributes, K is a *snapshot key* for R if the following conditions hold:

- $\forall a \in K(t[a] \neq \texttt{null})$
- $\forall t1, t2 \ (t1[VT, TT, AT] \cap t2[VT, TT, AT] \neq \emptyset$
$$\Rightarrow t1[K] \neq t2[K])$$

where $t[VT, TT, AT]$ denotes the temporal region (a cube) associated with the tuple t. ET is not involved in the definition of the snapshot key, because ET_i and ET_t denote two independent time instants (possibly related to different events), which do not identify any meaningful interval.

The model imposes some basic constraints on the relationships between the values of the various temporal dimensions: we cannot assign a value to ET if there exists no value for the corresponding VT and we cannot assign a value to AT if there exists no value for the corresponding TT.

The scenario reported in Example 4.1 is represented into the database composed of the two relations depicted in Tables 4.2.1 and 4.2.1. The relation PatTherapy stores information about patients (attribute PatId) and prescribed therapies (attributes Drug, DailyDose, and Unit), while the relation PatVisit stores information about patients and vital signs (attributes SBP and DBP). Both relations feature the four temporal dimensions VT, ET_i, ET_t, AT, TT. (PatId, Drug) and (PatId) are snapshot keys for the two relations, respectively. The special values *uc* and *now* denote current/available and still valid tuples, respectively.

As a final comment, let us consider some relevant information we are allowed to derive from these data in a decision making/decision evaluation perspective: at the current moment (surely after November 17, 2008), with regard to the content of the database, we are able to understand that, for example, patient p2 assumed atenolol from June 30, 2008 to July 11, 2008. The event initiating this therapy happened

Table 4.1 Database instance for patient visits and related therapies: the relation *PatTherapy*.

PatId	Drug	DailyDose	Unit	VT	
p2	Atenolol	5	mg	[2007Jun30, *now*]	....
p2	Lisinopril	10	mg	[2008Jul12, 2008Jul15]	....
p2	Atenolol	5	mg	[2007Jun30, 2008July11]	....
p2	Atenolol	50	mg/	[2008Jun30, 2008July11]	....

	ET_i	ET_t	AT	TT
....	2008Jun30		[2008Jun30, 2008Jul12]	[2008Jul02, 2008Jul12]
....	2008July12	2008Jul12	[2008Jul12, *uc*]	[2008Jul13, *uc*]
....	2008Jun30	2008Jul10	[2008Jul12, *uc*]	[2008Jul13, 2008Jul16]
....	2008Jun30	2008Jul10	[2008Jul12, *uc*]	[2008Jul17, *uc*]

Table 4.2 Database instance for patient visits and related therapies: the relation *PatVisit*.

PatId	SBP	DBP	VT	
p1	120	80	[2008May01, 2008May01]	..
p2	170	120	[2008Jun30, 2008Jun30]	....
p2	115	70	[2008July04, 2008Jul04]	....

	ET_i	ET_t	AT	TT
....	2008May01	2008May01	[2008Nov17, *uc*]	[2008Nov17 *uc*]
....	2008Jun28	2008Jun30	[2008Jun30, *uc*]	[2008Jul02, *uc*]
....	2008Jun30	2008Jul04	[2008July04, *uc*]	[2008Jul04, *uc*]

on June 30, while the event finishing the therapy happened on July 10, one day before the end of the therapy. All this information was available to the information system since July 12, and was inserted into the database on July, 17. From relation PatVisit we are able to understand that the high blood pressure of the patient on June 30 was initiated by an event happened on June 28, while the event terminating the high blood pressure episode happened on June 30: as a matter of fact, in this case the terminating event is the start of the therapy with atenolol, represented in relation PatTherapy.

Let us now consider how the availability time may become relevant when evaluating the quality of clinical decision making. If we consider the information known by the information system about the event terminating the atenolol prescription, we are able through the availability time to say that on July 12 the information system was aware that the patient's atenolol prescription ended the day before, i.e. on July 11. So, the decision of interrupting the therapy cannot be from the physician, who is part of the information system, as it corresponds to the terminating event time on July 10. On the other hand, we may assume that the prescription of atenolol for patient p2 was decided and administered immediately on June 30 as the physician (i.e., the information system) was aware of the high blood pressure of the patient.

It is worth noting that the same conclusion could not be reached if we did not consider the availability time: indeed, without availability time, we are allowed to use only transaction time, to try to estimate the moment at which data are known by the information system. In this case, transaction time for high blood pressure starts on July 02, 2008: using in a rough way this time value, we could think that the start of the therapy was decided without knowing anything about the patient's high blood pressure. It is clear that this second (wrong) interpretation could lead to a wrong evaluation of the decision making process, which seems in this case not to be based on any clinical evidence related to the patient's vital signs.

4.2.2 Querying Data with Multiple Temporal Dimensions

Let us now move to the main aspects of query languages supporting multiple temporal dimensions. In the following we will consider the temporal query language T4SQL [94], supporting all the four introduced temporal dimensions. This proposal takes into account and extends the expressiveness of well-known temporal query languages such as TSQL2 and the temporal part of SQL3 [133]. The main features of T4SQL include the following semantics: *current*, which considers only current tuples; *sequenced*, which corresponds to the homonymous SQL3 semantics; *atemporal*, which is equivalent to the SQL3 non-sequenced one; and the original *next*, which allows one to link consecutive states when evaluating a query. As a matter of fact, both *current*, *sequenced*, and *next* are special cases of the *atemporal* semantics; nevertheless, they allow one to express meaningful classes of queries in a much more compact way.

T4SQL queries receive input relations (via the FROM clause of a statement) with the four temporal dimensions: queries return relations with at most the four temporal dimensions. Relations without all the four temporal dimensions are prior converted to *complete*[2] relations according to either some default rules or other alternative rules, suitably defined according to the given application domain. T4SQL has both the constants and the standard temporal data types as in SQL92 and the PERIOD data type as in SQL3.

In the following, the only data type used to describe an instant is DATE; the PERIOD data type is defined by instants of type DATE, only; interval data types (i.e., representing durations) are *day* and *year-month*. These assumptions do not limit the expressiveness of the query language, which can be easily extended to manage the TIME and TIMESTAMP data types.

The values associated to a specific temporal dimension can be referenced by the following functions, only:

VALID(T) returns the VT of a tuple of the relation R;

TRANSACTION(T) returns the TT of a tuple of R;

AVAILABLE(T) returns the AT of a tuple of R;

---

[2] A temporal relation is complete if it has all the four temporal dimensions VT, TT, AT, and ET(i.e., ET_i and ET_t).

INITIATING_ET(T) returns ET_i for a tuple of R;
TERMINATING_ET(T) returns ET_t for a tuple of R.

The syntax of T4SQL is very close to that of TSQL, with some extensions. The (incomplete) BNF (Backus Naur Form) grammar is:

```
[SEMANTICS <sem> [ON] <dim> [[TIMESLICE] <ts_exp>]
        {, <sem> [ON] <dim> [[TIMESLICE] <ts_exp>]}]
SELECT  <sel_element_list>
  [WITH <w_exp> [AS] <dim> {, <w_exp> [AS] <dim>}]
  [TGROUPING]  [WEIGHTED]
FROM <tables>
[WHERE <cond>]
[WHEN  <t_cond>]
[GROUP BY <group_element_list>]
[HAVING <g_cond>]
<group_element_list> ::= <group_element>
        {, <group_element>}
<group_element> ::= <attribute> |
        <temp_attribute> USING <part_size>
<sem>:= ATEMPORAL | CURRENT | SEQUENCED |
        NEXT[(<duration>)] [THROUGH <att_list>]
<dim>:=VALID | TRANSACTION | AVAILABILITY |
        INITIATING_ET | TERMINATING_ET
```

4.2.2.1 The Clause SEMANTICS

T4SQL enables the user to specify different semantics for every temporal dimension, delegating the management of the temporal dimensions to the underlying DBMS as follows:

```
SEMANTICS <sem> [ON] <dim> [[TIMESLICE] <ts_exp>]
        {, <sem> [ON] <dim> [[TIMESLICE] <ts_exp>]}
```

where <sem> is the type of the required semantics and <dim> is the temporal dimension where the semantics is applied to. The tokens ON and TIMESLICE are optional and aim at increasing the readability of the query. The item <ts_exp> is a constant (p) either of type PERIOD (for dimensions VT, AT, and TT) or of type DATE (for dimensions ET_i and ET_t): focusing on the more important case of a period timeslice, the constant p is such that, if d is the considered temporal dimension, a generic tuple t, belonging to the relation r, is considered in the query only if $t(d) \cap p \neq \emptyset$. Additionally, the value assumed by t over the temporal dimension d is changed to $t(d) \cap p$.

A temporal dimension can be only once inside the SEMANTICS clause: one unique interpretation can be associated to every single temporal dimension.

As an example, consider the relation PatTherapy and the need of extracting data about all the patients that had a therapy with atenolol and the respective period during which the fact happened. In a relational DBMS where temporal dimensions are managed explicitly by the user, the query considering all the temporalities may look like the following:

```
SELECT    PatId, VT
FROM      PatTherapy
WHERE     Drug = 'Atenolol' AND
          END(TT) = DATE 'uc' AND
          END(AT) = DATE 'uc'
```

After having defined the suitable semantics, T4SQL manages all the temporal dimensions on behalf of the user, who does not have to explicitly consider all these temporal dimensions. The obtained code is more readable, less complex, and possibly, with a reduced number of errors. The previous query is expressed in T4SQL as follows:

```
SEMANTICS SEQUENCED ON VALID
SELECT    PatId
FROM      PatTherapy
WHERE     Drug = 'Atenolol'
```

The temporal dimension of VT is interpreted according to the use of the keyword SEQUENCED, while the temporal dimensions of TT and AT are automatically managed to consider only current and available tuples, respectively.

If instead we want to consider all the patients that assumed atenolol during June 2008, the TIMESLICE semantics helps us and the T4SQL query is the following:

```
SEMANTICS SEQUENCED ON VALID TIMESLICE
          PERIOD '[2008-06-01 - 2008-06-30]'
SELECT    PatId
FROM      PatTherapy
WHERE     Drug = 'Atenolol'
```

where the upper and lower bounds of a constant of type PERIOD are depicted by the symbols '[' and ']'.

When processing a T4SQL statement, the SEMANTICS token is processed first, well before the FROM clause, because all the statement needs to be interpreted according to the specified semantics. We now consider in detail the four types of available semantics: ATEMPORAL, CURRENT, SEQUENCED, and NEXT.

ATEMPORAL Semantics

If the ATEMPORAL semantics is adopted, the corresponding attribute(s) is dealt with as atemporal (timeless), providing the user with the highest level of freedom in managing temporal dimensions, even though any support by the system is disabled. The

ATEMPORAL semantics is exactly the same as the *non-sequenced* semantics discussed in Section 3.5 of Chapter 3.

If we want to retrieve all the patients where the therapy "Atenolol" has never been observed, the query does not require any temporal dimension, but requires to span over the entire temporal axis. The resulting T4SQL query is the following:

```
SEMANTICS ATEMPORAL ON VALID
SELECT      PatId
FROM        PatTherapy
WHERE       PatId NOT IN (
   SEMANTICS ATEMPORAL ON VALID
   SELECT      PatId
   FROM        PatTherapy
   WHERE       Drug = 'Atenolol')
```

By default, duplicates are removed, as by the DISTINCT clause of SQL.

The result of a T4SQL query is a relation with at most four temporal dimensions: one result relation may also have no temporal dimension at all. The above result relation has no VT: this comes from the assumption that, if not explicitly mentioned, the temporal dimension related to an ATEMPORAL semantics is not included in the result relation. In the considered example, the query does not suggest any function that can suitably evaluate a VT to be returned, unless explicitly defined by the user. In this latter case, a temporal dimension will be included in the result relation. Thus, the freedom of expression empowered by an ATEMPORAL query enables the user to obtain the same result as it could be obtained by using some other semantics, even though the ATEMPORAL query is a much more complex one.

CURRENT Semantics

The CURRENT semantics, applied to a temporal dimension d, considers only the tuples where the value associated to d includes the current date.

The CURRENT semantics may assume a different meaning according to the temporal data type associated to the considered temporal dimension:

- if the specified data type is a period, as for VT, TT or AT, the query considers only the tuples satisfying the condition t(d) CONTAINS DATE 'now' or t(d) CONTAINS DATE 'uc';
- if the specified data type is a date, as for ET, the query considers all the tuples satisfying the condition t(d) = DATE 'now'.

As an example, we want to retrieve all the patients from PatVisit, as currently stored. The T4SQL query is:

```
SEMANTICS CURRENT ON TRANSACTION, ATEMPORAL ON VALID
SELECT      PatId
FROM        PatVisit
```

By the above query, considered tuples are only those whose TT contains the current date (timestamp). The result relation does not contain neither VT (it should have contained it if the semantics were not ATEMPORAL) nor TT. TT is not included because the semantics CURRENT does not save the temporal dimension, too. This behavior is due to the following considerations: i) compatibility of T4SQL with SQL92, as described next; ii) a query with a CURRENT semantics considers the current state of the database for a given temporal dimension (i.e., at the current date), and in such a situation there is no reason to associate a temporal dimension because the information is related to a specific date (DATE 'now' or DATE 'uc' depending on the considered temporal dimension).

The above query can be expressed by an ATEMPORAL semantics (the complexity of the query increases if the user has to explicitly manage all the temporal dimensions) as:

```
SEMANTICS ATEMPORAL ON TRANSACTION,
                    ATEMPORAL ON VALID
SELECT    PatId
FROM      PatVisit AS pv
WHERE     TRANSACTION(pv) CONTAINS DATE 'uc'
```

The CURRENT semantics is the only case where TIMESLICE cannot be used, as CURRENT can be seen as a TIMESLICE containing the current date, only. The above query can be rewritten as:

```
SEMANTICS ATEMPORAL ON TRANSACTION
        TIMESLICE PERIOD '[uc - uc]', ATEMPORAL ON VALID
SELECT    PatId
FROM      PatVisit
```

Thus, a query including the statement CURRENT ON VALID TIMESLICE DATE '2008-07-01' will raise an error.

SEQUENCED Semantics

The SEQUENCED semantics forces a *time point by time point* evaluation of the statement, with exactly the same semantics as in [384] . This evaluation on a given temporal dimension d considers all the tuples of relations in the FROM clause where there exists a date i belonging to all the periods of the dimension d.

As an example, we consider the semantics SEQUENCED over VT. For every date over the time axis, we select all and only those tuples where VT contains the considered date. This semantics is very useful when we want to perform a historical analysis, considering only the information valid at a given date (while changing that date).

As a difference from the ATEMPORAL and CURRENT semantics, the SEQUENCED semantics returns in the result relation the temporal dimension specified in the query. By default, the returned value is evaluated for every tuple as follows:

let $T_1, \cdots, T_n$ be the relations used to evaluate the result, d be the temporal dimension to which the SEQUENCED semantics is applied, and T_r be the result relation. For every tuple $t_r \in T_r$ and for every $t_i \in T_i$, where $i \in [1..n]$, where any t_i participates in the evaluation of the result tuple t_r, we have that $t_r(d) = \bigcap_{i=1}^{n} t_i(d)$.

In order to understand this choice, let us consider VT. Every tuple t_r, belonging to the result relation, is valid when all the tuples involved in its definition are valid. The VT of every result tuple is the intersection of the VT of the involved tuples. If, according to the current state of the database, we want to retrieve for every patient the visits whose VT overlaps the VT of the prescribed therapies, the query is:

```
SEMANTICS SEQUENCED ON VALID, CURRENT ON TRANSACTION
SELECT     SBP, DBP, PatId
FROM       PatVisit AS pv, PatTherapy AS pt
WHERE      pv.PatId = pt.PatId
```

When the query involves one relation only, the SEQUENCED semantics considers all the tuples of that relation and associates to the result tuples the value of the considered temporal dimension. If we want to retrieve the patients who had high blood pressure, considering the information that the database in its current state believed correct on July 8, 2008, the query is:

```
SEMANTICS SEQUENCED ON VALID,
    CURRENT ON TRANSACTION, ATEMPORAL ON AVAILABLE
    TIMESLICE PERIOD '[0000 07 08    2008 07 08]'
SELECT    PatId
FROM      PatVisit
WHERE     SBP > 150 AND DBP > 100
```

The result is a temporal relation with one explicit attribute (PatId) and one implicit attribute (VT) managed by the system. In this case, too, the query can be translated to a different one with the ATEMPORAL semantics, as follows:

```
SEMANTICS ATEMPORAL ON VALID,
    ATEMPORAL ON TRANSACTION, ATEMPORAL ON AVAILABLE
    SELECT    PatId WITH VALID(pv) AS VALID
    FROM      PatVisit AS pv
    WHERE     SBP > 150 AND DBP > 100 AND
              TRANSACTION(pv) CONTAINS DATE 'uc' AND
              AVAILABLE(pv) CONTAINS DATE '2008-07-08'
```

The particular form of the SELECT clause will be described in Section 4.2.2.2. Generally, any query with a SEQUENCED semantics can be transformed to a different query defined by an ATEMPORAL semantics.

NEXT Semantics

The NEXT semantics enables the user to retrieve information about the same object as observed in two subsequent dates over the order of the temporal dimension. If we consider the example of patients and therapies, one may want to identify for every patient the time elapsed between two subsequent administrations of the same therapy or between two subsequent administrations of different therapies.

As main feature, the NEXT semantics considers two tuples related to the same entity and the two tuples must be subsequent in the order of the selected temporal dimension. Thus, the query in its basic form has to consider together two tuples with the same *snapshot* key (see Section 4.2.2).

The logical operation performed by the NEXT semantics is equivalent to a join between two instances of the same relation in the FROM clause: in its simpler form, this operation removes the couples of tuples that are not related to the same snapshot key. Next, the query selects the tuples where the data from the second tuple are subsequent to the data from the first tuple, according to the order of the selected temporal dimension. Due to the constraint over the snapshot key, we have a total order and, in the cases of VT, TT, and AT, the successor (if any) is unique, if we exclude the adoption of the ATEMPORAL semantics for some among the other dimensions. Moreover, in the case of ET_i and ET_t, we deal with temporal values of DATE type (instantaneous), while for the other dimensions we deal with values of type PERIOD.

By default, the NEXT semantics refers to the entire temporal axis. The query starts looking for the successor from the final date of the interval of the considered tuple till the very last instant representable by the system. Sometimes one may want to limit the upper bound of the interval considered by the query: to do that, a suitable integer parameter of the NEXT semantics identifies the duration of the period within which the successor of the tuple must be found. The complete syntax for the NEXT semantics is defined as:

```
NEXT[(<duration>)] [THROUGH <att_list] [ON]
      <dimension> [[TIMESLICE] <ts_exp>]
```

where <duration> is the width of the temporal interval where the successor must be found. The integer value refers to the smallest granularity of the considered temporal data type: for instance, for the VT of type PERIOD, the integer parameter has the day granularity. <dimension> is the dimension where the semantics must be applied to.

The interval over which the query evaluation looks for the successor tuple is [e, e+dur+1], where e is the final instant of the period associated to the considered temporal dimension for PERIOD data types (or the instant associated to the considered temporal dimension for DATE data types) and dur is the given duration.

Sometimes, it could be a strong limitation to consider next tuples only by considering the snapshot key: in the considered scenario, for example, it could be interesting to query the database not only on successive administrations of the same drug to patients but also on successive therapies to patients, no matter what the administered drug is. The clause THROUGH allows the user to explicitly set the attributes to consider when joining subsequent tuples in the evaluation of the query.

Example 4.2. Starting from the PatTherapy relation, we want to retrieve for every patient the period elapsed between two subsequent administrations disregarding the fact that therapies were related to a single drug. The corresponding T4SQL query is:

```
SEMANTICS NEXT THROUGH PatId ON VALID
SELECT    PatId, (BEGIN(VALID(NEXT(t)))-END(VALID(t))) DAY
FROM      PatTherapy AS t
```

In the SELECT clause of the example, the function NEXT has the parameter defined by a tuple variable (t), to reference the successor tuple of t, as obtained by the NEXT semantics. Thus, the BEGIN(VALID(NEXT(t))) refers to the lower bound of the VT for the successor tuple.

Example 4.3. We want to retrieve all the patients who had a therapy with 'Atenolol' moving from a dosage of 50 mg/day to a dosage of 70mg/day. The query in the T4SQL query language is the following:

```
SEMANTICS NEXT(0) ON VALID
SELECT    PatId
FROM      PatTherapy AS t
WHERE     t. Drug = 'Atenolol' AND
          t.DailyDose = 50 AND t.Unit = 'mg' AND
          NEXT(t).DailyDose = 70 AND
          NEXT(t).Unit = 'mg'
```

Temporal relations obtained by applying the NEXT semantics to the temporal dimension d, do not include that temporal dimension d. In fact T4SQL does not want to associate any temporal dimension to an object obtained starting from two tuples with disjoint values. The user can always specify by the WITH clause, described below, how to evaluate the temporal dimension to be included in the result relation.

Default Values

If no information or no temporal dimension is specified for the SEMANTICS clause, default values apply. Compatibility of T4SQL towards SQL92 drives the choices for default values.

In T4SQL any atemporal query[3], performed over the temporal relation corresponding to the atemporal query, produces an identical result as a query performed over the atemporal relation by a non-temporal DBMS. This definition is equivalent

---

[3] A query is named *atemporal* or *snapshot* if no construct is defined to manage any temporal dimension.

to the definition of *upward compatibility* as in [384]. Obviously, a query compatible with SQL92 cannot include the SEMANTICS clause.

In T4SQL, a SQL92-like query over a temporal relation evaluates only the information known to the DBMS and valid at query execution time: the query returns an atemporal relation evaluated according to the semantics of SQL92. This requires that the default semantics for VT and TT is CURRENT: in fact, the query considers only the information associated to the current state of the database with the same semantics as in SQL92. Additionally, no temporal dimension is returned. Due to similar reasons, the default semantics for AT is CURRENT.

A more accurate analysis is needed for ET. T4SQL cannot assume as default values a semantics CURRENT, or T4SQL will consider only the tuples where starting date, final date, and the current date coincide. As an example, we consider a tuple whose VT is [2009-01-01, 2009-12-31], and the current date is June 1, 2009: we also assume to have a co-active relation. In this case, a tuple valid at the current date will not be considered as both boundaries of ET differ from the current date 2009-06-01. Due to these reasons, the default semantics for ET is ATEMPORAL, as this does not influence the selection of the tuple and does not return the temporal dimension in the result relation.

Example 4.4. From the PatTherapy relation, we want to retrieve patients who are taking 'Lisinopril'. The SQL92 query has no condition over temporal aspects and it is:

```
SELECT    PatId
FROM      PatTherapy
WHERE     Drug = 'Lisinopril'
```

The result is an atemporal relation considering only information in the database and valid at the current time. The equivalent T4SQL code is:

```
SEMANTICS CURRENT ON VALID,
          CURRENT ON TRANSACTION,
          CURRENT ON AVAILABLE,
          ATEMPORAL ON INITIATING_ET,
          ATEMPORAL ON TERMINATING_ET
SELECT    PatId
FROM      PatTherapy
WHERE     Drug = 'Lisinopril'
```

4.2.2.2 The Clause SELECT

The SELECT clause of SQL92 performs a projection operation over the attributes to be included into the result relation. T4SQL manages both temporal and atemporal relations and thus it has to manage explicit attributes (i.e., those included in the SELECT clause), as well as temporal dimensions.

T4SQL introduces the token WITH in the SELECT clause, to separate the specification of explicit attributes from that of implicit attributes. Any statement before the token WITH is evaluated according to the SQL92 semantics for projection: any statement following the token WITH computes the temporal dimension(s) to be included in the result relation. The syntax for the SELECT clause in T4SQL is:

```
SELECT  <sel_element_list>
  [WITH <w_exp> [AS] <dim> {, <w_exp> [AS] <dim>}]
```

where <w_exp> computes a period (if the temporal dimension is associated to a PERIOD data type) or a date (if the temporal dimension is associated to a DATE data type), and <dim> is a temporal dimension. The optional token AS increases the readability of the code.

If no temporal dimension is specified for the result relation, the following default temporal dimensions are applied:

- ATEMPORAL Semantics: the temporal dimension is not included in the result relation;
- CURRENT Semantics: the temporal dimension is not included in the result relation;
- SEQUENCED Semantics: the temporal dimension is included in the result relation and included values are the intersection of the temporal attributes of the tuples involved in determining the result;
- NEXT Semantics: the temporal dimension is not included in the result relation.

Example 4.5. We want to retrieve the patients who had high diastolic blood pressure (i.e., more than 100 mmHg), received the drug *'Lisinopril'*, and had a measure of normal diastolic blood pressure within 5 days from the beginning of the therapy. Every element in the result relation must come with the VT, which is included within the beginning of the high diastolic blood pressure and the end of the therapy. The resulting T4SQL query is:

```
SEMANTICS SEQUENCED ON VALID
SELECT   PatId WITH PERIOD (BEGIN(VALID(pv1)),
             END(VALID(pt))) AS VALID
FROM     PatVisit AS pv1, PatVisit AS pv2, PatTherapy AS pt
WHERE    pv1.DBP > 100 AND pv2 < 100 AND
         pt. Drug = 'Lisinopril' AND
         pp.PatId = pt.PatId AND
         VALID(pv1) BEFORE VALID(pv2) AND
         (END(VALID(pt2)) - BEGIN(VALID(pt))) DAY
                             < INTERVAL '5' DAY
```

Here, the result relation has the VT attribute. The value of this temporal dimension, however, does not come from the intersection of the VT of the tuples included into the result, as the token WITH modifies the included temporal dimension.

Coalescing

In a temporal relation, all the tuples have to satisfy the *snapshot key* constraint. As the projection of the clause SELECT may produce a temporal relation violating the above constraint, we may thus have two tuples which have the same values for atemporal attributes and intersecting temporal dimensions (if any). To cope with this situation, the *coalescing* operator fuses the tuples with overlapping values of temporal dimensions and with the atemporal attributes having the same corresponding values [35, 132].

4.2.2.3 The Clause FROM

The FROM clause of SQL92 identifies the relations used to find the attributes and/or to perform join operations. In SQL92 several JOIN criteria can be defined: INNER join, which is the default value, LEFT OUTER, RIGHT OUTER and FULL OUTER joins. The token ON specifies the selection conditions of the FROM clause. T4SQL adopts the same FROM clause as SQL92. Additionally, the token JOIN of SQL92 has been replaced by the token TJOIN (*temporal join*), when some join conditions are temporal.

Relations in T4SQL are filtered to consider only the information needed for the specified interpretations. Thus, user-defined conditions are augmented by those coming from the semantics applied to every temporal dimension. According to the considered semantics, T4SQL behaves as follows:

* if the CURRENT semantics is specified, T4SQL considers only the tuples whose temporal dimension includes the current date;
* if the SEQUENCED semantics is specified, T4SQL considers only the tuples from the relations specified by the clause FROM with overlapping temporal dimensions;
* if the NEXT semantics is specified, T4SQL considers the tuples having a successor and the successor itself, only.

Example 4.6. We want to select, according to the current state of the database, all the patients, reporting also their vital signs, who had a visit during the assumption of a therapy that ended more than 10 days after the visit. The T4SQL query is:

```
SEMANTICS SEQUENCED ON VALID
SELECT    PatId, Drug, DBP, SBP
FROM      PatTherapy AS pt TJOIN
          PatVisit AS pv ON
          (pt.PatId = pv.PatId AND
          (END(VALID(pt)) - BEGIN(VALID(pv))) DAY >
              INTERVAL '10' DAY )
```

4.2.2.4 The Clauses WHERE and WHEN

In SQL92, the WHERE clause evaluates selection predicates over tuples from the relations in the FROM clause. The WHERE clause of T4SQL extends the SQL92 clause possibly including some temporal conditions. As temporal conditions may turn out to be very complex, the user can optionally separate temporal conditions from atemporal conditions by using the WHEN clause. The semantics of WHEN is very similar to that of WHERE, but WHEN includes temporal conditions only. Consider again the query of Example 4.6. It can be defined by using the clause WHEN as follows:

```
SEMANTICS SEQUENCED ON VALID
SELECT    PatId, Drug, DBP, SBP
FROM      PatTherapy AS pt, PatVisit AS pv
WHERE     pt.PatId = pv.PatId
WHEN      (END(VALID(pt)) - BEGIN(VALID(pv))) DAY >
          INTERVAL '10' DAY
```

The WHEN clause can be replaced with an equivalent statement in the WHERE clause, with a reduced readability.

4.2.2.5 The Clauses GROUP BY and HAVING

In a temporal query language, the GROUP BY clause can be used to implement a *temporal grouping*, rather than a punctual grouping over atemporal attributes. The selection of the tuples to be grouped together (we have one group for every partitioning element) is performed according to the value of the considered temporal attribute(s) of the tuple, according to the periods associated with the partition, and according to the temporal comparison operator used in the temporal grouping.

Let P be a period associated with a particular element of the partition, P_t be the value (that is, time period) assumed by the tuple t on the considered temporal attribute and Op be the temporal comparison operator. The tuple will belong to the group associated with the considered element of the partition if the condition $P \, Op \, P_t$ holds. To limit the number of new terms specifying the comparison operator for the temporal grouping, the default value is the INTERSECT operator. Let Op be the INTERSECT operator. For every tuple t and every element of the partition, if P_t INTERSECT $P = true$, then t belongs to the group associated with the element of the partition and the value of its temporal attribute (with respect to the considered group) is $P \cap P_t$.

The result of temporal grouping can be explained as follows. The time period P_t associated with the temporal attribute of the tuple can be viewed as a set of dates. Whenever P_t and the time period P of the element of the partition satisfy the requested condition, e.g., whenever they intersect, the tuple belongs to the group induced by the element of the partition and the time period associated with the temporal attribute is restricted to the set of dates that belong to P.

The token USING distinguishes between a classic (atemporal) grouping and a temporal grouping. Implemented temporal partitions are SECOND, MINUTE, HOUR, DAY, MONTH and YEAR. For instance, a temporal grouping can be represented by "GROUP BY <temp> USING MONTH", where <temp> refers to a temporal dimension of a table in the FROM clause.

In case of temporal grouping, it is only possible to use a constant value of type PERIOD, representing a specific partition, to identify a group. For homogeneity with SQL92, in T4SQL the grouping attributes could be included in the clause SELECT, while the temporal grouping is performed by the token TGROUPING(...). This token returns as many attributes as the parameters, possibly renamed as specified by the user.

T4SQL adopts the same semantics as SQL92 for grouping. For the functions MAX, MIN, AVG and SUM the token WEIGHTED is introduced by T4SQL. The token WEIGHTED must precede the aggregation function it is applied to: the token computes the weighted function over the dimension of the temporal period of every tuple. Let us assume that $t_1 \ldots t_n$ are the tuples belonging to the group G, $t_1(d) \ldots t_n(d)$ are the values of the tuple over the temporal attribute d according to which we performed the temporal grouping, and $d(G)$ is the duration of the considered grouping partition. Moreover, let $t_1(a) \ldots t_n(a)$ be the values of the tuples for attribute a which is the parameter for the weighted aggregate function. The resulting function is computed as follows:

- WEIGHTED MAX(a) $= max_{i=1}^{n} \{t_i(a) * \frac{duration(t_i(d))}{d(G)}\}$;
- WEIGHTED MIN(a) $= min_{i=1}^{n} \{t_i(a) * \frac{duration(t_i(d))}{d(G)}\}$;
- WEIGHTED SUM(a) $= \sum_{i=1}^{n} (t_i(a) * \frac{duration(t_i(d))}{d(G)})$;
- WEIGHTED AVG(a) $= \frac{\sum_{i=1}^{n} (t_i(a) * \frac{duration(t_i(d))}{d(G)})}{n}$.

The HAVING clause of T4SQL is exactly the same as that of SQL92, enhanced by the possibility of using weighted functions.

Example 4.7. Let us assume that we want to retrieve for every year the average duration of prescribed therapies. The corresponding T4SQL query is the following:

```
SELECT    TGROUPING(VALID(t) AS YearPeriod),
          AVG(CAST(INTERVAL(VALID(t) DAY))
          AS INTEGER)
FROM      PatTherapy AS t
GROUP BY  VALID(t) USING YEAR
```

Example 4.8. We want to compute from the table PatVisit the average level per month of DBP for each patient, returning only data for patients having assumed a (weighted) average quantity of atenolol per month more than 2 mg per day. The T4SQL query is:

```
SEMANTICS SEQUENCED ON VALID
SELECT    PatId, AVG(DBP),
```

```
            TGROUPING(VALID(v) AS VALID)
FROM        PatVisit AS v, PatTherapy AS t
WHERE       Drug = 'Atenolol'
GROUP BY    PatId, VALID(v) USING MONTH
HAVING      WEIGHTED AVG(DailyDose) > 2
```

It is worth noting in this case that the weighted average allows us to consider the real average assumption of atenolol per day (without the weighted average any daily quantity of atenolol is considered in the temporal grouping and in the average operation, without considering how many days the patient assumed the drug).

4.3 Granularity and Indeterminacy in Clinical Databases

Since the beginning of 1990s, results from research in temporal relational databases have been applied and extended for the management of temporal clinical data [111, 112, 123, 124, 80, 82, 113]: particularly, several work in the medical informatics field focused on the issue of temporal granularity and indeterminacy in modeling and querying clinical data. The proposed research considered suitable (and general) extensions both to the relational model and query languages and to object-oriented models and query languages. In the following we will describe two well known methodologies adopting the relational model and the object-oriented one, respectively. In particular, the object-oriented approach will be treated as a complete case study, considering modeling issues, querying issues, some technological aspects, the real world clinical domain considered for the application, and the web-based architecture of the designed and implemented system.

4.3.1 A Temporally Extended Clinical Relational System

One of the main problems faced for temporal relational clinical databases is the seamless management of instant- and interval- valid times with different granularities; in several temporal database systems, valid times are homogeneous both in granularity and in instant/interval reference for all the tuples of a given relation. This is a limitation in the clinical domain: in a relation containing descriptions of pathologies, for example, instantaneous tuples (e.g. "cerebral stroke at 21:23 of May 16th, 1997") and interval-based tuples (e.g. "vision loss on February 13th 1997 from 18:35:15 to 18:45:28") must co-exist.

A widely known and general temporal data model proposed for the management of temporal clinical data is Chronus, proposed by Das and Musen in [111], and successively used in several projects dealing with the integrated management of data and knowledge in the clinical context [113, 327]. The main problem faced here is that clinical information can come with heterogeneous temporal data and with indeterminacy. Chronus distinguishes two kinds of medical temporal data: instantaneous data represent *events*, while interval-based data represent *states*. Instantaneous data and interval data are stored into *history tables*. For managing the indeterminacy of events and states, Das and Musen [111] define four different types of relational tuples: *event*, *start*, *body*, and *stop* tuples. Moreover, an extension is proposed for the relational algebra, to manage temporal information and temporal relational operations. As an example, let us consider the Chronus database depicted in Figure 4.1, which represents a simple clinical database containing some basic demographic data for the patients, some information about their blood pressures and heart rate during some visits, and some information on their therapies.

As shown in the figure, any Chronus relational table contains three extra-attributes, which have a predefined meaning: the attributes *Start_time* and *Stop_time* contain the starting timestamp and the ending timestamp for the fact represented into the tuple; moreover, tuples are typed according to the attribute *Type*, which distinguish instantaneous facts and interval-based facts.

As an example, the first tuple of table *Visit* in Figure 4.1b refers to the fact that patient Smith had values 130 and 90 for systolic blood pressure (SBP) and diastolic blood pressure (DBP), respectively, at a time point between 11 a.m. and noon of June 6, 1998: more precisely, the value *event* for the attribute *Type* says that the tuple represents an instantaneous fact, while the different values for *Start_time* and *Stop_time* indicate that the temporal location of the represented fact is not given at a precise timepoint, but is within the two given upper and lower bounds. As for states, three tuples of a Chronus table are needed: two tuple, having the values *start event* and *stop event*, respectively, for the attribute *Type* represent the indeterminacy of starting and ending points of the interval of validity of the considered state; a further tuple, with value *body* for the attribute *Type*, is needed to represent when the state is valid for sure. It is worth noting that in Chronus the tuple interval expressed through the attributes *Start_time* and *Stop_time* assumes different meanings according to the value of the attribute *Type*. For *event*, *start*, and *stop* tuples, the tuple interval represents an interval of uncertainty (IOU), i.e., the interval within which the represented instantaneous event occurs: such indeterminacy could derive, for example, from the need to represent data given at different granularities. For *body* tuples, the tuple interval represents the interval of certainty (IOC), i.e. the interval over which the represented state holds. As a special case, when clinical information is provided without indeterminacy, a state is represented through a single *body* tuple (as for the tuples of relation *Patient* and for the first tuple of relation *Therapy* in Figure 4.1c), while an event is represented through an *event* tuple having the same values for both attributes *Start_time* and *Stop_time*.

The Chronus data model is also completed by an algebra, called *historical algebra*, which allows one to compare intervals and instants, to perform operations on time intervals and instants, and on the related tuple [111]. Among the operations supported by the proposed algebra, we mention here temporal selection and projection, catenation (i.e, coalescing), and different kinds of temporal joins. Further work on Chronus has been done to improve the representation of temporal uncertainty of clinical information by associating a probability distribution function to each IOU describing the possible temporal location of a time point [284].

Start_time	Stop_time	Type	PatId	Name
97/11/10/11/00	99/9/25/11/00	body	SM1	Smith
95/3/4/12/00	99/12/6/17/00	body	RS1	Rossi
97/6/11/11/00	97/6/15/11/00	body	HB3	Hubbard

a) the relation *Patient*

Start_time	Stop_time	Type	PatId	SBP	DBP
98/6/6/11/00	98/6/6/11/59	event	SM1	130	90
98/11/10/10/00	98/11/10/10/09	event	SM1	120	70
99/7/20/12/45	99/7/20/12/59	event	RS1	150	110
97/6/12/00/00	97/6/12/11/59	event	HB3	80	60

b) the relation *Visit*

Start_time	Stop_time	Type	PatId	drug	dosage
98/5/12/00/00	98/5/22/23/59	body	SM1	thiazide diuretics	30 mg once a day
99/7/20/00/00	99/7/20/23/59	start event	RS1	aspirin	120 mg daily
99/7/21/00/00	99/7/31/23/59	body	RS1	aspirin	120 mg daily
99/8/1/00/00	99/8/31/23/59	stop event	RS1	aspirin	120 mg daily
99/7/21/16/00	99/7/21/16/59	start event	RS1	heparin	18 units/kg/hr
99/7/21/17/00	99/7/26/16/59	body	RS1	heparin	18 units/kg/hr
99/7/26/17/00	99/7/26/17/59	stop event	RS1	heparin	18 units/kg/hr

c) the relation *Therapy*

Fig. 4.1 Relations of the example clinical database.

In general, several needs have been identified in designing a relational database system for temporal clinical data:

- compatibility with the flat relational model and with SQL; a large amount of clinical data is stored in conventional relational databases. Temporal queries on these data have to be performed. Temporal query languages and related data models have to consider also flat data, containing some user-defined temporal dimension.
- addition and enhancement of some specific clauses, functions and predicates of temporal query languages. We often need to identify the clinical state of patients: to this end it is necessary to query temporal clinical data by specifying complex conditions on data based on temporal proximity, temporal order, and complex temporal relationships on collected data. For example, it is important to have the capability to observe data by a window, having a predefined duration, moving on the time axis (e.g. "find the patients having had the systolic blood pressure below 110 mmHg for ten days").

With respect to these needs and to the requirements related to the clinical domain, there are two main approaches: the first one is based on the proposal of an extended SQL syntax, which is compatible also for querying standard SQL tables; the second one is related to the design and implementation of software modules for providing physicians with a more simple temporal query language, which is in turn based on SQL.

According to the first approach, Das and Musen propose an extension of SQL, based on the algebra proposed for the Chronus data model, called Time Line SQL (TLSQL) [111]. To give an example of how to express temporal queries by TLSQL, let us consider the following scenario.

Example 4.9. Considering tables *Patient*, *Visit*, and *Therapy* depicted in Figure 4.1, we want to retrieve all the patients' visits that occurred during a therapy lasting for sure more than 7 days. As result of the query, the system must return, for each patient's visit, the surname of the patient and the systolic blood pressure measured in that visit. The query expressed by TLSQL is:

```
SELECT Name, SBP
FROM   Visit V, Therapy T, Patient P
WHEN   [V.Start_time, V.Stop_time]  DURING
              [T.Start_time, T.Stop_time] AND
       DURATION([T.Start_time, T.Stop_time]) > 7 DAYS AND
       V.Type = "event" AND T.Type = "body"
WHERE  V.PatId = T.PatId AND P.PatId = T.PatId
```

In the above query we may observe that TLSQL uses the square brackets as interval constructor, to make the comparison between temporal dimensions more compact. On the other side, as the query asks for sure relationships, a suitable condition is required to deal only with tuples of body type (i.e., representing an IOC).

4.3.2 An Object-Oriented Approach for Temporal Clinical Data

Clinical data is a good example of complex information, that can be suitably modeled and managed by object-oriented technologies [217, 82]. Indeed, medical records are complex documents, composed by different kinds of multimedia data, often involving strong temporal aspects: unstructured and structured text, coded data, numerical parameters, static images (radiographies, CTs), dynamic images (angiographic films, echographies), sounds (cardiac phonoscopies), bio-signals (ECGs, EEGs), graphical and vocal comments from clinical reports [51, 308]. Considering all this heterogeneous data, a lot of information involves temporal aspects: to monitor patient conditions, for example, several parameters, e.g., SBP, DBP, HR, are periodically monitored and historical data, related to past therapies, symptoms and so on, has to be considered too [387, 82, 292]. Combi et al. [82] extended an object-oriented data model and the related query language to deal with temporal clinical data: Granular Clinical History - Object SQL (GCH-OSQL) was proposed as a query language for temporal clinical databases, taking into account different and mixed temporal granularities. Goralwalla and colleagues adapted an existent object database model to the management of time-oriented data, and have applied it to the modeling of pharmacoeconomic clinical trials [158]. The broad set of types supported by the adopted object data model enables, for example, a modeling of branching timelines, corresponding, for instance, to the evaluation of different pharmacological treatments. In the following, we will consider some details of GCH-OSQL and of the related object-oriented temporal data model GCH-OODM.

4.3.2.1 The Temporal Data Model GCH-OODM

GCH-OODM is an object-oriented data model, extended to consider and manage the valid time of information. The model focuses on the capability of managing valid time expressed by different and mixed granularities and/or with indeterminacy.

GCH-OODM supports the main features of object-oriented data models, as applied to databases: besides the already mentioned object identity and encapsulation, the data model supports also single inheritance, polymorphism, management of complex objects, persistence (for further details on these basic concepts, see Chapter 3 and the related bibliography).

GCH-OODM Types

Besides the usual types (string, int, real) GCH-OODM uses some collection types: set<t>, bag<t>, list<t>, array<t>. Each of these types is a type generator, in respect with the type t in the angle brackets. GCH-OODM uses, to model the temporal dimension of information, some predefined data types: the type hierarchy el_time, instant, duration, interval; the collection type t_o_set, and some of its specializations, by which set of temporal objects are modeled. GCH-OODM relies on a three-valued logic, modeled by the type bool3, to manage uncertainty coming from comparison between temporal dimensions expressed at different granularities/indeterminacies.

The basic time domain **T**, called also *time axis*, is isomorphic to the natural numbers with the usual ordering relation $\leq$. As discussed in Section 2.2 of Chapter 2, a granularity is defined as a mapping from an index set, i.e. the set of points related to the given granularity, to the powerset of the time domain. The considered index sets are isomorphic to integers; different notations for different index sets based on that for dates and durations, e.g. YY/MM/DD/HH:Mi:SS for seconds, YY/MM/DD for days, n_1 y n_2 m for durations expressed using months and years.

The set *Gran* of granularity mappings is related to granularities of the Gregorian calendar (years, months, days, hours, seconds). Granularity mappings consider granularities for both anchored and unanchored time spans [159, 26, 160]: for example, the granularity of months can be used for expressing a certain period in a year (*October, 1999*), as well as for expressing a duration (*for three months*). In this direction, the mappings Y, M, D ... represent the usual granularities of the Gregorian calendar (they manage leap years, months with 28, 29, 30, or 31 days, and so on). On the other side, *mean_Y*, *mean_M*, ... provide regular mappings, that will be used in modeling duration, i.e. unanchored time spans, based on the (astronomical) mean length of a year. To identify the ith granule of a granularity, we will use the symbols ⟨.⟩.

The type el_time allows one to model time points on the basic time axis, named elementary instants. Each elementary instant is the basic unit of time supported by the temporal DBMS. By the type el_time properties of integers are extended to the time axis. This way, both time points and spans between time points are modeled in a homogeneous way: time points are identified on the basic time axis by their distance from the origin of the axis. The type el_time provides, then, functions both to manage the absolute location of time points on the time axis - i.e., calendar-related functions able to deal with leap years, months having 28, 29, 30, or 31 days - and to manage time spans - i.e., functions able to perform operations on time spans by the adoption of concepts like the mean month - and also to compute sum and difference operations on time points/time spans. For clarity reasons two different formats are used, to specify time points, i.e. anchored time spans, and distances between time points, i.e. unanchored time spans. The calendric notation YY/MM/DD/HH:Mi:SS allows the specification of a time point. The notation Y yy M mm D dd H hh Mi min S ss is used to identify a distance between time points (Y, M, D, H, Mi, and S stand for values related to the corresponding time unit). The time point 98/6/6/0:0:0, for example, identifies the first second of June 6, 1998; the time span 6 min 32 ss identifies a duration lasting 6 minutes and 32 seconds (we will omit to specify 0 yy 0 mm 0 dd 0 hh, but only for time units coarser than the coarsest time unit having a non-zero value).

The presence of different granularities/indeterminacies leads to manage relations between intervals possibly having, besides the two logical values *True* or *False*, a logical value *Undefined*. It is not always possible to establish with certainty the truth or the falsehood of relations existing between intervals. Let us consider the two sentences "In July 1998 the patient suffered from headache for eight days", and "In July 1998 the patient had fever for 17 days". While we can affirm for sure that for the patient the fever lasted more than the headache, we cannot answer with *True* or *False* to the question whether the patient suffered from headache before having fever. Both these answers could be wrong, because we haven't enough information (which are the starting and the ending day of the two symptoms).

GCH-OODM uses a three-valued logic, in which the values T: *True*, F: *False*, and U: *Undefined* are present. The usual logical connectives AND, OR, NOT, IMPLIES, ..., and the logical quantifiers *EXISTS* (∃), and *FOR EACH* (∀) have been extended to consider the third truth value *Undefined*. The adopted three-valued logic derives from Kleene's logic, where the third truth value U is related to situations, about which it is not possible to know the truth or falsehood [295]. In comparison with the Kleene's logic, the new logical connectives T(), U() and F() explicitly manage each of the three truth values. The interpretation of the logical connectives, depending on the values of the formulas A and B, is described by the following truth tables. In GCH-OODM formulas may consist in: a) methods returning a logical value (managed by the type bool3); b) comparison operations between objects returned by suitable methods and/or suitable typed constants; or c) composition by the logical connectives of formulas of type a) or b).

Table 4.3 Truth table for the formula A AND B

A B	(A AND B)
T T	T
F T	F
U T	U
T F	F
F F	F
U F	F
T U	U
F U	F
U U	U

Table 4.4 Truth table for formulas NOTA and T(A)

A	(NOTA)	T(A)
T	F	T
F	T	F
U	U	F

The meaning of the other logical connectives can be defined by the above defined ones: A OR B stands for NOT((NOT A) AND (NOT B)); F(A) stands for T(NOT A); U(A) stands for NOT (T(A) OR T(NOT A)). This three-valued logic is managed by the predefined data type bool3.

Instants, durations, and intervals

An interval, expressed in a heterogeneous way like in examples of the previous section, is represented by its starting instant, duration and ending instant. Obviously some constraints exist among the values of starting instant, ending instant and duration of an interval: if we are specifying instants and duration at different granularities or with indeterminacy, given the values of two of the three entities characterizing an interval, the value of the third entity depends on the granularities/indeterminacy of both the given values. This is the reason for which, only using both starting and ending instants and duration, it is possible to express an interval with different and heterogeneous granularities or with indeterminacy.

Instants and duration, expressed at different granularities/indeterminacy, are based on a discrete time axis, where points, named elementary instants, are represented by the finest time unit considered by the model. In our data model that of seconds is the unit of measure of the time axis. It is also possible to define duration, having values of one or more orders of magnitude lower than seconds (by the symbol ϵ), and unknown (by the label unknown).

The type *instant* allows us to represent a time point, identified either by the granule, i.e. a set of contiguous chronons, containing it or by the period on the time axis containing it. This type uses, by the methods *inf()* and *sup()*, two objects of type el_time, to represent the lower and upper bound of the granule, in which the generic time point is located. A granule can be expressed by different time units, e.g. by the format YY/MM/DD or YY/MM, while the period may be specified by two time points, i.e. the upper and the lower bound of the period, by the format <YY/MM/DD/HH:Mi:SS, YY/MM/DD/HH:Mi:SS>, if we have to model explicit indeterminacy. The instant 98/6/6, for example, may coincide with anyone of the time points included between the two bounds 98/6/6/0:0:0 and 98/6/6/23:59:59, represented by two objects of el_time type: the notation 98/6/6 will be equivalent to <98/6/6/0:0:0, 98/6/6/23:59:59>. The instant <98/7/12/12:30:0, 98/7/12/12:36:59> specifies a time point between 12:30 and 12:36 of July 12, 1998.

The type *duration* allows us to model a generic duration, specified at arbitrary granularity. This type uses, by the methods *inf()* and *sup()*, two objects of type el_time, to represent the lower and upper distances between chronons, between which the value of the given duration is included. A duration is expressed by an ordered sequence of elements, composed by an integer followed by a granularity specifier (from years to seconds, yy, mm, dd, hh, min, ss): e.g., 7 yy, 2 yy 3 mm 23 dd. The duration 13 dd, for example, stands for a time span between 13 dd 0 hh 0 min 0 ss and 13 dd 23 hh 59 min 59 ss. A duration may also be expressed by specifying the lower and upper distances, e.g. <23 dd 14 hh 6 min 23 ss, 24 dd 8 hh 51 min 12 ss>, for explicit indeterminacy. Suitable methods allow the expression of relations and of operations, like sum or differences, on instances of the types instant and duration.

A generic interval, i.e. a set of contiguous time points, is modeled by the type interval. The methods *start()*, *end()* and *dur()* allow us to identify, respectively, the starting instant, the ending instant and the duration of the interval. Suitable methods of the type interval allow us to establish temporal relations between two intervals, specified by different and not predefined granularity and/or indeterminacy. Relations between intervals are a superset of the 13 Allen's relations and they can be divided in granularity-related relations and granularity-independent relations or in relations based on the location of intervals on time axis and duration-related relations [82, 88]. Using methods *start()*, *end()*, and *dur()* is not redundant to identify an interval: e.g., the interval x having the methods $x.start()$ and $x.end()$ returning respectively 98/7/7 and 98/7/9, could have the method $x.dur()$ returning <1 dd 0 hh 1 ss, 2 dd 23 hh 59 min 59 ss> (i.e. a duration between one day and one second and three days less one second) or 48 hh 0 min 0 ss. This last interval, having the duration (48 hours) specified at a granularity level (seconds) finer than that used in specifying starting (ending) instant, cannot be expressed, for example, in TSQL2 [387]. To explicitly specify an interval x, the following notations have been introduced:

notation 1. ⟨YY⟩, or ⟨YY/MM⟩, or ⟨YY/MM/DD⟩, and so on, when the interval x is a granule of the Calendar, e.g. ⟨1994/10⟩; this notation is used to model intervals given by sentences like "the year 2000", "January '03". By these different notations we refer to intervals given as granules at one of the granularities of the Gregorian Calendar.

notation 2. ⟨x.start(), x.end()⟩ when starting and ending instants are given, e.g. ⟨98/6/6, 99/3/12/13⟩; this notation is used to model intervals given by "from ... to

..." sentences. This is the usual way to express intervals in temporal databases (but using the same granularity both for the starting instant and for the ending one).

notation 3. ⟨x.start(), x.dur()⟩ when starting instant and duration are given, e.g. ⟨98/6/6, 3 h⟩; this notation is used to model intervals given by "from ... for ... " sentences.

notation 4. ⟨x.dur(), x.end()⟩ when ending instant and duration are given, e.g. ⟨33 h, 96/10⟩; this notation is used to model intervals given by "for ...to ... " sentences.

notation 5. ⟨in, x.dur()⟩, where in is a granule; this notation allows one to express intervals given by "in ... for ...", e.g., ⟨⟨98/6⟩, 7mi33s⟩.

notation 6. ⟨x.start(), x.dur(), x.end()⟩ when both starting instant, duration, and ending instant are given, e.g. ⟨97/8/9, <5 h 5 mi 2 s, 24 h 8 mi 3 s>, 97/8/10⟩ or ⟨<96/4/3/12/30/10, 96/4/3/23/30/0>, <4 h 6 mi 3 s, 24 h 35 mi 2 s>, 96/4/4⟩; this is the more general notation, allowing one to express also all the intervals expressible by the previous notations.

Some constraints and relations exist between the starting instant, the ending one, and the duration of an interval: for example, given an interval by specifying its starting instant and duration (notation 3.), the ending instant can be computed by adding the given duration to the starting instant [88].

Temporal and atemporal types

GCH-OODM distinguishes *temporal types* and *atemporal types*. Objects instances of temporal types (hereinafter temporal objects) have an associated valid interval. By these temporal objects we are able to represent information for which it is important to know the time during which the information is true in the modeled world. For example, the type modeling the concept of pathology (or therapy) has to consider the interval, during which the pathology was present (or the therapy was administered). The method *validInterval()* returns the interval of validity of an object. By the valid interval it is possible to verify temporal relations between objects instances of temporal types. Temporal objects may have many temporal properties; these properties, defined by suitable methods, are represented in their turn by temporal objects, having their own valid interval.

Objects instances of atemporal types (hereinafter atemporal objects) model information, not having an associated temporal dimension. By these objects we are able to represent information, for which the temporal dimension is not interesting (let us think of an object modeling demographic data of a patient, about which we do not want to record the history: address, profession, ...). An atemporal object can, however, have properties represented by temporal objects. For example, the object modeling a hospital division, without any valid interval, may have a property related to the history of the heads of the division, modeled by temporal objects.

Both in temporal and atemporal types we distinguish, then, (a) *temporal methods*, modeling temporal features, returning temporal objects, and (b) *atemporal methods* returning atemporal objects. In the following some examples are given both for temporal and atemporal types and for temporal and atemporal methods.

In GCH-OODM temporal properties are modeled by temporal objects, which can be composed by a set of temporal objects. Several temporal constraints can be defined by GCH-OODM. It is possible, for example, to model the constraints on the valid time of objects and properties: being o.p() an object modeling a temporal property of a temporal object o, the following relation must hold:

$$T(o.p().validInterval().IN(o.validInterval()))$$

In GCH-OODM the temporal properties of a temporal object are not constrained like the above defined one. GCH-OODM allows us to model many other constraints existing between the valid interval of an object and the valid interval of a temporal property. Let us think, for example, of a temporal property o.prev() of a temporal object o; o.prev() is an object having a valid interval that must precede the valid interval of the object o. In this case the following condition holds:

$$T(o.validInterval().AFTER(o.prev().validInterval()))$$

For this kind of constraints that may exist between temporal objects, GCH-OODM allows us to make explicit the existence of constraints, during the design of the methods of temporal types for the considered database.

The predefined type t_o_set (temporal_object_set) allows the construction and the management of sets of temporal objects. To the instances of the type t_o_set it is possible to apply the usual operations on sets: insertion, deletion, intersection, union, difference, existence of an element, emptiness, contained-in relation. Some methods are defined to verify the existence, at a given granularity (if needed), of temporal relations between objects belonging to an instance of the type t_o_set, characterized also by some atemporal features. Let us consider for example the following methods, related to an instance I of the type t_o_set. Let p, q be two logical expressions, x and y two temporal objects, X an assigned granularity. To explain the meaning of the following methods, we will use the method subset: *I.subset(p)* returns the subset of temporal objects belonging to I and satisfying the expression p. The method *OCCURS(p)* allows us to establish if in the set I there is an object satisfying the condition p.

$$I.OCCURS(p) \equiv I.subset(p) \neq \emptyset$$

The method *CONTEMPORARY(p, q, X)* allows us to establish if in the set I there are two temporal objects, satisfying, respectively, the logical expressions p and q, and having the valid intervals equal at a predefined granularity X.

$$I.CONTEMPORARY(p,q,X) \equiv \exists x \in I.subset(p),$$
$$\exists y \in I.subset(q)(x.valid\_time().CONTEMPORARY(y.valid\_time(),X))$$

Several specializations of the type t_o_set can be defined, to manage only some types of temporal objects. This is the usual way to specialize types defined by type generators: t_o_set<c> is a specialization of t_o_set<c'> iff the temporal type c is a specialization of the temporal type c. We can specialize in an orthogonal way the type t_o_set to consider also temporal constraints among managed temporal objects. Some specializations of the type t_o_set allow the management of particular features of sets of temporal objects. For example, in GCH-OODM it is possible to model temporal properties in a way similar to what is done by time-varying properties in [430], by constraining the temporal objects representing properties to compose a temporal sequence of temporally non-intersecting elements.

Further specializations allow us to verify some logical consistency of the stored information. In a set of diagnoses related to the same patient, for example, there cannot exist two diagnoses having two overlapping valid intervals, and contradicting or implying one another.

The example database

To show both the modeling features of GCH-OODM and the query capabilities of GCH-OSQL, we will consider in the following the database schema represented in Figure 4.2 and the database instance depicted in Figure 4.3. The database schema, obviously far from the real clinical data complexity, considers both temporal and atemporal types. A hierarchy of types is described, using a UML-like graphical syntax. It refers to a clinical database, where data about patients are stored: data are related to the therapies prescribed for a patient and to the vital signs (e.g., blood pressures) collected during follow up visits. The atemporal type patient has two temporal properties, modeled by the methods *therapySet()* and *visitSet()*, returning instances of the temporal type t_o_set. These instances are specialized to manage sets of temporal objects of, respectively, therapy and visit types. Methods *drug()* or *SBP()*, for example, are atemporal.

4.3.2.2 The Temporal Query Language GCH-OSQL

The temporal extension to SQL syntax concerns the part needed for database querying. No specific syntax is provided for update, insert and delete operations, to preserve information hiding [56, 388, 57]. The specific programming language of the adopted object-oriented DBMS directly manages these operations.

The temporal extension includes the addition of the TIME-SLICE and MOVING WINDOW clauses in the original SELECT statement; the temporal dimension of objects may be referred to in the WHERE and SELECT clauses.

A GCH-OSQL query may be expressed as in the following, where, as usually, square brackets mean that the clause is optional:

SELECT <*type methods or path expressions*>
FROM <*classes*>
[WHERE <*atemporal and temporal conditions*>]
[TIME-SLICE <*time interval*>]
[MOVING WINDOW <*duration*>]

The query returns data retrieved through methods listed in the SELECT clause, from instances in the database of types listed in the FROM clause, satisfying the

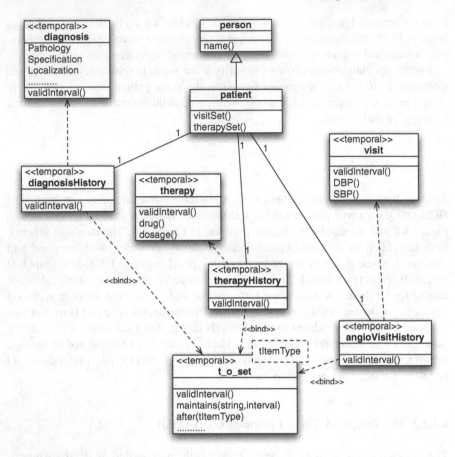

Fig. 4.2 The schema of the example clinical database.

conditions imposed through the optional clauses WHERE, TIME-SLICE, MOVING WINDOW. Retrieved objects are those for which the specified conditions result in TRUE or UNDEFINED logical values. Therefore, objects who might satisfy the specified conditions are also included in the result. According to the object-oriented approach we exposed before, object attributes are referred to through methods listed in the clauses, hiding implementation details from users. When a method is specified in a clause, the related code is executed. An object can be reached through a path expression (implicit join): it consists of a sequence of methods separated by a "." (see the following examples).

OID	name()
p1	Smith
p2	Hubbard
p3	Rossi

a) objects of type person

OID	therapySet()	visitSet()
p1	tos1	tos2
p2	tos3	tos4
p3	-	tos5

b) objects of type patient

OID	validInterval()	Object type	OID_set
tos1	$\langle 98/5/11, 98/5/23\rangle$	therapy	{t1}
tos2	$\langle\ 98/6/6/11, <98/11/10/10/15/0, 98/11/11/15:58>\ \rangle$	visit	{v1, v2}
tos3	$\langle 99/7/20, 99/8\rangle$	therapy	{t2, t3}
tos4	$\langle\ <99/7/20/12/45/0, 99/7/20/13/0/59>, 99/7/20/13/30\ \rangle$	visit	{v3}
tos5	$\langle 97/6/12, 97/6/12\rangle$	visit	{v4}

c) objects of type t_o_set

OID	validInterval()	SBP()	DBP()
v1	$\langle 98/6/6/11, 30\ min\rangle$	130	90
v2	$\langle 98/11/10/10, 15\ min\rangle$	120	70
v3	$\langle\ <30\ min\ 0\ ss, 45\ min\ 0\ ss>, 99/7/20/13/30\rangle$	150	110
v4	$\langle 97/6/12, 97/6/12\rangle$	80	60

d) objects of type visit

OID	validInterval()	drug()	dosage()
t1	$\langle 98/5/11, 98/5/23\rangle$	thiazide diuretics	30 mg once a day
t2	$\langle 99/7/20, 99/8\rangle$	aspirin	120 mg daily
t3	$\langle 99/7/21/16, 5\ dd\rangle$	heparin	18 units/kg/hr

e) objects of type therapy

Fig. 4.3 The instance of the example clinical object-oriented database. In showing an instance of the example database, we use a table for objects of the same type. Each object, corresponding to a row of the table, is represented by its OID and by the values of its methods. If methods return complex objects, the OIDs of the returned objects are contained in the corresponding column. The inheritance of the type patient from the type person is represented by using the same OIDs in the corresponding parts. The table for the objects of the type t_o_set has two special columns containing, respectively, the type name and the OIDs of the managed temporal objects (*object_type* and *OID_set*). Values of valid intervals are given according to the notations described in Section 4.3.2.1. Values of valid intervals for objects of the type t_o_set are evaluated as detailed in [82] and are the minimal intervals obtained by merging all the valid intervals of the contained temporal objects.

The SELECT FROM WHERE clauses

In the SELECT clause, either object methods or path expressions can be listed, with a comma between them. In this clause only methods related to data reading are allowed: it is not possible to use in this clause updating methods, that have side effects on the state of the database. In the following, we assume that all the GCH-OODM types for time-related concepts are suitably represented as strings. For more complex types, we explicitly use the method *display()* in the clause, to underline that a suitable string-based representation of any complex object is explicitly required for the final result of a GCH-OSQL query.

Example 4.10. The query "Find all the vital signs measured during visits; display all data about visits and also the name of visited patients" will be expressed in the following way:

```
SELECT P.name(), P.visitSet().display()
FROM patient P
```

Objects in the database containing the data we are interested in are instances of types listed in the FROM clause. To each listed type an object variable is associated. An object variable is represented by an alphanumeric string, whose first character can not be a digit: it is used to refer to object instances of the related type in the database. In GCH-OSQL object variables have to be specified for each type. In Example 4.10 the object variable P is declared for the type `patient`.

In the WHERE clause the logical conditions which express the constraints that must be satisfied by the selected objects are specified. Complex constraints may be composed by simpler conditions, using the logical connectives AND, OR, NOT, and the connectives MUSTBE, MAYBE, MUST_NOTBE, translating the T(), U(), F() GCH-OODM connectives, respectively. Conditions involving temporal relations are expressed in the WHERE clause through methods of types `instant`, `duration`, `interval`, and `t_o_set`.

Example 4.11. The query "Find all the patients having had systolic blood pressure below 130 mmHg and display the patient name, the starting instant and the drug of therapies assigned to these patients" will be expressed in the following way:

```
SELECT P.name(), T.validInterval().start(), T.drug()
FROM patient P, therapy T, visit V
WHERE P.therapySet().HAS_MEMBER(T) AND
      P.VisitSet().HAS_MEMBER(V) AND V.SBP()< 130
```

Example 4.12. "Find the patients having had a therapy with thiazide diuretics and with aspirin, while aspirin-based therapy surely occurred before that with diurectis". This query can be expressed in the following two ways:

```
•   SELECT P.name()
    FROM  patient P, therapy T1, therapy T2
    WHERE P.therapySet().HAS_MEMBER(T1) AND
     P.therapySet().HAS_MEMBER(T2) AND
     T1.drug() = ''thiazide diuretics'' AND
     T2.drug() = ''aspirin'' AND
     MUSTBE(T2.validInterval().BEFORE(T1.validInterval()))
```

- ```
  SELECT P.name()
  FROM patient P
  WHERE MUSTBE
       P.therapySet().BEFORE(''drug() = 'aspirin' '',
                             ''drug = 'thiazide diuretics' '')
  ```

The second, more compact, expression of the query is based on the method BEFORE of type t_o_set; this method, as OCCURS and others verifying further temporal relationships, is able to parse the passed string and, this way, to verify complex properties on the temporal objects belonging to the temporal set.

Alike other proposals, GCH-OSQL has no further clauses (as WHEN or WHILE [64, 333]), that would allow one to separately express the temporal part of the query. This choice allows one to express the query constraints in a seamless way, without forcing the user to divide the select condition. This choice, moreover, avoids some anomalies, as, for example, expressing some temporal constraints in the WHEN clause and some others in the WHERE clause [333].

The TIME-SLICE and MOVING WINDOW clauses

The TIME-SLICE clause allows the user to query along the temporal dimension of objects, considering only those objects in the database whose valid time is contained in the interval specified in the clause. Also objects whose valid time could be contained in the specified interval are selected. In this clause it is possible to use the MUST and MAY keywords. Using the MUST keyword, only those objects whose valid time is certainly contained in the interval specified in the clause are selected, while using the MAY keyword only those objects are selected for which it is uncertain that their valid time is contained in the specified interval.

In the TIME-SLICE clause the time interval may be expressed in many different ways:

- by the FROM..TO keywords, in order to define an interval by its starting and ending instants: e.g., FROM 1994/12/11 TO 1994/12/23/11:00. It is also possible to specify only the FROM or the TO keywords.
- by the FROM..FOR keywords, in order to define an interval by its starting instant and its duration: e.g., FROM 1994/12/11 FOR 2 mm.
- by the FOR..TO keywords, to define an interval by its duration and its ending instant: e.g., FOR 3 dd TO 1995/4.
- by the AT keyword, to define an interval as a single granule: e.g., AT 1996/5.

Example 4.13. The query "Find all therapies administered from October 2, 1994 to November 12th, 1996 in the afternoon; display the drug and the interval of validity of the selected therapies" will be expressed in the following way:

```
SELECT T.validInterval(), T.drug()
FROM therapy T
TIME-SLICE FROM 1994/10/2 TO
```
```

<1996/11/12/12:0:0, 1996/11/12/17:0:0>

It is worth noting that the capability of defining the TIME-SLICE interval by, respectively, FROM..TO, FROM..FOR, and FOR..TO keywords is not redundant: for example, the clause TIME-SLICE FROM 1994/6/23 TO 1994/6/25 will consider, on the chronon time axis, an interval having a duration between 24 hours plus 1 second and 72 hours less 1 second; the clause TIME-SLICE FROM 1994/6/23 FOR 48 hh 0 min 0 ss will consider, on the chronon time axis, an interval having starting and ending instants expressed exactly like in the previous clause, but having a duration of exactly 48 hours.

On the other hand, using the MOVING WINDOW clause, objects stored in the database are examined through a temporal window, of the width specified in the clause, moving along the temporal axis. The constraints expressed in the other clauses are checked only on the database objects visible through that window.

Example 4.14. "Show the name of patients having had diastolic blood pressure greater than 120 and therapies with diuretics in a period of fifteen days"; this query is expressed as:

```
SELECT P.name()
FROM patient P
WHERE P.visitSet().OCCURS(''DBP()>120'') AND
    P.therapySet().OCCURS(''drug() LIKE 'diuretics' '')
MOVING WINDOW 15 dd
```

In the MOVING WINDOW clause through the MUST or MAY keywords, only, respectively, certain or uncertain situations can be considered.

Some closing queries

To have a comprehensive idea of GCH-OSQL, let us consider some more complex queries on our example database.

Example 4.15. "Find those patients having had, within a period of four months, a diastolic blood pressure measure less than 60 mmHg before another one more than 100 and other measurements of vital signs surely before the measure of less than 60. Return their names, the vital signs of visits they had surely before the DBP measurement of less than 60, and the therapies holding during these visits. Consider only the period starting from about noon, October 9, 1997, and lasting 25 months"; this query is expressed as:

```
SELECT P.name(), V1.SBP(), V1.DBP(), T3.drug(), T3.dosage()
FROM patient P, visit V1, visit V2, visit V3, therapy T
WHERE P. visitSet().HAS_MEMBER(V1) AND
  P.visitSet().HAS_MEMBER(V2) AND
  P.visitSet().HAS_MEMBER(V3) AND
```

```
MUSTBE(V1.validInterval().BEFORE(V2.validInterval())) AND
V2.validInterval().BEFORE(V3.validInterval()) AND
V2.DBP()<60 AND V3.DBP()>100 AND
V1.validInterval().DURING(T.validInterval())
TIME-SLICE FROM <1997/10/9/11:30:0, 1997/10/9/12:30:0>
              FOR 25 mm
MOVING WINDOW 4 mm
```

Example 4.16. "Find those patients having had a therapy surely during the period composed by the second half of July and the first ten days of August, 1998. Return their names, their therapies, and their previous vital signs of visits having a time span of less than 15-20 minutes. Consider only therapies and vital signs having a temporal distance less than 30-40 days"; this query is expressed as:

```
SELECT P.name(), T.drug(), T.validInterval(),
              V.SBP(), V.DBP()
FROM patient P, therapy T, visit V, therapy T
WHERE P.therapySet().HAS_MEMBER(T) AND
  P.visitSet().HAS_MEMBER(V) AND
  MUSTBE T.validInterval().DURING((1998/7/15, 1998/8/10)) AND
  V.validInterval().BEFORE(T.validInterval()) AND
  V.validInterval().duration().
              SMALLER(<15 min 0 ss, 20 min 0 ss>)
MOVING WINDOW <30 dd 0 hh 0 min 0 ss, 40 dd 0 hh 0 min 0 ss>
```

4.3.2.3 Query Processing

Different logical steps can be identified in the evaluation of a GCH-OSQL query.

1. Evaluation of the contents of the FROM clause: for each object variable there is a corresponding internal variable ranging on OIDs of objects of the related type. Candidate solutions for the query can be represented as tuples of OIDs, ranging on the corresponding type, as defined in the FROM clause.
2. Evaluation of the atemporal conditions (i.e. conditions not involving methods of the types el_time, instant, duration, interval, t_o_set or of types inheriting from them) expressed in the WHERE clause: among all the candidate solutions only the solutions for which the corresponding objects satisfy atemporal conditions are retained.
3. Evaluation of the temporal part of the content of the WHERE clause: methods related to the time-modeling types are considered. By these methods it is possible to consider relations between, for example, intervals given at different granularity/indeterminacy in a seamless way. Candidate solutions are those coming from the previous step, for which the temporal part of the condition returns the truth

values True or Undefined. By the connectives MUSTBE (MAYBE) only the truth value True (Undefined) is considered.

4. Evaluation of the content of the TIME-SLICE clause: each valid interval of temporal objects of each candidate solution is compared with the interval specified in the TIME-SLICE clause. Only candidate solutions are retained, having temporal objects with the valid interval contained in the TIME-SLICE interval (i.e., satisfying the relation modeled by the DURING method). The keyword MUST (MAY) allows us to consider only objects, for which the relation is sure (possible).

5. Evaluation of the clause MOVING WINDOW: the candidate solutions coming from the previous steps are finally "viewed" through a window moving along the time axis. Among the candidate solutions, only those are selected, for which there is a time window having the duration specified in the clause MOVING WINDOW containing the valid intervals of all their temporal objects. To do that, an instance of the type t_o_set is created for each candidate solution, managing all its temporal objects. If the valid interval of the t_o_set instance has a duration smaller than the duration expressed in the MOVING WINDOW clause (i.e., satisfying the relation modeled by the method SMALLER of type duration), the corresponding candidate solution is considered for the final query result. The keyword MUST (MAY) allows us to consider only objects, for which the relation is sure (possible).

6. Application of the methods defined in the clause SELECT to the suitable objects of the final candidate solutions.

4.3.2.4 The Clinical Database

GCH-OODM and GCH-OSQL have been used in the definition and development of a clinical database, containing data coming from patients who underwent a coronary-artery angioplasty [82]. These patients suffer from an insufficient supply of blood to the coronary arteries due to a partial (or total) obstruction of some coronaric vessels. Coronary revascularization is performed by inflations of a balloon, placed on a suitable catheter: the catheter causes a dilation of the stenotic area of the vessel and lets more blood through. This operation, also known as PTCA (Percoutaneous Transluminal Coronary Angioplasty), receives more and more consensus in the clinical field, and in a lot of cases is a valid alternative to by-pass surgery operations. Patients who have undergone this kind of operation are periodically followed up, to prevent sufferance from new stenoses or re-stenoses.

The object-oriented database containing data about this kind of patients is composed of different data categories:

- patient ID data;
- auxiliary demographic data;
- data related to risk factors;
- data related to current and previous therapies;
- data related to current and previous diagnoses;

- data related to follow-up visits.

The clinical database contains some temporal objects, modeled through the temporal types therapy, relative to previous and current therapies, diagnosis, relative to previous and current pathologies, angio_visit, relative to vital signs recorded in a visit, as heart rate, systolic and diastolic blood pressure, and others. The same clinical database allows also the integration among alphanumeric data and related images, to manage relationships between data about observed stenosis and angio-cardiographic images displaying stenoses, as detailed in [308].

The database is patient-oriented: the type patient allows the access to all patient data. Every object of the patient type can have multiple object instances of the temporal types therapy, diagnosis, angio_visit, managed through multiple instances of the t_o_set type (or its subtypes). Figure 4.4 shows the type diagram according to a UML-based graphical notation.

To show the GCH-OSQL expressiveness about a situation of clinical relevance, consider the following query: retrieve the name of all the patients that, after having had normal values of blood pressure (DBP values between 100 and 60 and SBP between 150 and 100) for three weeks and more, suffered from angina, followed by PTCA intervention in 36 months. Only the period starting from Winter, 1988 must be considered.

```
SELECT P.surname(), P.name()
FROM patient P, diagnosis A, angio_visit B
WHERE P.Dia_Set().HAS_MEMBER(A) AND
  P.visitSet().HAS_MEMBER(B) AND
  B.angio_exam().exam_type() = ''PTCA''
  AND A.pathology()=''angina'' AND
  MUSTBE A.validInterval().BEFORE(B.validInterval()) AND
  P.visitSet().MAINTAINS(''SBP()>100 AND SBP()<150'',
        ⟨21 dd, A.validInterval().start()⟩) AND
  P.visitSet().MAINTAINS(''DBP()>60 AND DBP()<100'',
        ⟨21 dd, A.validInterval().start()⟩)
TIME SLICE FROM <87/12/21/0:0:0, 88/3/20/23:59:59>
MOVING WINDOW 36 mm
```

In this GCH-OSQL query some particular features may be noticed: i) different time granularities (years, months, days) and indeterminacy (Winter 1988) are explicitly used; ii) the MUSTBE connective allows us to verify by sure the precedence condition between angina occurrence and angioplasty intervention; iii) for temporal relations the query uses both methods of the interval type (BEFORE is a method of this type) and t_o_set type (MANTAINS is a method of that type); iv) through methods of the t_o_set type it is possible to verify some complex conditions involving objects contained in a set (in this case that at the end of a time period of at least 21 days of normal blood pressure, an angina episode has happened); v) the condition expressed through the MANTAINS method includes also the patients for which a period of 21 days of normal blood pressure has been possible.

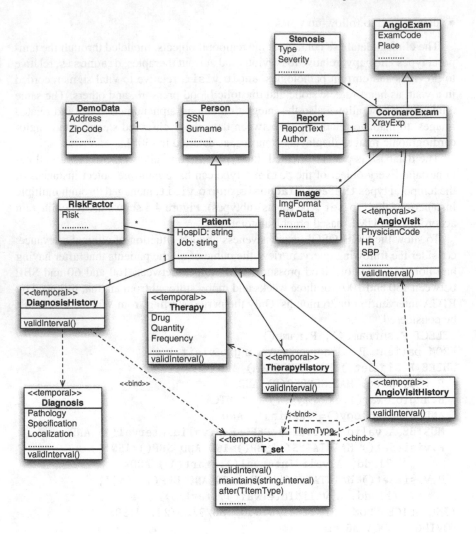

Fig. 4.4 The type diagram of the temporal object-oriented database for PTCA patients.

4.3.2.5 System Implementation

GCH-OODM and GCH-OSQL are characterized by the presence of some prototypical implementations, applied to a management system of clinical histories [83, 415, 82, 317, 318]. Initially, a prototype of GCH-OSQL and a graphical interface have been implemented on a Sun workstation, in the OpenLook graphical environment; the prototype was based on the ONTOS object-oriented database management system [229]. ONTOS has been used in the implementation of the clinical database too, to which GCH-OSQL has been applied, related to the follow-up of patients after a coronary angioplasty intervention, as described in the previous section [83]. Tests performed on the prototype confirmed the capability of the system to store, represent, and query the database about complex temporal features of data. A second prototype system was designed and implemented using the OODBMS Ode [15, 243, 242]. Finally, the Ode-based prototype has been extended to allow a web-based interaction with the system [318]. In this last prototype, named KHOSPAD (Knocking at the Hospital for PAtient Data) the designed and adopted data model extends GCH-OODM to deal with views [318]: views allow different users, e.g., physicians, nurses, technicians, to access the same temporal objects stored into the database in several different ways, according to the needs and the authorizations related to the role the user has in the healthcare organization. For example, when aiming at improving the quality of the process of patient care concerning general practitioner-patient-hospital relationships for the population of PTCA follow-up patients, the general practitioner has to deal with complex history data, to assess the efficacy of current patient therapies. This data is acquired during hospitalization and in the follow-up visits and managed by a DBMS in the cardiology division. Views allow the general practitioner to access history data in a compact and ad-hoc way, with respect to the way hospital physicians access the same data [318].

The overall system architecture of KHOSPAD is depicted in Figure 4.5. It is composed of different modules. The global architecture of the system is a web based client-server one: HTML pages and applets compose the client; the server consists of the Ode database management system extended with classes for managing both the GCH-SQL language, the GCH-OODM model, and the user view based access to the database. The modules *GCH-OSQL query manager* and *User-oriented view schemas* are accessed by the applets through the web server via suitable CGI applications and allow the application to access clinical data at a high abstraction level: indeed, these modules provide a user-oriented temporally-oriented data access, based on GCH-OODM, and a fully fledged temporal object-oriented query language. The module *Clinical database schema* contains the description of types of the clinical database (e.g., types `patient, therapy`); The module *GCH-OODM classes* contains the description of types related to the temporal data model (e.g. types `t_o_set, interval, el_time`). The module *Classes for view definition* contains the description of types for the view specification. All these modules are, finally, based on classes provided by the Ode DBMS: indeed, modules *Ode classes* and *Support classes* allow the above mentioned modules to perform usual database services (persistency, recovery, concurrency, query, data insertion and deletion, and so on) on the clinical database stored through the Ode DBMS.

4.4 Further Research Directions

Temporal reasoning systems and temporal data-maintenance systems are often independent efforts, even though they usually contribute towards the same goal. For example, time-oriented decision-support systems often do not adopt any kind of a

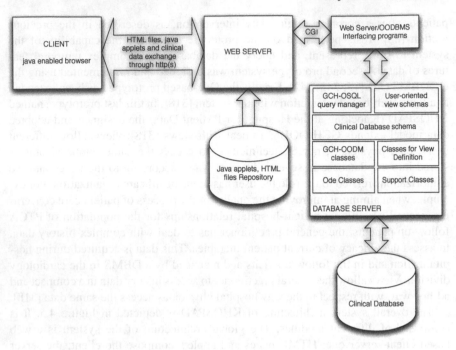

Fig. 4.5 The architecture of KHOSPAD.

formal temporal data model or a temporal query language to manage stored time-oriented clinical data.

We suggest that currently, after several years of research on the topics described in previous sections, new and more powerful solutions could be derived from a merging of different approaches.

The temporal-abstraction task and the management of temporal granularity seem to be a meeting point between research efforts originating in the artificial-intelligence and in the database communities, at least as these efforts have been applied to medical domains. Furthermore, as previously pointed out, the issue of the appropriate time model is always a pertinent one. Thus, several research themes, most of which are relevant to the community of general computer scientists, will be important, in our opinion, for next-generation time-oriented systems in medicine.

- **Adoption of advanced data models**. The adoption of advanced data models, such as the object-oriented data model and the EER data model, will improve the capability of describing real world clinical entities at high abstraction levels. Thus, the focus may shift to more domain-specific inference actions.
- **Maintenance of clinical raw data and abstractions**. Several recent systems allow not only the modeling of complex clinical concepts at the database level, but also the maintenance of certain inference operations at that level. For example, active databases can store and query also derived data; these data are obtained

by the execution of rules that are triggered by external events, such as the insertion of patient related data. Furthermore, integrity constraints based on temporal reasoning could be evaluated at the database level, for example to validate data during their acquisition.

- **Merging the functions of temporal reasoning and temporal maintenance**. By combining these two functions within one architecture, sometimes called a temporal mediator, a transparent interface is created to a database, a knowledge base, or both. An example of ongoing research is the Tzolkin temporal-mediation module [114], which is being developed within the EON guideline-based-therapy system [278]. The Tzolkin module merges Shahar's temporal-abstraction system, RÉSUMÉ [357], with Das's temporal-maintenance system, Chronus [111], into a unified temporal-mediator server. The Tzolkin server answers complex temporal queries using both the time-oriented patient database and the domain-specific temporal-abstraction knowledge base, but hides the internal division of computational tasks from the user (or from the calling process). Many questions will still have to be answered, such as how does a temporal mediator decide which computational module to use for what temporal queries, and will provide interesting issues for future research.

- **Resolution of conflicts between temporal-reasoning and temporal data-base systems within hybrid architectures**. Currently, it is common to have temporal-reasoning systems working purely within a short-term, random-access memory, while the temporal-maintenance system stores and retrieves data and abstractions using a long-term storage device such as an external database. As a result, multiple conflicts might arise, especially when systems need to be accessed concurrently by multiple users. One problem is the inherent non-monotonicity of temporal abstractions, which might be retracted when additional data arrives (whose valid time is either the present or the past). This problem is solved, for instance, in [357], by the use of a logical truth-maintenance system (TMS). However, integrating a temporal-abstraction system with an external database (as might happen in a temporal-mediator architecture such as mentioned in this section) might create inconsistency problems: the temporal-abstraction system might update its old conclusions as newly-available data arrive; but a standard database system, not having the benefit of the dependency links and the TMS mechanism, will also keep the old, incorrect conclusions. In addition, arrival of new data to the patient database should be reported to the temporal-abstraction module. Thus, we need to investigate whether the short-term, random-access temporal-reasoning fact base and the long-term external database should be tightly coupled (each update is reflected immediately in the other database), loosely coupled (updates are sent intermittently to the other database) or not coupled at all. Several protocols for connecting and mutually updating the internal and external databases are theoretically possible. The choice among these protocols might depend on the properties of the specific medical domain, and the capabilities of the external database (e.g., object-oriented databases handle links among entities better); adding a transaction time to the patient's electronic record, while keeping the valid time (i.e., using a bitemporal database), would obviously be very helpful.

In addition, the capabilities of active and of deductive databases might provide several advantages, similar to a TMS [134]. In any case, the problem deserves further research.

- **Providing efficient storage protocols for hybrid architectures**. Finally, another issue, closely related to the conflict-resolution problem, is whether some, all, or none of the temporal-reasoning conclusions should be saved in the external, long-term database. Given that many abstractions are only intermediate, and that other abstractions might be changed by data arriving in the future (possibly even data with a past valid-time stamp, or data that exert some influence on the interpretation of the past), it might be advisable not to save any abstractions, due to their logically defeasible nature. However, it is obviously useful, from an efficiency point of view, to cache key conclusions for future use, either to respond to a direct query or to support another temporal-reasoning process. The caching is especially important for saving high-level abstractions, such as "nephrotic syndrome," that have occurred in the past, are unlikely to change, and are useful for interpreting the present. Such abstractions might be available for querying by other users (including medical decision-support programs), who do not necessarily have access to the temporal-abstraction module or to the domain's full temporal-abstraction knowledge base. One option that might be worth investigating is an episodic use of "temporal checkpoints" beyond which past abstractions are cached, available for querying but not for modification.
- **Temporal clinical data warehousing and mining**. An interesting issue faced by the research community is to collect temporal clinical data from different sources, to clean and merge them, and to analyze and mine highlighting interesting temporal patterns and association rules. Indeed, both temporal patterns and temporal association rules may provide insights on on clinical data, allowing to distinguish important relationships between vital signs, therapies, and the clinical paths/evolutions of patient's state [338].
- **Extraction of temporal information from unstructured clinical data**. A common important source for information about the longitudinal clinical course of a patient is a text-based narrative, such as discharge summaries and progress notes. The challenge in this case is to reconstruct the implicit temporal database underlying the narrative [437]. In some cases, this implicit database can only be guessed at, judging by the text. In other cases, researchers have validated the reconstructed temporal predicates by having access to the original quantitative timestamped patient records [438]. An example of highly useful application is the detection of adverse drug events from physicians' clinical notes [422].

Work on each of the new research areas we listed would contribute towards the important goal of integrating temporal data-maintenance and temporal-reasoning systems in medical domains, and thus lead to both a better understanding and to a better solution of important problems in management and reasoning about time-oriented clinical data.

Summary

In this chapter we have overviewed some specific aspects related to the management, modeling and querying of temporal clinical data. In particular we considered two different aspects: first, we considered multiple temporal dimensions of clinical data; then, we considered the issue of multiple granularities and indeterminacy in modeling and querying clinical data. Both these issues have been recognized as important for the medical domain by the scientific community, though of general interest. After having mentioned several clinical domains where the management of temporal data have been considered, we discussed the multi-dimensionality of clinical data: besides the well-known concepts of valid and transaction times, we discussed some other dimensions, namely the event time and the availability time, which are useful in the medical domain to be able to correctly consider clinical data. The chapter discussed, using some simple examples, both the modeling and the querying issues when multiple temporal dimensions are considered. Then, we discussed how to model and query clinical data given at different granularities or with indeterminacy. Both the relational approach and the object-oriented one have been discussed. The object-oriented approach has been described also with respect to a real clinical application, dealing with data from cardiology follow-up patients, and with regard to some architectural features, related to a web-based clinical temporal database system prototype. The main purpose of this chapter was to allow the reader to become aware of some domain specific, yet of general interest, issues when dealing with temporal clinical data and designing and implementing software tools for the management of medical data having complex and multi-faceted temporal features.

Bibliographic Notes

Apart from the specific references that are mentioned in this chapter, several survey papers covered topics related to the management of temporal clinical data. In particular, we mention here: the survey by Combi and Shahar [364] where both temporal reasoning and temporal data maintenance in medicine are considered; the position paper by Adlassnig et al. [5], where some specific research directions are discussed for temporal clinical databases, together with other research topics on temporal reasoning in medicine; the paper by Dorda et al. [127], where the authors summarize in 20 issues the most important lessons learnt in 25 years of development and use of temporal query systems in real and challenging clinical settings.

Problems

4.1. In Table 4.5 (relation *pat_sympt*), information about the symptoms of two patients, Mary (P_id=1) and Sam (P_id=2), are reported.

P_id	symptom	VT	ET_i	ET_t		TT
1	headache	[97Oct1, ∞)	97Sept5	null	[97Oct7, 97Oct15)	[97Oct10, 97Oct15)
2	vertigo	[97Aug8, 97Aug15)	97Aug7	97Aug12	[97Sept3, 97Oct17)	[97Oct15, 97Oct21)
2	vertigo	[97Aug10, 97Aug15)	97Aug7	97Aug12	[97Oct19, ∞)	[97Oct21, ∞)
1	headache	[97Oct1, 97Oct14)	97Sept5	97Oct9	[97Oct15, 97Oct20)	[97Oct15, 97Oct21)
1	headache	[97Oct1, 97Oct14)	97Sept15	97Oct9	[97Oct20, ∞)	[97Oct21, ∞)

Table 4.5 Database instance of patient symptoms: the relation *pat_sympt*.

1. What would be the result of determining which data is available to the physician on October 18, 1997, according to the database contents on October 20, 1997?
2. What would be the T4SQL query corresponding to the previous question?
3. Describe and discuss at least two possible scenarios consistent with data collected into the table.
4. Even on the base of the provided example, discuss the difference between availability and transaction times.

4.2. What is the difference between granularity and indeterminacy? Provide some medically-oriented examples of situations where the management of indeterminacy and granularity is needed.

4.3. Consider both the Chronus data model and GCH-OODM: what are the main differences with regards to the expression of clinical data involving uncertainty?

4.4. Read the following clinical scenario.

Dr. Jones has examined Ms. Smith in her clinic on May 7, 1997, at noon (Ms. Smith came since she had symptoms of a urinary-tract infection). Dr. Jones recorded a fever of 102° Fahrenheit at that time. About 15 minutes later, she has drawn a blood sample and sent it to the laboratory, and asked Ms. Smith to provide her with a urine sample. Dr. Jones proceeded to immediately test the urine for traces of white blood cells (WBC) (result: "highly positive"). She has also sent the urine sample to the laboratory, asking to culture it. The day after, Dr. Jones called the laboratory and recorded in her notes that the WBC count in Ms. Smith's blood sample was 11, 500/cc. Subsequently, the fever was measured as 98.6° on May 8, between 1:00 a.m. and 1:37am. On May 10, 11:33 a.m., Dr. Jones added also that she just learned that the urine culture had been positive for E. Coli bacteria, and that the bacteria were sensitive to Sulfisoxazole. Dr. Jones called Ms. Smith two hours later and told her to start taking that drug (dose: 1gr, frequency: four times a day), which she has previously prescribed for her, but asked that she wait before using it. Afterwards, on May, Dr. Jones got another call from the laboratory; it seems that there was an error in the first report; in fact, the bacteria were resistant to Sulfisoxazole but were sensitive to Ampicilin. Dr. Jones promptly called Ms. Smith and asked her to stop the Sulfisoxazole and to start an Ampicilin regimen (dose: 500mg, frequency: four times a day). She recorded immediately the details of the administrations of the two drugs in her notebook on May 17.

1. What are the temporal aspects of this scenario that cannot be captured by a GCH-OODM temporal database? Please, provide a motivation to your answer.
2. Create a database schema in GCH-OODM and a related temporal database replacing Dr. Jones' notebook, which will describe all the parameters and events mentioned in the scenario that can be captured by GCH-OODM.

4.5. Provide the results of all the example queries of the chapter, considering the databases provided for multidimensional temporal clinical data and for temporal data with different granularities and indeterminacy, respectively.

Chapter 5
Abstraction of Time-Oriented Clinical Data

Overview

The chapter aims to give a comprehensive and critical review of current approaches to the common task of abstraction of time-oriented data in medicine, or *temporal abstraction*. Temporal-data abstraction constitutes a central requirement that presently receives much and justifiable attention. The role of this process is especially crucial in the context of time-oriented clinical monitoring, therapy planning, and exploration of clinical databases. General theories of time typically used in artificial intelligence do not fully address the requirements for temporal abstraction in medical reasoning (see Chapter 2).

The reader is expected to gain the following from this chapter:

- An understanding of basic types of data abstraction, both atemporal and temporal types, and of the desired computational behavior of a process that creates meaningful abstractions, and overall an appreciation of the significance of interval-based abstractions.
- A familiarization with early approaches in dealing with time-oriented medical data in the context of rudimentary knowledge discovery in terms of temporal-causal relations (Rx system), and in the context of producing intelligent summaries of patient data (systems IDEFIX and TOPAZ), as well as a familiarization with early and more recent approaches in the context of time-oriented monitoring of patients (systems VM, TCS, TUP, TrenDx, M-HTP and VIE-VENT).
- An understanding of the new concept of a temporal mediator, a system that aims to provide facilities both for temporal reasoning and temporal maintenance, thus forming a linkage between a database system (storing patient data and their abstractions) and a decision-support system, while decoupling the temporal-data abstraction process from the specifics of the decision-support process.
- A detailed understanding of Shahar's generic Knowledge-Based Temporal Abstraction (KBTA) ontology and associated method and the RÉSUMÉ system that implements it.

C. Combi et al., *Temporal Information Systems in Medicine*,
DOI 10.1007/978-1-4419-6543-1_5, © Springer Science+Business Media, LLC 2010

Structure of the Chapter

The chapter is structured as follows. Section 5.1 gives an exposition on various, key types of temporal data abstraction, namely merge (or state) abstraction, persistence abstraction, trend abstraction, and periodic abstraction. Obviously there are other types of temporal-data abstraction, given the specifics of different domains. The same section overviews the desired computational behavior of a method that creates meaningful abstractions from time-stamped data in medical domains. Sections 5.2 and 5.3 present different approaches to temporal data abstraction. More specifically Section 5.2 outlines data abstraction for knowledge discovery and presents three pioneering attempts: the Rx system that examined a time-oriented clinical database, and produced a set of possible causal relationships among various clinical parameters; the IDEFIX system that created an intelligent summary of a patient's current status using an electronic medical record; and the later TOPAZ system that also produced intelligent summaries of patient records, exploiting an integrated interpretation model approach. Section 5.3 focuses on approaches whose scope of application is primarily time-oriented monitoring. Here again a number of representative approaches are presented and discussed, starting from the classic system VM. This was one of the first knowledge-based systems that included an explicit representation of time. It was designed to assist care providers managing patients on ventilators in intensive-care units. The other approaches presented are: the TCS system that aims to decouple time-oriented reasoning from the domain-specific inference procedures, emphasizing the derivation of state abstractions; the TUP system that represents qualitative and quantitative relations among temporal intervals, and maintains and propagates the constraints posed by these relations through a constraint network of temporal (or any other) intervals; the TrenDx system that focuses on using efficient general methods for representing and detecting predefined temporal patterns in raw time-stamped data; the M-HTP system that monitors heart-transplant patients; and the VIE-VENT system, that validates data and plans therapy for artificially ventilated newborn infants. It is restated that the presented systems do not provide an exhaustive list of all approaches so far, but merely a representative sample. Section 5.4 presents a novel new idea, that of a temporal mediator, a system that merges temporal reasoning and temporal maintenance. Temporal-data maintenance includes the capability to store and retrieve also the different temporal dimensions distinguished in time-oriented data (transaction time, valid time, user-defined time and more recently other specialized times such as decision-time, event time, and availability time [93, 94]). The mediator serves as an intermediate layer of processing between client applications and databases, and as a result it is tied neither to a particular application, domain, or task, nor to a particular database. This important concept is illustrated through the IDAN architecture. Section 5.5 presents in detail Shahar's ontology for knowledge-based temporal-data abstraction (KBTA), a comprehensive, domain-independent ontology specific only to the task of temporal abstraction. This ontology and its associated method, which is also presented, have been implemented in the RÉSUMÉ system. The chapter concludes by summarizing what has been done and suggesting issues that need further exploration.

Keywords

Temporal abstraction, temporal reasoning, merge abstraction, state abstraction, persistence abstraction, trend abstraction, periodic abstraction, computational characteristics of an abstraction process, knowledge discovery, patient record summarization, Rx, IDEFIX, TOPAZ, time-oriented monitoring, VM, Temporal Control Structure (TCS), Temporal-Utilities Package (TUP), TrenDx, M-HTP, VIE-VENT, temporal-abstraction mediator, Knowledge-Based Temporal-Abstraction (KBTA) ontology and method, IDAN, RÉSUMÉ.

5.1 Temporal-Data Abstraction

Medical knowledge-based systems involve the application of medical knowledge to patient-specific data with the goal of reaching diagnoses or prognoses, deciding the best therapy regime for the patient, or monitoring the effectiveness of some ongoing therapy and if necessary applying rectification actions. Medical knowledge, like any kind of knowledge, is expressed in as general a form as possible, for example, in terms of associations or rules, causal models of pathophysiological states, behavior (evolution) models of disease processes, patient management protocols and guidelines, etc. Data on a specific patient, on the other hand, comprise numeric measurements of various parameters (such as blood pressure, body temperature, etc.) at different points in time. The record of a patient gives the history of the patient (past operations and other treatments), results of laboratory and physical examinations as well as the patient's own symptomatic recollections.

To perform any kind of medical problem solving, patient data have to be "matched" against medical knowledge. For example, a forward-driven rule is activated if its antecedent can be unified against patient information; similarly, a patient management protocol is activated if its underlying preconditions can be unified against patient information, etc. The difficulty encountered here is that often the abstraction gap between the highly specific, raw patient data, and the highly abstract medical knowledge does not permit any direct unification between data and knowledge. The process of data abstraction aims to close this gap; in other words, it aims to bring the raw patient data to the level of medical knowledge in order to permit the derivation of diagnostic, prognostic or therapeutic conclusions. Hence data abstraction can be seen as an auxiliary process that aids the problem solving process per se. However it is a critical auxiliary process since the success of some medical knowledge-based system can depend on it; data abstraction involves low level processing, but this processing could be of a more "intelligent" and computationally demanding nature, than that of the higher level reasoning process.

The significance of a data abstraction process in the context of a knowledge-based system was first perceived by Clancey in his seminal proposal on heuristic classification [75]. In Clancey's work, data abstraction is used as the stepping stone towards the activation of nodes on a solution hierarchy. Such nodes, especially at the high levels of the hierarchy, are associated with triggers, where a trigger is a conjunction of observable items of information. In heuristic classification, data abstraction is applied in an event-driven fashion with the aim of mapping raw case data to the level of abstraction used in the expression of triggers, in order to enable the activation of triggers (i.e., their unification against data).

Obviously, a knowledge-based system that does not possess any data abstraction capabilities would require its user to express the case data at the level of abstraction corresponding to its knowledge. Such a system puts the onus on the user to perform the data abstraction process. This approach has limitations. Firstly the user, often a non-specialist, is burdened with the task of not only observing, measuring, and reporting data, but also of interpreting such data for the special needs of the particular problem solving. Secondly, manual abstraction is prone to errors and inconsistencies even for domains where it can be considered "doable". There are, however, many domains where the sheer amount of raw data renders such a thing practically impossible. In short, the usefulness of a medical knowledge-based system that does not possess data abstraction capabilities is substantially reduced. For instance, in clinical domains, a final diagnosis is not always the main goal. What is often needed is a coherent intermediate-level interpretation of the relationships between data and events, and among data, especially when the overall context (e.g., a major diagnosis) is known. The goal is then to abstract the clinical data, which often is acquired or recorded as time-stamped measurements, into higher-level concepts, which often hold over time periods. These concepts should be useful for one or more tasks (e.g., planning of therapy or summarization of a patient's record). Thus, the goal is often to create, from time-stamped input data, interval-based temporal abstractions, such as "bone-marrow toxicity grade 2 or more for 3 weeks in the context of administration of a prednisone/azathioprine protocol for treating patients who have chronic graft-versus-host disease, and complication of bone-marrow transplantation" and more complex patterns, involving several intervals (Figure 5.1).

5.1.1 Types of Data Abstraction

The purpose of data abstraction, in the context of medical problem solving, is therefore the intelligent interpretation of the raw data on some patient, so that the derived abstract data are at the level of abstraction corresponding to the given body of knowledge. Abstract data are useful since they can be unified against knowledge.

There are different types of data abstraction. Some are rather simple and others quite complicated. The types discussed below are more for illustration; they are not meant to provide an exhaustive classification. This is due to the rather open-ended nature of data abstraction and the multitude of ways basic types can be combined to yield complex types. The common feature of all these types, even the very simple ones, is that their derivation is knowledge-driven; hence data abstraction is itself a knowledge-based process. The use of knowledge in the derivation of abstractions is the feature that distinguishes data abstraction from statistical data analysis, e.g.,

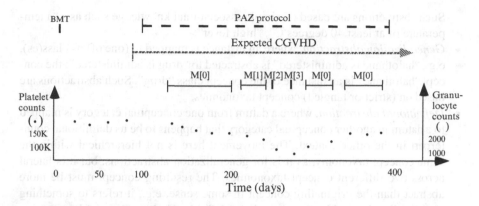

Fig. 5.1 Temporal abstraction of platelet and granulocyte values during administration of a prednisone/azathioprine (PAZ) clinical protocol for treating patients who have chronic graft-versus-host disease (CGVHD). The time line starts with a bone-marrow-transplantation (BMT) external event. The platelet- and granulocyte-count parameters and the PAZ and BMT external events (interventions) are typical inputs. The abstraction and context intervals are typically part of the output. · = platelet counts; Δ = granulocyte counts; dashed line with bars = event; striped arrow = open context interval; full line with bars = closed abstraction interval; M[n] = myelotoxicity (bone-marrow-toxicity) grade n.

the derivation of trends through time-series analysis. Data abstraction is knowledge-based and heuristic while statistical analysis is "syntactic" and algorithmic.

Before listing the types of data abstraction it is necessary to say a few words about the nature of raw patient data. Their highly specific form has already been stressed. In addition they can be noisy and inconsistent. For some domains, e.g., intensive care monitoring, the data are voluminous, while for other domains they are grossly incomplete, e.g., for medical domains dealing with skeletal abnormalities. Different medical parameters can have very different sampling frequencies and hence different time units (granularities) arise. Thus for some parameter there could be too much and very specific data, while for another parameter only very few and far between recordings. In either case, data abstraction tries to ferret out the useful (abstract) information, safeguarding against the possibility of noise; in the first case it tries to eliminate the detail while in the second case to fill the gaps, two orthogonal aims. Since noise is an unavoidable phenomenon a viable data abstraction process should perform some kind of data validation and verification which also makes use of knowledge [179].

Simple types of data abstraction are atemporal and often involve a single datum, which is mapped to a more abstract concept. The knowledge underlying such abstractions often comprises concept taxonomies or concept associations. Examples of simple data abstractions are:

- *Qualitative abstraction*, where a numeric expression is mapped to a qualitative expression, e.g., "a temperature of 41 degrees C" is abstracted to "high fever".

Such abstractions are based on simple associational knowledge such as <"a temperature of at least 40 degrees C", "high fever">

• *Generalization abstraction*, where an instance is mapped to (one of) its class(es), e.g., "halothane is administered" is abstracted to "drug is administered"; the concept "halothane" is an instance of the concept class "drug". Such abstractions are based on (strict or tangled) concept taxonomies.

• *Definitional abstraction*, where a datum from one conceptual category is mapped to a datum in another conceptual category that happens to be its definitional counterpart in the other context. The movement here is not hierarchical within the same concept taxonomy, as it is for generalization abstractions, but it is lateral across two different concept taxonomies. The resulting concept must be more abstract than the originating concept in some sense, e.g., it refers to something more easily observable. An example of definitional abstraction is the mapping of "generalized platyspondyly" to "short trunk". "Generalized platyspondyly" is a radiological concept, the observation of which requires the taking of a radiography of the spine; platyspondyly means flattening of vertebrae and generalized platyspondyly means the flattening of all vertebrae. "Short trunk" is a clinical concept, the observation of which does not require any special procedure. The knowledge driving such abstractions consists of simple associations between concepts across different categories.

In all the above types of data abstraction time is implicit. The abstractions refer to the same times, explicitly or implicitly, associated with the raw data. Thus in an atemporal situation, where everything is assumed to refer to 'now', we have the general implication $holds(P,D) \Longrightarrow holds(P,abs(D))$, where predicate $holds$ denotes that datum D holds for patient P now and function abs embodies any of the above types of simple abstraction. Predicate $holds$ can be extended to have a third argument giving an explicit time, thus having $holds(P,D,T_1) \Longrightarrow holds(P,abs(D),T_2)$. (In the case of a basic abstraction, such as a definitional one, $T_1 = T_2$; in the case of general temporal patterns, or inducement of a context [e.g., following some drug administration] $T1 \neq T2$ is common). When time becomes an explicit and inherent dimension of patient data, and medical knowledge, it plays a central role in data abstraction, hence the name *temporal data abstraction*.

The dimension of time adds a new aspect of complexity to the derivation of (temporal) abstractions. In the simple types of (atemporal) data abstraction discussed above, often it is just a single datum which is mapped to a more abstract datum, although several data (with implicitly the same temporal dimension) might be mapped to one abstract concept. In temporal abstractions, however, it is a cluster of (time-stamped) data that is mapped to an abstract temporal datum. Atemporal data abstraction is "concept abstraction", going from a specific concept to a more abstract concept. Temporal data abstraction is both "concept abstraction" and "temporal abstraction". The latter encompasses different notions, such as going from discrete time-points (used in the expression of raw patient data) to continuous (convex) time-intervals or (nonconvex) collections of time-intervals (used in the expression of medical knowledge), or moving from a fine time granularity to a grosser time granularity, etc. Temporal data abstraction can therefore be decomposed into concept abstraction, i.e., atemporal data abstraction, followed by temporal abstraction. The reverse sequence is not valid since the (concrete) concepts involved have to be mapped to more abstract concepts, to facilitate temporal abstractions.

Temporal data abstraction entails temporal reasoning, both of a commonsense nature (e.g., intuitive handling of multiple time granularities and temporal relations such as before, overlaps, disjoint, etc.), as well as of a specialist nature dealing with persistence semantics of concepts, etc. Examples of important types of temporal abstraction are (here a datum is assumed to be an association between a property and a temporal aspect, which often is a time-point at a given time-unit; a simple property is a tuple comprising a subject [parameter or concept] and a list of attribute value pairs):

- *Merge abstraction*, where a collection of data, all having the *concatenable* property [376] (which enables concatenating several intervals, over which the same type of proposition holds, into a longer one) and whose temporal aspects collectively form a (possibly overlapping) chain are abstracted to a single datum with the given property, whose temporal aspect is the maximal time-interval spanning the original data. For example three consecutive, daily, recordings of fever can be mapped to the temporal abstraction that the patient had fever for a three-day interval. Merge abstraction is also known as state abstraction, since its aim is to derive maximal intervals over which there is no change in the state of some parameter.

- *Persistence abstraction*, where again the aim is to derive maximal intervals spanning the extent of some property; here, though, there could be just one datum on that property, and hence the difficulty is in filling the gaps by "seeing" both backwards and forwards in time from the specific, discrete, recording of the given property. For example, if it is known that the patient had headache in the morning, can it be assumed that he also had headache in the afternoon and/or the evening before? Also if the patient is reported to have gone blind in one eye in December 1997 can it be assumed that this situation persists now? In some temporal reasoning approaches the persistence rule is that some property is assumed to persist indefinitely until some event (e.g., a therapy) is known to have taken place and this terminates the persistence of the property. This rule is obviously unrealistic for patient data, since often symptoms have a finite existence and go away even without the administration of any therapy. Thus, persistence derivation with respect to patient data can be a complicated process, drawing from the persistence semantics of properties. These categorize properties into finitely or infinitely persisting, where finitely persisting properties are further categorized into recurring and non-recurring. In addition, ranges for the duration of finitely persisting properties may be specified, in the absence of any external factors such as treatments. Thus blindness could be classified as an infinitely persistent property, chickenpox as a finitely persisting but not a recurring property, and flu as a finitely persisting, recurring, property. Persistence derivation is often context-sensitive, where contexts can also be dynamically derived (abstracted) from the raw data [347]. Within different contexts, clinical propositions can have different persistence semantics. As already said, persistence can be either forward in time, to the future, or backwards in time, to the past. Thus, a certain value of hemoglobin measured at a certain time point might indicate that with high probability the value was true at least within the previous day and within the next 2 days [347]. For example, if it is known that the patient with the headache took aspirin at noon, it can be inferred that the persistence of headache lasted most probably (that is, above a certain probability threshold) up to about 1pm, and that with high-enough probability there was no headache up to 3pm. This is based on the derivation of the time interval spanning the persistence of the effectiveness of the event of aspirin administration; e.g., relevant knowledge may dictate that this starts about 1 hour after the occurrence of the event and lasts for about 2 hours. Such time intervals defining the persistence of the effectiveness of treatments are referred to as context intervals. Qualitative abstractions (see above) can also be context-sensitive.

- *Trend abstraction*, where the aim is to derive the significant changes and the rates of change in the progression of some parameter. Trend abstraction entails merge and persistence abstraction in order to derive the extents where there is no change in the value of the given parameter. However the difficulty is in subsequently joining everything together (which may well involve filling gaps), deciding the points of significant change and the directions of change. Again this type of abstraction is driven by knowledge. Most of the current work in temporal data abstraction concerns trend abstraction, where often the medical domain under examination involves especially difficult data such as very noisy and largely incomplete data [233]. The problem is often compounded by a fast rate of arriving sampled data, such as in the intensive care unit domain, which necessitates initial smoothing of the raw data before extracting meaningful trends [339].

- *Periodic abstraction*, where repetitive occurrences, with some regularity in the pattern of repetition, are derived, e.g., headache every morning, for a week, with increasing severity. Such repetitive, cyclic occurrences are not uncommon in medical domains. A periodic abstraction is expressed in terms of a repetition element (e.g., headache), a repetition pattern (e.g., every morning for a week) and a progression pattern (e.g., increasing severity) [207]. It can also be viewed as a set of local constraints on the time and value of the repeating element (single time interval), and a set of global constraints on each pair of elements or even on the whole element set [58, 59]. The repetition element can be of any order of complexity (e.g., it could itself be a periodic abstraction, or a trend abstraction), giving rise to very complex periodic abstractions. The period spanning the extent of a periodic occurrence is nonconvex by default; i.e., it is the collection of time intervals spanning the extents of the distinct instantiations of the repetition element, and the collection can include gaps. Periodic abstraction uses the other types of data abstraction and it is also knowledge driven. Relevant knowledge can include acceptable regularity patterns, means for justifying local irregularities, etc. The knowledge-intensive, heuristic, derivation of periodic abstractions is currently largely unexplored although its significance in medical problem solving is widely acknowledged.

Table 5.1 summarizes the discussed types of data abstraction. As already mentioned these types can be combined in a multitude of ways yielding complex abstractions. As already explained, data abstraction is deployed in some problem solving system and hence the derivation of abstractions is largely done in a directed fashion. This means that the given system, in exploring its hypothesis space, predicts various abstractions which the data abstraction process is required to corroborate against the raw patient data; in this respect the data abstraction process is goal-driven. However, for the creation of the initial hypothesis space the data abstraction process needs to operate in a non-directed, or event-driven fashion (as already discussed with respect to Clancey's seminal proposal [75]). In the case of a monitoring system, data abstraction, which is the heart of the system, in fact operates in a largely event-driven fashion. This is because the aim is to comprehensively interpret all the data covered by the moving time window underlying the operation of the monitoring system, i.e., to derive all abstractions, of any degree of complexity, and on the basis of such abstractions the system decides whether the patient situation is static, or it is improving or worsening.

Non-directed data abstraction repeatedly applies the methods for deriving the different types of data abstraction, until no more derivations are possible. Data abstraction, operating under such a mode, can be used in a stand alone fashion, i.e., in direct interaction with the user rather than with a higher level reasoning engine; in such a case the derived abstractions should be presented to the user in a visual form. Visualization is also of relevance when a data abstraction process is not used in a stand alone fashion; since the overall reasoning of the system depends critically on the derived abstractions a good way of justifying this reasoning is the presentation of the relevant abstractions in a visual form.

In summary, the ability to automatically create interval-based abstractions of time-stamped clinical data has multiple implications:

1. *Data summaries* of time-oriented electronic data, such as patient medical records, have an immediate value to a human user, such as to a care provider scanning a long patient record for meaningful trends [128]. These summaries can be presented in a visual format as we shall discuss in detail in Chapter 8, but can also be used to generate natural language summaries of large amounts of data, as was effectively achieved in the neonatal intensive care unit domain [316].
2. Temporal abstractions support *recommendations* by intelligent decision-support systems, such as diagnostic and therapeutic systems [212].
3. Abstractions support *monitoring* of plans (e. g., therapy plans) during execution of these plans (e. g., application of clinical guidelines [278]). More will be elaborated on this aspect when clinical guidelines are discussed in Chapter 7.
4. Meaningful time-oriented contexts enable generation of *context-specific abstractions*, maintenance of *several interpretations* of the same data within different contexts, and certain hindsight and foresight inferences [347].
5. Temporal abstractions are helpful for *explanation* of recommended actions by an intelligent system.
6. Temporal abstractions are a useful representation for the intentions of designers of clinical guidelines, and enable real time and retrospective critiquing and

Table 5.1 Example Data Abstraction Types

Atemporal Types
• *Qualitative Abstraction*: Converting numeric expressions to qualitative expressions. • *Generalization Abstraction*: Mapping instances into classes. • *Definitional Abstraction*: Mapping across different conceptual categories.
Temporal Types
• *Merge (or State) Abstraction*: Deriving maximal intervals for some concatenable property from a group of time-stamped data for that property. • *Persistence Abstraction*: Applying (default) persistence rules to project maximal intervals for some property, both backwards and forwards in time and possibly on the basis of a single datum. • *Trend Abstraction*: Deriving significant changes and rates of change in the progression of some parameter. • *Periodic Abstraction*: Deriving repetitive occurrences, with some regularity in the pattern of repetition.

quality assessment of the application of these guidelines by care providers [355]. This aspect has been emphasized within the Asgaard project and will be described in detail in Chapter 7.

7. Domain-specific, meaningful, interval-based characterizations of time-oriented medical data are a prerequisite for effective *visualization* and dynamic exploration of these data by care providers and researchers [351, 352, 350]. Visualization and exploration of information in general, and of large amounts of time-oriented medical data in particular, is essential for effective decision making. Examples include deciding whether a patient had several episodes of bone marrow toxicity of a certain severity and duration, caused by therapy with a particular drug; or deciding if a certain therapeutic action has been effective. Different types of care providers require access to different types of time-oriented data, which might be distributed over multiple databases.

However, there are several points to note with respect to the *desired* computational behavior of a method that creates meaningful abstractions from time-stamped data in medical domains:

1. The method should be able to accept as input both *numeric* and *qualitative* data. Some of these data might be at *different levels of abstraction* (i.e., we might be given either raw data or higher-level concepts as primary input, perhaps abstracted by the care provider from the same or additional data). The data might

also involve different forms of temporal representation (e.g., time *points* or time *intervals*).

2. The output abstractions should also be available for query purposes *at all levels of abstraction*, and should be created as time *points* or as time *intervals*, as necessary, aggregating relevant conclusions together as much as possible (e.g., "extremely high blood pressures for the past 8 months in the context of treatment of hypertension"). The outputs generated by the method should be controlled, sensitive to the goals of the abstraction process for the task at hand (e.g., only particular types of output might be required). The output abstractions should also be sensitive to the context in which they were created.

3. Input data should be used and incorporated in the interpretation even if they arrive *out of temporal order* (e.g., a laboratory result from last Tuesday arrives today). Thus, the past can change our view of the present. This phenomenon has been called a *view update*[359]. Furthermore, new data should enable us to reflect on the past; thus, the present (or future) can change our interpretation of the past, a property referred to as *hindsight* [336].

4. Several possible interpretations of the data might be reasonable, each depending on additional factors that are perhaps unknown at the time (such as whether the patient has AIDS); interpretation should be specific to the context in which it is applied. All reasonable interpretations of the same data relevant to the task at hand should be available automatically or upon query.

5. The method should leave room for some *uncertainty* in the input and the expected data *values*, and some uncertainty in the *time* of the input or the expected temporal pattern.

6. The method should be generalizable to other clinical domains and tasks. The domain-specific assumptions underlying it should be explicit and as declarative as possible (as opposed to procedural code), so as to enable reuse of the method without rebuilding the system, *acquisition* of the necessary knowledge for applying it to other domains, *maintenance* of that knowledge, and *sharing* that knowledge with other applications in the same domain.

5.2 Approaches to Temporal Data Abstraction

This section discusses a number of specific approaches to temporal data abstraction. As already mentioned temporal data abstraction is primarily used for converting the raw data on some patient to more useful information. However, it can also be used for discovering new knowledge. The latter role of temporal data abstraction is briefly addressed in this section that also overviews one of the first programs developed for this purpose, Rx. This program aimed to discover possible causal relationships by examining a time-oriented clinical database. The principal role of temporal data abstraction, summarization of patient records, is addressed more extensively, through two pioneering systems, IDEFIX and TOPAZ. Further approaches of this type, used specifically in the context of patient monitoring, are addressed in Section 5.3

5.2.1 Data Abstraction for Knowledge Discovery

Data abstraction and more specifically temporal data abstraction can be utilized for the discovery of medical knowledge. Data is patient specific, while knowledge is patient independent, it consists of generalizations that apply across patients. Machine learning for medical domains aims to discover medical knowledge by inducing generalizations from the records of representative samples of patients. Trying to induce such generalizations directly from the raw patient data, which are recorded at the level of "the systolic blood pressure reading was 9 at 10 am on March 26th 1998", is infeasible, since at this level finding the same datum in more than one patient's record is highly unlikely. True generalizations can be more effectively discovered by comparing the patient profiles at a high level of abstraction, in terms of derived data abstractions such as periodic occurrences, trends and other temporal patterns. Different raw data can yield the same abstractions, even if they differ substantially in volume. The number of derived abstractions is relatively constant across patients with the same medical situation, and of course this number is considerably smaller than the number of raw data. Temporal data abstractions reveal the essence of the profile of a patient, hide superfluous detail, and last but not least eliminate noisy information. Furthermore, the temporal scope of abstractions such as trends and periodic occurrences are far more meaningful and prone to adequate comparison than the time-points corresponding to raw data. In addition, temporal abstractions incorporate domain-specific knowledge (e.g., meaningful ranges) that might not be learnable from the raw data itself. If the same complex abstraction, such as a nested periodic occurrence, is associated with a significant number of patients from a representative sample, it makes a strong candidate for knowledgehood.

Current machine learning approaches do not attempt to first abstract, on an individual basis, the example cases that constitute their training sets, and then to apply whatever learning technique they employ for the induction of further generalizations. Strictly speaking every machine learning algorithm performs a kind of abstraction over the entire collection of cases; however it does not perform any abstraction on the individual cases. Cases tend to be atemporal, or at best they model time (implicitly) as just another attribute. Data abstractions on the selected cases are often manually performed by the domain experts as a preprocessing step. Such manual processing is prone to non uniformity and inconsistency, while the automatic extraction of abstractions is uniform and objective.

One of the goals behind the staging of a series of international workshops called IDAMAP (Intelligent Data Analysis in Medicine and Pharmacology) is to bring together the machine learning and temporal data abstraction communities interested in medical problems [238]. The efforts have led to significant progress in a new area usually referred to as *temporal data mining*, in which interval-based abstractions are used as features for a supervised or non supervised data mining process that either characterizes the data as containing meaningful temporal patterns, or associates such patterns with a given outcome [338, 272, 276]. Another approach enables the user to interactively explore a visual representation of the abstractions and their association [219].

5.2.2 Discovery in Time-Oriented Clinical Databases: Blum's Rx Project

Rx [31] was a program that examined a time-oriented clinical database, and produced a set of possible causal relationships among various clinical parameters. Rx used a *discovery module* for automated discovery of statistical correlations in clinical databases. Then, a *study module* used a medical knowledge base to rule out spurious correlations by creating and testing a statistical model of a hypothesis. Data for Rx were provided from the American Rheumatism Association Medical Information System (ARAMIS), a chronic-disease time-oriented database that accumulates time-stamped data about thousands of patients who have rheumatic diseases and who are usually followed for many years. The ARAMIS database evolved from the mainframe-based Time Oriented Database (TOD) [147]. Both databases incorporate a simple three-dimensional structure that records, in an entry indexed by the patient, the patient's visit, the clinical parameter, and the value of that parameter, if entered on that visit. The TOD was thus a *historical* database [382] (see Chapter 3).

The representation of data in the Rx program included *point events*, such as a laboratory test, and *interval events*, which required an extension to TOD to support diseases, the duration of which was typically more than one visit. The medical knowledge base was organized into two hierarchies: *states* (e.g., disease categories, symptoms, and findings) and *actions* (drugs).

The Rx program determined whether interval-based complex states, such as diseases, existed by using a hierarchical *derivation tree*: Event A can be defined in terms of events B_1 and B_2, which in turn can be derived from events C_{11}, C_{12}, C_{13} and C_{21}, C_{22}, and so on. When necessary, to assess the value of A, Rx traversed the derivation tree and collected values for all A's descendants [31].

Due to the requirements of the Rx modules, in particular those of the study module, Rx sometimes had to assess the value of a clinical parameter when it was not actually measured, a so called *latent* variable. One way to estimate latent variables was by using *proxy variables* that are known to be highly correlated with the required parameter. An example is estimating what was termed in the Rx project the *intensity* of a disease during a visit when only some of the disease's clinical manifestations had been measured.

The main method used to access data at time points when a value for them did not necessarily exist used *time-dependent database access functions*. One such function was *delayed-action (variable, day, onset-delay, interpolation-days)*, which returned the assumed value of *variable* at *onset-delay* days before *day*, but not if the last visit preceded *day* by more than *interpolation-days* days. Thus, the dose of prednisone therapy, 1 week before a certain visit, was concluded on the basis of the dose known at the previous visit, if that previous visit was not too far in the past. A similar *delayed-effect* function for states used interpolation if the gap between visits was not excessive. The *delayed-interval* function, whose variable was an interval event, checked that no residual effects of the interval event remained within a given carryover time interval. Other time-dependent database-access functions included functions such as *previous-value* (*variable, day*), which returned the last value before *day*; *during* (*variable, day*), which returned a value of *variable* if *day* fell within an episode of *variable*; and *rapidly_tapered* (*variable, slope*), which returned the interval events in which the point event *variable* was decreasing at a rate greater than *slope*. All these functions and their intelligent use were assumed to be supplied by the user. Thus, Rx could have a modicum of control over *value uncertainty* and *persistence uncertainty*.

In addition, to create interval events, Rx used a parameter-specific *intraepisode gap* to determine whether visits could be joined, and an *interepisode definition* using the medical knowledge base to define clinical circumstances under which two separate intervals of the parameter could *not* be merged. The intraepisode gap was not dependent on clinical contexts or on other parameters.

5.2.3 Summarization of On-line Medical Records

5.2.3.1 De Zegher-Geets' IDEFIX Program for Medical-Record Summarization

De Zegher Geets' IDEFIX program [435, 436], had goals similar to an earlier program developed by Downes [128]-namely, to create an intelligent summary of the patient's current status, using an electronic medical record. IDEFIX used the ARAMIS project's database (in particular, for patients who had systemic lupus erythematosus (SLE)). This program updated the disease likelihood by using essentially a Bayesian odds-update function. IDEFIX used probabilities that were taken from a probabilistic interpretation of the INTERNIST-1 [266] knowledge base, based on Heckerman's work [176]. However, IDEFIX dealt with some of the limitations of Downs' program, such as the assumption of infinite persistence of the same abnormal attributes, and the merging of static, general, and dynamic, patient-specific, medical knowledge. IDEFIX also presented an approach for solving a problem closely related to the persistence problem-namely, that older data should be used, but should not have the same weight for concluding higher-level concepts as do new data. IDEFIX used weighted severity functions, which computed the severity of the manifestations (given clinical cut-off ranges) and then the severity of the state or disease by a linear-combination weighting scheme. (Temporal evidence, however, had no influence on the total severity of the abnormal state.) Use of clinical, rather than purely statistical, severity measures improved the performance of the system-the derived conclusions were closer to those of human expert care providers looking at the same data.

The IDEFIX medical knowledge ontology included abnormal primary attributes (APAs), such as the presence of protein in the urine; abnormal states, such as nephrotic syndrome; and diseases, such as SLE-related nephritis. APAs were derived directly from ARAMIS attribute values. IDEFIX inferred abnormal states from APAs; these states were essentially an intermediate-level diagnosis. From abnormal states and APAs, IDEFIX derived and weighted evidence to deduce the likelihood and severity of diseases, which were higher-level abnormal states with a common etiology. IDEFIX used two strategies. First, it used a goal-directed strategy, in which the program sought to explain the given APAs and states and their severity using the list of known complications of the current disease (e.g., SLE). Then, it used a data-driven strategy, in which the system tried to explain the remaining, unexplained APAs using a cover-and-differentiate approach based on odds-likelihood ratios.

De Zegher-Geets added a novel improvement to Downs' program by using time-oriented probabilistic functions (TOPFs). A TOPF was a function that returned the conditional probability of a disease D given a manifestation M, $P(D|M)$, as a function of a time interval, if such a time interval was found. The time interval could be the time since M was last known to be true, or the time since M started to be true, or any other expression returning a time interval. Figure 5.2 shows a TOPF for the conditional probability that a patient with SLE has a renal complication (lupus nephritis) as time passes from the last known episode of lupus nephritis. A temporal predicate that used the same syntax as did Downs' temporal predicates, but which could represent higher-level concepts, was used to express the temporal interval for which IDEFIX looked. For instance, PREVIOUS. ADJACENT. EPISODE(LUPUS. NEPHRITIS) looked for the time since the last episode of lupus nephritis. Thus, *as time progressed, the strength of the (probabilistic) connection between the disease and the manifestation could be changed in a predefined way*. For instance, as SLE progressed in time, the probability of a complication such as lupus nephritis increased as a logarithmic function (Figure 5.2). TOPFs were one of four functions: linear increasing, exponential decreasing, exponential increasing and logarithmic. Thus, only the type and coefficients of the function had to be given, simplifying the knowledge representation.

Note that TOPFs were used to compute only *positive* evidence; *negative* evidence likelihood ratios were constant, which might be unrealistic in many domains. The derivation of diseases was theoretically based on derived states, but in practice depended on APAs and states. In addition, TOPFs did not depend on the *context* in which they were used (e.g., the patient is also receiving a certain therapy) or on the *value* of the manifestation (e.g., the severity of the last lupus-nephritis episode). TOPFs were not dependent on the *length* of time for which the manifestation was true (i.e., for how long a manifestation, such as the presence of lupus nephritis, existed).

TOPFs included an implicit strong assumption of *conditional independence* among related diseases and findings (some of which was alleviated by grouping together of related findings as disjunctions). Knowledge about APAs included an *expected time of validity* attribute, but it was also, like TOPFs, independent of the clinical context.

The *goal* of the IDEFIX reasoning module was to explain, for a particular patient visit, the various manifestations for that visit, taking as certain all previous data. There was no explicit intention of creating *interval-based abstractions*, such as "a 6-month episode of lupus nephritis" for the purposes of enabling queries by a care provider or by another program; such conclusions were apparently left to the care

provider who, using the graphical display module, looked at all the visits[1]. Therefore, such intervals were not used explicitly by the reasoning module.

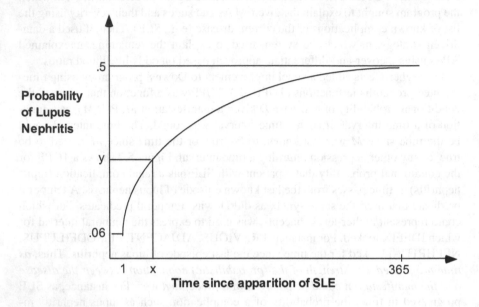

Fig. 5.2 A time-oriented probabilistic function (TOPF) associated with the predicate "previous
episode of lupus nephritis." Modified from [435].

5.2.3.2 Kahn's TOPAZ System: an Integrated Interpretation Model

Kahn has suggested using more than one temporal model to exploit the full power
of different formalisms of representing medical knowledge. Kahn [199] has implemented a temporal-data summarization program, TOPAZ, based on three temporal
models (Figure 5.3):

1. A numeric model represented quantitatively the underlying processes, such as
   bone-marrow responses to certain drugs, and their expected influence on the patient's granulocyte counts. The numeric model was based on differential equa-
   tions expressing relations among hidden patient-specific parameters assumed
   by the model, and measured findings. When the system processed the initial
   data, the model represented a *prototypical-patient model* and contained general,

---

[1] In fact, the graphical module originally assumed infinite persistence of states, and concatenated
automatically adjacent state or disease intervals, regardless of the expected duration of each state;
it was modified by the introduction of an *expected-length* attribute that was used only for display
purposes.

population-based parameters. That model was specialized for a particular patient, thus turning it into an *atemporal patient-specific model*-by addition of details such as the patient's weight. Finally, the parameters in the atemporal patient-specific model were adjusted to fit actual patient-specific data that accumulate over time (such as response to previous therapy), turning the model into a *patient-specific temporal model* [198].

2. A symbolic interval-based model aggregated intervals that were clinically interesting in the sense that they violated expectations. The model encoded abstractions as a hierarchy of symbolic intervals. The symbolic model created these intervals by comparing population-based model predictions to patient-specific predictions (to detect surprising observations), by comparing population-based model parameters to patient-specific parameter estimates (for explanation purposes), or by comparing actual patient observations to the expected patient-specific predictions (for purposes of critiquing the numeric model). The abstraction step was implemented by context-specific rules.

3. A symbolic state-based model generated text paragraphs that used the domain's language, from the interval-based abstractions, using a representation based on augmented transition networks (ATNs). The ATNs encoded the possible summary statements as a network of potential interesting states. The state model transformed interval-based abstractions into text paragraphs.

In addition, [196] designed a temporal-maintenance system, TNET, to maintain relationships among intervals in related contexts and an associated temporal query language, TQuery [197]. TNET and TQuery were used in the context of the ONCOCIN project [406] to assist care providers who were treating cancer patients enrolled in experimental clinical protocols. The TNET system was extended to the ETNET system, which was used in the TOPAZ system. ETNET [196] extended the temporal-representation capabilities of TNET while simplifying the latter's structure. In addition, ETNET had the ability to associate interpretation methods with ETNET intervals; such intervals represented contexts of interest, such as a period of lower-than-expected granulocyte counts. ETNET was not only a *temporal-reasoning* system, but also a flexible *temporal-maintenance* system. Kahn noted, however, that ETNET could not replace a database-management system, and suggested implementing it on top of one.

TOPAZ used different formalisms to represent different aspects of the complex interpretation task. TOPAZ represents a landmark attempt to create a hybrid interpretation system for time-oriented data, comprising three different, integrated, temporal models.

The numeric model used for the representation of the prototypical (population-based) patient model, for the generation of the atemporal patient-specific model, and for the fitting of the calculated parameters with the observed time-stamped observations (thus adjusting the model to a temporal patient-specific model), was a complex one. It was also highly dependent on the domain and on the task at hand. In particular, the developer created a complex model just for predicting *one* parameter (granulocytes) by modeling *one* anatomical site (the bone marrow) for patients who had *one* disease (Hodgkin's lymphoma) and who were receiving treatment by

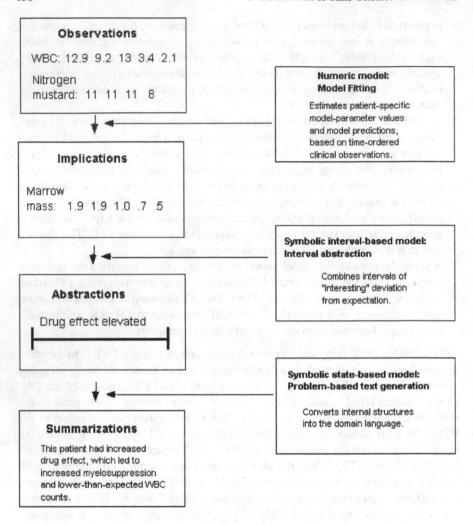

Fig. 5.3 Summarization of time-ordered data in the TOPAZ system. Three steps were taken: (1) estimation of system-specific model features from observations, using the numeric model, (2) aggregation of periods in which model predictions deviate significantly from system observations, using the symbolic interval-based model, and (3) generation of text by presentation of "interesting" abstractions in the domain's language, using the symbolic state-based model. (Source: modified from [198] (pp. 16 and 118)).

one particular form of chemotherapy (MOPP, a clinical protocol that administers nitrogen mustard, vincristine, procarbazine, and prednisone). Even given these considerable restrictions, the model encoded multiple simplifications. For instance, all the drugs were combined into a pseudodrug to represent more simply a combined myelosuppressive (bone-marrow-toxicity) effect. The model represents the decay of the drug's *effect*, rather than the decay of the actual drug *metabolites*. This modeling simplification was introduced because the two main drugs specifically toxic to the bone-marrow target organ had similar myelosuppressive (toxic to the bone marrow) effects. As Kahn notes, this assumption might not be appropriate even for other MOPP toxicity types for the same patients and the same protocols; it certainly might not hold for other cancer-therapy protocols, or in other protocol-therapy domains. In fact, it is not clear how the model can be adjusted to fit even the rather related domain of treatment of chronic Graft-Versus-Host Disease (GVHD) patients. Chronic GVHD patients suffer from similar, but not quite the same, effects due to myelosuppressive drug therapy, as well as from multiple-organ (e.g., skin and liver) involvement due to the chronic GVHD disease itself; such effects might complicate the interpretation of other drug toxicities.

In addition, many clinical domains seem to defy complete numeric modeling, e.g., the domain of monitoring children's growth. Similarly, in many other clinical domains, the parameter associations are well known, but the underlying physiology and pathology are little or incompletely understood, and cannot be modeled with any reasonable accuracy. Quantitative modeling is especially problematic in data-poor domains, where measurements are taken once a week or once a month.

TOPAZ used the patient-specific predictions, not the actual observed data, for comparisons to the expected population data. The reason for this choice was that data produced for patient-specific predictions (assuming a correct, complete, patient-specific model) should be *smoother* than actual data and should contain fewer spurious values. However, using predictions rather than observed data might make it more difficult to detect changes in patient parameters. Furthermore, the calculated, patient-specific expected values do not appear in the generated summary and therefore would not be saved in the patient's medical record. It is therefore difficult to produce an explanation to a care provider who might want a justification for the system's conclusions, at least without a highly sophisticated text-generating module.

The ETNET system was highly expressive and flexible. It depended, however, on a model of unambiguous time-stamped observations. This assumption was also made (at least as far as the input, is concerned) in Russ' TCS system (see Section 5.3.2) and in Shahar's RÉSUMÉ system (see Section 5.5). In addition, TOPAZ did not handle well vertical (value) or horizontal (temporal) uncertainty, and, as Kahn remarks, it is in general difficult to apply statistical techniques to data-poor domains.

5.3 Time-Oriented Monitoring

Most clinical monitoring tasks require measurement and capture over time of numerous patient data, often on electronic media. Care providers who have to make diagnostic or therapeutic decisions based on these data may be overwhelmed by the number of data if the care providers' ability to *reason* with the data does not scale up to the data-storage capabilities. Thus, support of automated monitoring is a major task involving reasoning about time in medical applications.

Most stored clinical data include a time stamp in which the particular datum is valid; an emerging pattern over a stretch of time has much more significance than an isolated finding or even a set of findings. Experienced care providers are able to combine several significant contemporaneous findings, to abstract such findings into clinically meaningful higher-level concepts in a context-sensitive manner, and to detect significant trends in both low-level data and abstract concepts.

Thus, it is desirable to provide short, informative, context-sensitive summaries of time-oriented clinical data stored on electronic media, and to be able to answer queries about abstract concepts that summarize the data. Providing these abilities benefits both a human care provider and an automated decision-support tool that recommends therapeutic and diagnostic measures based on the patient's clinical history up to the present. Such concise, meaningful summaries, apart from their immediate value to a care provider, support an automated system's further recommendations for diagnostic or therapeutic interventions, provide a justification for the system's or for the human user's actions, and monitor not just patient data, but also therapy plans suggested by the care provider or by the decision-support system. A meaningful summary cannot use only *time points*, such as dates when data were collected; it must be able to characterize significant features over *periods* of time, such as "5 months of decreasing liver enzyme levels in the context of recovering from hepatitis."

Many of the temporal-abstraction methodologies mentioned above are intended, in part, to support the automated monitoring task. One of the problems encountered in monitoring is that of time-oriented validation of clinical data. Several methodologies have been proposed or used, including intricate schemes for detecting inconsistencies that can only be revealed over time [179].

5.3.1 Fagan's VM Program: a State-Transition Temporal-Interpretation Model

Fagan's VM system was one of the first knowledge-based systems that included an explicit representation for time. It was designed to assist care providers managing patients on ventilators in intensive-care units [140]. VM was designed as a rule-based system inspired by MYCIN, but it was different in several respects: VM could reason explicitly about time units, accept time-stamped measurements of patient parameters, and calculate time-dependent concepts such as rates of change. In addition, VM relied on a state-transition model of different intensive-care therapeutic situations, or contexts (in the VM case, different ventilation modes). In each context, different expectation rules would apply to determine what, for instance, is an ACCEPTABLE mean arterial pressure in a particular context. Except for such state-specific rules, the rest of the rules could ignore the context in which they were applied, since the context-specific classification rules created a context-free, "common denominator," symbolic-value environment. Thus, similar values of the same parameter that appeared in meeting intervals (e.g., IDEAL mean arterial pressure) could be joined and aggregated into longer intervals, even though the meaning of the value could be different, depending on the context in which the symbolic value was determined. The fact that the system changed state was inferred by special rules, since VM was not connected directly to the ventilator output.

Another point to note is that the VM program used a classification of expiration dates of parameters, signifying for how long VM could assume the correctness of the parameter's value if that value was not sampled again. The expiration date value was used to fill a GOOD-FOR slot in the parameter's description. Constants (e.g., gender) are good (valid) forever, until replaced. Continuous parameters (e.g., heart rate) are good when given at their regular, expected sampling frequency unless input data are missing or have unlikely values. Volunteered parameters (e.g., temperature) are given at irregular intervals and are good for a parameter- and context-specific amount of time. Deduced parameters (e.g., hyperventilation) are calculated from other parameters, and their reliability depends on the reliability of these parameters.

VM did not use the MYCIN certainty factors, although they were built into the rules. The reason was that most of the uncertainty was modeled within the domain-specific rules. Data were not believed after a long time had passed since they were last measured; aberrant values were excluded automatically; and wide (e.g., ACCEPTABLE) ranges were used for conclusions, thus already accounting for a large measurement variability. Fagan notes that the lack of uncertainty in the rules might occur because, in clinical contexts, care providers do not make inferences unless the latter are strongly supported, or because the intensive-care domain tends to have measurements that have a high correlation with patient states.

VM could not accept data arriving out of order, such as blood-gas results that arrive after the current context has changed, and thus could not revise past conclusions. In that sense, VM could not create a valid *historical* database (see Chapter 3), although it did store the last hour of parameter measurements and all former conclusions; in that respect, VM maintained a *rollback* database of measurements and conclusions.

5.3.2 Temporal Bookkeeping: Russ' Temporal Control Structure

Russ designed a system called the temporal control structure (TCS), which supports reasoning in time-oriented domains, by allowing the domain-specific inference procedures to ignore temporal issues, such as the particular time stamps attached to values of measured variables [336, 337].

The main emphasis in the TCS methodology is creating what Russ terms as a state abstraction: an abstraction of continuous processes into steady-state time intervals, when all the database variables relevant for the knowledge-based system's reasoning modules are known to be fixed at some particular value. The state-abstraction intervals are similar to VM's states, which were used as triggers for VM's context-based rules. TCS is introduced as a control-system buffer between the database and the rule environment. The actual reasoning processes (e.g., domain-specific rules) are activated by TCS over all the intervals representing such steady states, and thus can reason even though the rules do not represent time explicitly. That ignorance of time by the rules is allowed because, by definition, after the various intervals representing different propositions have been broken down by the control system into steady-state, homogeneous subintervals, there can be no change in any of the parameters relevant to the rule inside these subintervals, and time is no longer a factor.

The TCS system allows user-defined code modules that reason over the homogeneous intervals, as well as user-defined data variables that hold the data in the database. Modules define inputs and outputs for their code; Russ also allows for a memory variable that can transfer data from one module to a succeeding or a preceding interval module (otherwise, there can be no reasoning about change). Information variables from future processes are termed oracles; variables from the past are termed history.

The TCS system creates a process for each time interval in which a module is executed; the process has access only to those input data that occur within that time interval. The TCS system can chain processes using the memory variables. All process computations are considered by the TCS system as black boxes; the TCS system is responsible for applying these computations to the appropriate variables at the appropriate time intervals, and for updating these computations, should the value of any input variable change. Figure 5.4 shows a chain of processes in the TCS system.

The underlying temporal primitive in the TCS architecture is a time *point* denoting an exact date. Propositions are represented by point variables or by interval variables. Intervals are created by an abstraction process that employs user-defined procedural Lisp code inside the TCS modules to create steady-state periods, such as a period of stable blood pressure. The abstraction process and the subsequent updates are data driven. Variables can take only a single value, which can be a complex structure; the only restriction on the value is the need to provide an equality predicate.

A particularly interesting feature of TCS is the truth-maintenance capability of the system, that is, the abilities to maintain dependencies among data and conclusions in every steady-state interval, and to propagate the effects of a change in past or present values of parameters to all concerned reasoning modules. Thus, the TCS system creates a *historical* database that can be updated at arbitrary time points, in which all the time-stamped conclusions are valid. Another interesting property of Russ's system is the ability to reason by hindsight, that is, to reassess past

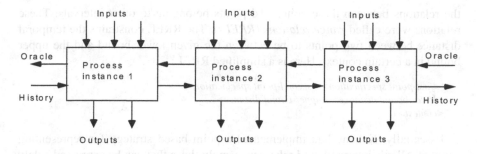

Fig. 5.4 A chain of processes in the TCS system. Each process has in it user-defined code, a set of predefined inputs and outputs, and memory variables connecting it to future processes (oracle variables) and to past processes (history variables).

conclusions based on new, present data [336]. This process is performed essentially by information flowing through the memory variables backward in time.

5.3.3 Kohane's Temporal Utilities Package

Kohane [224, 223] has written a general-purpose *temporal-utilities package (TUP)* for representing qualitative and quantitative relations among temporal intervals, and for maintaining and propagating the constraints posed by these relations through a *constraint network* of temporal (or any other) intervals. The use of constraint networks is a general technique for representing and maintaining a set of objects (called the *nodes* of the network) such that, between at least some pairs of nodes, there are links (known as *arcs*) which represent a relation that must hold between the two nodes. Updating a constraint network by setting the values of certain nodes or arcs to be fixed propagates the changes to all the other nodes and arcs.

Kohane's goal was mainly to represent and reason about the complex, sometimes vague, relations found in clinical medicine, such as "the onset of jaundice follows the symptom of nausea within 3 to 5 weeks but before the enzyme-level elevation." When such relations exist, it might be best not to force the patient or the physician to provide the decision-support system with accurate, unambiguous time-stamped data. Instead, it may be useful to store the relation, and to update it when more information becomes available. Thus, a relation such as "2 to 4 weeks after the onset of jaundice" might be updated to "3 to 4 weeks after the onset of jaundice" when other constraints are considered or when additional data, such as enzyme levels, became available. Such a strategy is at least a partial solution to the issue of *horizontal* (temporal) uncertainty in clinical domains, in which vague patient histories and unclear disease evolution patterns are common.

The TUP system used a point-based temporal ontology. Intervals were represented implicitly by the relations between their start points and end points, or by the relations between these points and points belonging to other intervals. These relations were called *range relations* (*RRELs*). The RREL constrains the temporal distance between two points to be between the given lower bound and the upper bound in a certain context. Here is a simplified RREL:

(*<first-point specification> <second-point specification>*
<lower-bound distance> <upper-bound distance>
<context>)

Essentially, Kohane had implemented a point-based strategy for representing some of Allen's interval-based relations, namely those that can be expressed solely by constraints between two points. For instance, to specify that interval *A* precedes interval *B*, it is sufficient to maintain the constraint that "the end of *A* is between +INFINITY and +ε before the start of *B*." This point-based, restricted temporal logic has been shown in other studies to be computationally sound and complete in polynomial time, since point-based constraints can be propagated efficiently through the arcs of the constraint network. However, such a restricted logic cannot capture *disjunctions* of the type "interval *A* is either before or after interval *B*," since no equivalent set of constraints expressed as conjunctions using the end points of A and B can express such a relation. Whether such relations are needed often, if at all, in clinical medicine is debatable.

Kohane tested the TUP system by designing a simple medical expert system, *temporal-hypothesis reasoning in patient history* (*THRIPHT*). The THRIPHT system accepted data in the form of RRELs and propagated newly computed upper and lower bounds on temporal distances throughout the TUP-based network. The diagnostic, rule-based algorithm (in the domain of hepatitis) waited until all constraints were propagated, and then queried the TUP system using temporal predicates such as, "Did the patient use drugs within the past 7 months, starting as least 2 months before the onset of jaundice?" [223].

5.3.4 Haimowitz and Kohane's TrenDx System

A very different system, which had demonstrated initial encouraging results, is Haimowitz and Kohane's TrenDx temporal pattern-matching system [170]. TrenDx focuses on using efficient general methods for representing and detecting predefined temporal patterns in raw time-stamped data.

Trend templates (TTs) describe typical clinical temporal patterns, such as normal growth development, or specific types of patterns known to be associated with functional states or disease states, by representing these patterns as *horizontal* (temporal) and *vertical* (measurement) constraints. The TrenDx system has been developed mainly within the domain of pediatric growth monitoring, although examples from other domains have been presented to demonstrate its more general potential. For example, the growth TT (Figure 5.5) declares several predefined events, such as PUBERTY ONSET; these events are constrained to occur within a predefined temporal range, e.g., PUBERTY ONSET must occur within 10 to 13 years after birth. Within that temporal range, height should vary only by $\pm\delta$.

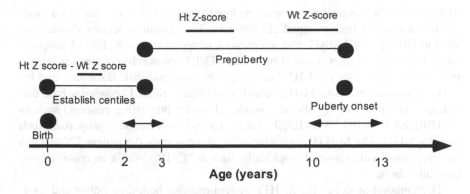

Fig. 5.5 A portion of a trend template (TT) in TrenDx that describes the male average normal growth as a set of functional and interval-based constraints. All Z scores are for the average population. The Birth landmark, assumed to denote time 0, is followed by an uncertain period of 2 to 3 years in which the child's growth percentiles are established, and in which the difference between the Ht Z score and the Wt Z score are constrained to be constant. During an uncertain period of prepuberty ending in the puberty onset landmark sometime between the age of 10 and 13, the Ht Z score and the Wt Z score are both constrained to be constant. = landmark or transition point; = constant value indicator; Ht = height; Wt = weight; Z score indicates number of standard deviations from the mean. (Source: adapted from [171](p. 45)).

TrenDx has the rather unique ability to match *partial* patterns by maintaining an agenda of candidate patterns that *possibly* match an evolving pattern. Thus, even if TrenDx gets only one point as input, it might (at least in theory) still be able to return a few possible patterns as output. As more data points are known, the list of potential matching patterns and their particular instantiation in the data is modified. This continuous pattern-matching process might be considered a goal-directed approach to pattern matching.

A TT indeed provides a powerful mechanism for expressing the dynamics of some process, in terms of the different phases comprising it, the uncertainty governing the transitions from one phase to the next, the significant events marking these transitions and various constraints on parameter-values associated with the different phases. However, the abstraction levels are not explicit, and there is no decoupling between an intermediate level of data interpretation (derivation of abstractions) and a higher level of decision making. Data interpretation involves the selection of the TT instantiation that matches best the raw temporal data (this procedure solves the problems of noise detection and positioning of transitions). The selected TT instantiation is the final solution; thus temporal data abstraction and diagnostic-monitoring reasoning per se are tangled up into single process. This makes the overall reasoning more efficient, but it limits the reusability of the approach.

5.3.5 Larizza et al.'s Temporal-Abstraction Module in the M-HTP System

M-HTP [235] is a system for monitoring heart-transplant patients that has a module for abstracting time-stamped clinical data. The system generates abstractions such as HB-DECREASING, and maintains a temporal network (TN) of temporal intervals, using a design inspired by Kahn's TNET temporal-maintenance system (see above). Like TNET, M-HTP uses an object-oriented visit taxonomy and indexes parameters by visits. M-HTP also has an object-oriented knowledge base that defines a taxonomy of significant-episodes-clinically interesting concepts such as DIARRHEA or WBC_DECREASE. Parameter instances can have properties, such as MINIMUM. The M-HTP output includes intervals from the patient TN that can be represented and examined graphically, such as "CMV_viremia_increase" during particular dates.

The temporal model of the M-HTP system includes both time points and intervals. The M-HTP system uses a temporal query language to define the antecedent part of its rules, such as "an episode of decrease in platelet count that *overlaps* an episode of decrease of WBC count *at least for* 3 days *during* the past week implies suspicion of CMV infection".

5.3.6 Miksch et al.'s VIE-VENT System

Miksch et al. [262] have developed VIE-VENT, a system for data validation and therapy planning for artificially ventilated newborn infants. The overall aim is the context-based validation and interpretation of temporal data, where data can be of different types (continuously assessed quantitative data, discontinuously assessed quantitative data, and qualitative data). The interpretation contexts are not dynamically derived, but they are defined through schemata with thresholds that can be dynamically tailored to the patient under examination. The context schemata correspond to potential treatment regimes; which context is actually active depends on the current regime of the patient. If the interpretation of data suggests an alarming situation, the higher level reasoning task of therapy assessment and (re)planning is invoked which may result in changing the patient's regime, thus switching to a new context. Context switching should be done in a smooth way and again relevant thresholds are dynamically adapted to take care of this. The data abstraction process per se is fairly decoupled from the therapy planning process. Hence this approach differs from Haimowitz and Kohane's approach where the selection and instantiation of an interpretation context (trend template) represents the overall reasoning task. In VIE-VENT the data abstraction process does not need to select the interpretation context, as this is given to it by the therapy planning process.

The types of knowledge required are classification knowledge and temporal dynamic knowledge (e.g., default persistences, expected qualitative trend descriptions, etc.). Everything is expressed declaratively in terms of schemata that can be dynamically adjusted depending on the state of the patient. First quantitative point-based data are translated into qualitative values, depending on the operative context. Smoothing of data oscillating near thresholds then takes place. Interval data are then transformed to qualitative descriptions resulting in a verbal categorization of the change of a parameter over time, using schemata for trend-curve fitting. The system deals with four types of trends: very short-term, short-term, medium-term and long-term.

Overall this approach is aimed at a specific type of medical applications, and so, unlike Shahar's knowledge-based temporal-abstraction method, to be discussed in detail in Section 5.5, the aim is not to formulate in generic terms a reusable kernel for temporal data abstraction.

5.4 Merging Temporal Reasoning and Temporal Maintenance: Temporal Mediators

As the reader recalls from Chapter 3, in addition to *reasoning* about time-oriented medical data, it is also necessary to consider the *management* of these data: insertion, deletion, and query, tasks often collectively referred to as *temporal-data maintenance*.

Recall that temporal-data maintenance includes the capability to store and retrieve also the different *temporal dimensions* which have been distinguished in time-oriented data [382]: (1) the *transaction time*, that is, the time at which data are stored in the database; (2) the *valid time*, that is, the time at which the data are true for the modeled real world entity; and (3) the *user-defined time*, whose meaning is related to the application and thus is defined by the user. *Bitemporal databases*, which represent explicitly both valid and transaction times, are the only representation mode that fulfils the necessary functional and legal requirements for time-oriented medical databases, although historical and rollback databases are currently most common. Another temporal dimension of information considered recently is the *decision-time* [148]: the decision time of a therapy, for example, could be different, both from the valid time during which the therapy is administered and from the transaction time, at which the data related to the therapy are inserted into the database.

Several systems allow not only the modeling of complex clinical concepts at the database level, but also the maintenance of certain inference operations at that level. For example, *active databases* [425] can also store and query derived data; these data are obtained by the execution of rules that are triggered by external events, such as the insertion of patient related data [48].

Furthermore, integrity constraints based on temporal reasoning [179] can often be evaluated at the database level, for example to *validate* clinical data during their acquisition. This validation, however, requires domain-specific knowledge (e.g., height is a monotonically increasing function, and should never decrease, at least for children; weight cannot increase by more then a certain number of pounds a day). As we shall see, ultimately, maintenance of both raw data and abstractions requires an integration of the temporal reasoning and temporal maintenance tasks.

When building a time-oriented decision-support application, one needs to consider the mode of integration between the application, the data-abstraction process (essentially, a temporal-reasoning task), and the temporal-maintenance aspect of the system. Both the temporal-reasoning and the temporal-abstraction processes are needed to support clinical tasks such as diagnosis, monitoring, and therapy.

Data abstraction is a critical auxiliary process. It is usually deployed in the context of a higher-level problem solving system, it is knowledge-based, and it operates in a goal- or event-driven fashion, or both. The knowledge used by the data abstraction process comprises both specialist knowledge and so called "world" knowledge, i.e., commonsense knowledge which is assumed domain and even task independent. The knowledge is organized on the basis of some *ontology*, which defines the classes of concepts, their properties, and the types of relations among them.

In this section we briefly discuss the mode of integration between a medical decision-support system and a temporal-data abstraction process. This mode can be described as *loosely coupled* or *tightly coupled* and denotes the level of generality, and thus degree of reusability, of the data abstraction process. A *loosely-coupled* process implies that the data abstraction process is domain independent (e.g., it can be integrated with any diagnostic system irrespective of its medical domain), task independent (e.g., it can be integrated with different reasoning tasks, such as diagnosis, monitoring, prognosis, etc., within the same medical domain), or both (e.g., it can be integrated with different reasoning tasks applied to different domains). A *tightly coupled* process, on the other hand, implies that the data-abstraction process is an embedded component of the problem solving system; thus, its usability outside that system is limited.

In general, one can conceive of three basic modes of integration among the temporal-data abstraction process, the temporal-data management process, a medical decision-support application (e.g., diagnosis, therapy), and a time-oriented database (Figure 5.6):

1. Incorporating the abstraction process within the database: the drawbacks include relying on the database management system's language, typically simpler than a programming language, and forming a tight coupling to the particular syntax and semantics of the database used;
2. Adding a data-management capability to the application system, assuming an inherent data-abstraction capability: the drawbacks include the problem of duplicating quite a few of the functions already inherent to a database management system, the lack of ability to take advantage of the sophisticated data-management and query-optimization techniques implemented within the database, and tight coupling to the particular application, without the ability to reuse the abstraction mechanisms elsewhere; and
3. Out-sourcing both the temporal-data management and the temporal-data abstraction capabilities, by encapsulating them within an intermediate *mediator* that is independent of either, thus enjoying all advantages of the specialized modules, with none of the above-mentioned drawbacks.

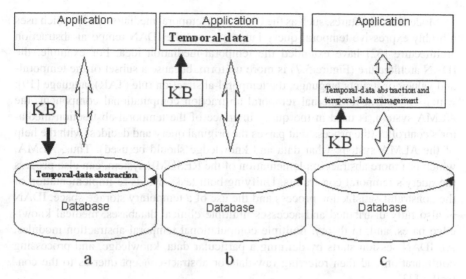

Fig. 5.6 Three modes of integrating the temporal-data abstraction and temporal-data management processes with a medical decision-support application (e.g., diagnosis, therapy) and a time-oriented database. (a) Incorporating the abstraction process within the database; (b) adding a data-management capability to the application system, assuming an inherent data-abstraction capability; (c) encapsulating the temporal-data management and the temporal-data abstraction capabilities within an intermediate *mediator* that is independent of either. KB = [temporal-abstraction] knowledge base.

The concept of a mediator has been proposed in the early 1990s [427]. It is called a *mediator* because it serves as an intermediate layer of processing between client applications and databases. As a result, the mediator is tied to neither a particular application, domain, or task, nor to a particular database [428]. By combining the functions of temporal reasoning and temporal maintenance within one architecture, which we refer to as a *temporal mediator* (or, more precisely, a *temporal-abstraction mediator*, since it should include also the data-abstraction capability), a transparent interface can be created to the patient's time-oriented database. An example of such an architecture was the Tzolkin temporal-mediation module [281], which supported the EON guideline-based-therapy system [278]. The Tzolkin module combined the RÉSUMÉ temporal-abstraction system [359], the Chronus temporal-maintenance system [111], and a controller into a unified temporal-mediation server. The Tzolkin server answered complex temporal queries, regarding either raw clinical data or their abstractions, submitted by care providers or clinical decision-support applications, hiding the internal division of computational tasks from the user (or from the clinical decision-support application). When users asked complex temporal queries including abstract terms that do not exist in the database, the Tzolkin controller loaded the necessary raw data from the database, used RÉSUMÉ to abstract the data, saved the results in a temporary database, and used Chronus to access the results and answer the original temporal query.

Modern architectures, such as the Chronus-2 temporal mediator [284], which uses a highly expressive temporal-query language, and the IDAN temporal-abstraction architecture [32] have extended the temporal-mediation idea. For example, the IDAN architecture (Figure 5.7) is more uniform, because a subset of the temporal- and value-constraints language, the temporal-abstraction rule (TAR) language [19], which is used in its internal temporal-abstraction computational component, the ALMA system, is used in the query interface of the temporal-abstraction media- tor's controller (the process that parses the original query and decides, with the help of the ALMA system, what data and knowledge should be used). Thus, ALMA, which is a more abstract implementation of the RÉSUMÉ system, can also process the query's temporal constraints. Unifying both tasks avoids re-implementation of the constraint-satisfaction process and the use of a temporary storage space. IDAN is also fully distributed and accesses multiple clinical databases, medical knowl- edge bases, and, in theory, multiple computational temporal-abstraction modules. An IDAN session starts by defining a particular data, knowledge, and processing configuration and then referring raw-data or abstract-concept queries to the con- troller [33].

IDAN is used by multiple applications. A typical example is the *KNAVE-II* ar- chitecture [361, 366], a distributed re-implementation of *KNAVE* [351, 352]. The KNAVE architecture supports interactive knowledge-based visual exploration of time-oriented clinical databases by sending queries to the IDAN controller, using the domain's ontology of raw data and of abstract concepts derivable from them, and displaying the resulting data, derived concepts, and knowledge. The combination of the IDAN mediator and the KNAVE-II interactive exploration module has been evaluated successfully, demonstrating ease of use by clinicians and reduced time to answer typical time-oriented clinical queries when compared to paper charts and an electronic spreadsheet [254, 255]. *DeGeL* [361, 360, 367], which is discussed in detail in Chapter 7 is a distributed framework that supports clinical-guideline spec- ification, retrieval, application, and quality assessment, by sending runtime queries about the current patient to the IDAN controller.

Thus, the ideal situation is to have a data abstraction process that is both domain- and task-independent, and implemented within a temporal-abstraction mediator. Whether this is fully achievable remains to be seen, although some significant steps have been taken in this direction. (For example, Shahar's knowledge-based temporal-abstraction method, which is described in detail in the next section, has been applied to several different medical domains and tasks, and even to several nonmedical domains and tasks.)

The looseness or otherwise of coupling between a data abstraction process and a problem solving system can be decided on the basis of the following questions:

1. Is the ontology underlying the specific knowledge domain independent? If so, then removing that knowledge and incorporating a knowledge-acquisition com- ponent that functions to fill the given knowledge base with the relevant knowl- edge from another domain will result in a traditional skeletal system for data abstraction, applicable to different domains for the same task.

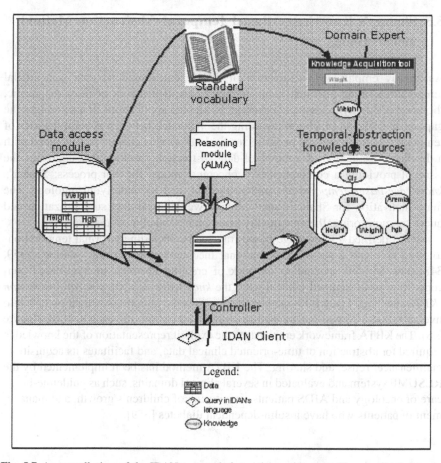

Fig. 5.7 An overall view of the IDAN temporal-abstraction architecture. User applications submit time-oriented raw or abstract-data queries to the temporal-abstraction mediator. The temporal mediator, using data from the appropriate local data-source, and temporal-abstraction knowledge from the appropriate domain-specific knowledge base, delegates the query, data and knowledge to the inference engine (the default implementation is the ALMA system) which processes the query, and returns the answer set. Diamonds denote queries, ovals denote knowledge elements and squares denote database relations.

2. Is the overall ontology task independent? If so, we can obtain a skeletal system for data abstraction, applicable to different tasks within the same domain.
3. Is the specific knowledge task independent? If so, the data abstraction process is already applicable to different tasks within the same domain.
4. Do generated abstractions constitute the system's main and final output? If so, the data abstraction process is strongly coupled to the problem solving system. In the spirit of the new generation of knowledge-engineering methodologies, the objective should be to form a *library* of generic data abstraction methods, with different underlying ontologies and computational mechanisms (e.g., [347, 337]).

5.5 Shahar's Knowledge-Based Temporal-Abstraction Method and Ontology

As already emphasized, abstraction of time-oriented medical data is a crucial temporal-reasoning task that is an implicit or explicit aspect of most diagnostic, therapeutic, quality-assessment or research-oriented applications. It also bridges the gap noted earlier in this book regarding the mismatch between general theories of temporal reasoning and the needs of medical tasks. This chapter is concluded with a detailed presentation of a temporal-data abstraction ontology that addresses the issue of providing a comprehensive conceptual model for that process: Shahar's *knowledge-based temporal-abstraction method*, and its underlying ontology. The main motivation for Shahar's methodology was clinical-data summarization and query for monitoring, therapy, quality assessment, and clinical research.

Shahar defined a knowledge-based framework, including a formal temporal ontology [347] and a set of computational mechanisms using that ontology [359, 348, 349, 58, 59] specific to the task of creating abstract, interval-based concepts from time-stamped clinical data: the *knowledge-based temporal-abstraction* (*KBTA*) method. The KBTA method decomposes the temporal-abstraction task into five subtasks; a formal mechanism was proposed for solving each subtask (Figure 5.8). The KBTA framework emphasizes the explicit representation of the knowledge required for abstraction of time-oriented clinical data, and facilitates its acquisition, maintenance, reuse, and sharing. The KBTA method has been implemented by the RÉSUMÉ system and evaluated in several clinical domains, such as guideline-based care of oncology and AIDS patients, monitoring of children's growth, and management of patients who have insulin-dependent diabetes [359].

5.5.1 The Knowledge-Based Temporal-Abstraction Ontology

The KBTA theory defines the following set of entities:

1. The basic time primitives are *time stamps*, $T_i \in$ T. Time stamps are structures (e.g., dates) that can be mapped, by a time-standardization function $f_s(T_i)$, into an integer amount of any element of a set of predefined *temporal granularity units* $G_i \in \Gamma$ (e.g., DAY). A *zero-point* time stamp (the start of the positive time line) must exist. Time stamps are therefore either positive or negative shifts from the zero point measured in the G_i units. (Intuitively, the 0 point might be grounded in each domain to different absolute, "real-world," time points: the patient's age, the start of the therapy, the first day of the twentieth century.) The domain must have a time unit G_0 of the lowest granularity (e.g., SECOND); there must exist a mapping from any integer amount of granularity units G_i into an integer amount of G_0. (The time unit G_0 can be a task-specific choice.) A finite negative or positive integer amount of G_i units is a *time measure*.

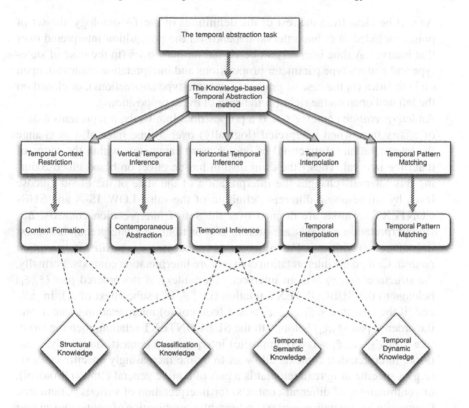

Fig. 5.8 The knowledge-based temporal-abstraction (KBTA) method. The TA *task* is decomposed by the KBTA *method* into five *subtasks*. Each subtask can be performed by one of five TA *mechanisms*. The TA mechanisms depend on four domain- and task-specific *knowledge types*. Rectangle = task; oval = method or mechanism; diamond = knowledge type; striped arrow = DECOMPOSED-INTO relation; full arrow = PERFORMED-BY relation; dashed arrow = USED-BY relation.

The special symbols +∞ and −∞ are both time stamps and time measures, denoting the furthest future and the most remote past, respectively. Any two time stamps must belong to either a precedence relation or an equivalence relation defined on the set of pairs of time stamps. The precedence relation corresponds to a temporal order; the equivalence relation denotes temporal equivalence for the domain. The −∞ time stamp precedes any other time stamp; the +∞ time stamp follows (is preceded by) all other time stamps. Subtraction of any time stamp from another must be defined and should return a time measure. Addition or subtraction of a time measure to or from a time stamp must return a time stamp.

2. A *time interval I* is an ordered pair of time stamps representing the interval's end points: [*I.start, I.end*]. *Time points T_i* are therefore represented as zero-length intervals where *I.start = I.end*. Propositions can be interpreted only over time intervals.

As will be clear from the rest of the definitions of the TA ontology, the set of points included in a time interval depends on the proposition interpreted over that interval. A time interval can be closed on both sides (in the case of state-type and pattern-type parameter propositions and interpretation contexts), open on both sides (in the case of gradient- and rate-type abstractions), or closed on the left and open on the right (in the case of event propositions).

3. An *interpretation context* $\xi \in \Xi$ is a proposition. Intuitively, it represents a state of affairs that, when interpreted (logically) over a time interval, can change the interpretation (abstraction) of one or more parameters within the scope of that time interval. Thus, "the drug insulin has an effect on blood glucose during this interval" changes the interpretation of the state of the blood-glucose level, by suggesting a different definition of the value LOW. IS-A and SUB-CONTEXT relations are defined over the set of interpretation contexts. *Basic* interpretation contexts are atomic propositions. An interpretation context in conjunction with one of its subcontexts can create a *composite interpretation context*. Composite interpretation contexts are interpretation contexts. Formally, the structure $< \xi_i, \xi_j >$ is an interpretation context, if the ordered pair (ξ_j, ξ_i) belongs to the SUBCONTEXT relation (i.e., ξ_j is a subcontext of ξ_i). In general, if the structure $< \xi_1, \xi_2, ... \xi_i >$ is a (composite) interpretation context, and the ordered pair (ξ_j, ξ_i) belongs to the SUBCONTEXT relation, then the structure $< \xi_1, \xi_2, ... \xi_i, \xi_j >$ is a (composite) interpretation context. Intuitively, composite interpretation contexts allow us to define increasingly specific contexts (e.g., a specific drug regimen that is a part of a more general clinical protocol), or combinations of different contexts, for interpretation of various parameters. Composite interpretation contexts represent a combination of *contemporaneous* interpretation contexts whose *conjunction* denotes a new context that has significance for the interpretation of one or more parameters. Finally, *generalized* and *nonconvex interpretation contexts* are special types of contexts that enable, when present, the joining of propositions formed within different context intervals. The first allows the generalization of propositions of similar type that hold in different but temporally adjacent contexts into a proposition that holds in a more general context. The second allows the aggregation of propositions of similar types that hold in similar but temporally disjoint contexts into a proposition that holds over a (temporally) nonconvex interval within a corresponding nonconvex context interval (e.g., blood-glucose abstractions before breakfast over several consecutive days).

4. A *context interval* is a structure $< \xi, I >$, consisting of an interpretation context ξ and a temporal interval I. Intuitively, a context interval represents an interpretation context interpreted over a time interval; the interpretation of one or more parameters is different within the temporal scope of that interval. Thus, the effects of chemotherapy form an interpretation context that can hold over several weeks, within which the values of hematological parameters might be abstracted differently.

5. An *event proposition* $e \in E$ (or an *event*, for short, when no ambiguity exists) represents the occurrence of an external volitional action or process, such as the administration of a drug (as opposed to a measurable datum, such as

temperature). Events have a series a_i of *event attributes* (e.g., dose) and a corresponding series v_i of *attribute values*. (Typically, events are controlled by a human or an automated agent, and thus neither are they measured data, nor can they be abstracted from the other input data.)

An IS-A hierarchy (in the usual sense) of *event schemata* (or event types) exists. Event schemata have a list of attributes a_i, where each attribute has a domain of possible values V_i, but do not necessarily contain any corresponding attribute values. Thus, an *event proposition* is an event schema in which each attribute a_i is mapped to some value $v_i \in V_i$. A PART-OF relation is defined over the set of event schemata. If the pair of event schemata (e_i, e_j) belongs to the PART-OF relation, then event schema e_i can be a *subevent* of an event schema e_j (e.g., a Clinical-protocol event can have several parts, all of them Medication events).

6. An *event interval* is a structure $< e, I >$, consisting of an event proposition e and a time interval I. The time interval I represents the duration of the event.

7. A *parameter schema* (or a *parameter*, for short) $\pi \in \Pi$ is, intuitively, a measurable aspect or a describable state of the world, such as a patient's temperature. Parameter schemata have various *properties*, such as a domain V_π of possible symbolic or numeric values, measurement units, and a measurement scale (which can be one of NOMINAL, ORDINAL, INTERVAL, or RATIO, corresponding to the standard distinction in statistics among types of measurement[2]). Not all properties need have values in a parameter schema. An IS-A hierarchy (in the usual sense) of parameter schemata exists. The combination $< \pi, \xi >$ of a parameter π and an interpretation context ξ is an *extended parameter schema* (or an *extended parameter*, for short). Extended parameters are parameters (e.g., blood glucose in the context of insulin action, or platelet count in the context of chemotherapy effects). Note that an extended parameter can have properties, such as possible domains of value, that are different from that of the original (non extended) parameter (e.g., in a specific context, a parameter might have a more refined set of possible values). Extended parameters also have a special property, a value $v \in V_\pi$. Values often are known only at runtime.

Intuitively, parameters denote either input (usually raw) data, or any level of abstraction of the raw data (up to a whole pattern). For instance, the Hemoglobin level is a parameter, the White-blood-cell count is a parameter, the Temperature level is a parameter, and so is the Bone-marrow-toxicity level (which is abstracted from Hemoglobin and other parameters).

The combination of a parameter, a parameter value, and an interpretation context-that is, the tuple $< \pi, v, \xi >$ (i.e., an extended parameter and a value)-is called a *parameter proposition* (e.g., "the state of Hemoglobin has the value LOW in

---

[2] Nominal-scale parameters have values that can be listed, but that cannot be ordered (e.g., color). Ordinal-scale parameters have values that can be ordered, but the intervals among these values are not meaningful by themselves and are not necessarily equal (e.g., military ranks). Interval-scale parameters have scale with meaningful, comparable intervals, although a ratio comparison is not necessarily meaningful (e.g., temperature measured on a Celsius scale). Ratio-scale parameters have, in addition to all these properties, a fixed zero point (e.g., height); thus, a ratio comparison, such as "twice as tall," is meaningful regardless of the height measurement unit.

the context of therapy by AZT"). A mapping exists from all parameter propositions and the properties of their corresponding parameter (or extended parameter) schema into specific property values.

Much of the knowledge about abstraction of higher-level concepts over time depends on knowledge of specific parameter-proposition properties, such as persistence over time of a certain parameter with a certain value within a particular
context. Different TA mechanisms typically require knowledge about different
parameter properties of the same parameter propositions.

Primitive parameters are parameters that play the role of raw data in the particular domain in which the TA task is being solved. They cannot be inferred
by the TA process from any other parameters (e.g., laboratory measurements).
They can appear only in the input of the TA task.

Abstract parameters are parameters that play the role of intermediate concepts
at various levels of abstraction; these parameters can be part of the output of the
TA task, having been *abstracted* from other parameters and events, or they may
be given as part of the input (e.g., the value of the state of Hemoglobin is MOD-
ERATE_ANEMIA). There is an ABSTRACTED-INTO relationship between
one or more parameters and an abstract parameter. Each pair of parameters that
belongs to an ABSTRACTED-INTO relation represents only one abstraction
step; that is, the ABSTRACTED-INTO relation is not transitive. It is also are-
flexive and antisymmetric.

Constant parameters are parameters that are considered atemporal in the context of the particular interpretation task that is being performed, so their values
are not expected to be time-dependent (e.g., the patient's gender, the patient's
address, the patient's father's height). There are few, if any, truly constant parameters. (Indeed, using explicit semantic properties, constants can be represented as fluents with a particular set of temporal-semantic inferential properties, such as infinite persistence into the past and future, thus removing, in effect,
the traditional distinction between temporal and atemporal variables.) It is often useful to distinguish between (1) *case-specific constants*, which are specific
to the particular case being interpreted and which appear in the runtime input
(e.g., the patient's date of birth), and (2) *case-independent constants*, which are
inherent to the overall task, and which are typically prespecified or appear in
the domain ontology (e.g., the local population's distribution of heights).

8. *Abstraction functions* $\theta \in \Theta$ are unary or multiple-argument functions from one or
   more parameters to an abstract parameter. The "output" abstract parameters can
   have one of several *abstraction types* (which are equivalent to the abstraction
   function used). We distinguish among at least three basic abstraction types:
   state, *gradient*, and *rate*. (Other abstraction functions and therefore types,
   such as *acceleration* and *frequency*, can be added). These abstraction types
   correspond, respectively, to a classification (or computational transformation)
   of the parameter's value, the sign of the derivative of the parameter's value, and
   the magnitude of the derivative of the parameter's value during the interval (e.g.,
   LOW, DECREASING, and FAST abstractions for the Platelet-count parameter).
   The state abstraction is always possible, even with qualitative parameters having

only a nominal scale (e.g., different values of the Skin-color parameter can be mapped into the state-abstraction value RED); the gradient and rate abstractions are meaningful for only those parameters that have at least an ordinal scale (e.g., degrees of physical fitness) or an interval scale (e.g., Temperature), respectively. The θ abstraction of a parameter schema π is a new parameter schema $\theta(\pi)$-a parameter different from any of the arguments of the θ function (e.g., STATE(Hemoglobin), which we will write as Hemoglobin_state). This new parameter has its own domain of values and other properties (e.g., scale), typically different from those of the parameters from which it was abstracted. It can also be abstracted further (e.g., GRADIENT(STATE(Hemoglobin))).

A special type of abstraction function (and a respective proposition type) is *pattern*: A function that creates a temporal pattern from temporal intervals, over which hold parameters, events, contexts, or other patterns (e.g., a QUIESCENT-ONSET pattern of chronic graft-versus-host disease). Patterns were previously defined as a special abstract-parameter type [347] but have since been recognized as an independent proposition type. Patterns have interval-based components. Local constraints on these components (e.g., duration) and global constraints among components (e.g., qualitative temporal relations) define the pattern. Patterns can be *linear* or *periodic*.

Statistics such as *minimum, maximum,* and *average value* are not abstraction types in this ontology. Rather, these statistics are *functions* on parameter *values* that return simply a *value* of a parameter, possibly during a time interval, often from the domain of the original parameter (e.g., the minimum Hemoglobin value within a time interval I can be 8. 9 gr. /100cc, a value from the domain of Hemoglobin values), rather than a parameter *schema*, which can have a new domain of values (e.g., the Hemoglobin_state can have the value INCREASING).

9. A *parameter interval* is a tuple $< \pi, v, \xi, I >$, where $< \pi, v, \xi >$ is a parameter proposition and I is a time interval. If I is in fact a time point (i.e., $I.start = I.end$), then the tuple can be referred to as a *parameter point*. Intuitively, a parameter interval denotes the value v of parameter π in the context ξ during time interval I. The value of parameter π at the beginning of interval I is denoted as $I.start.\pi$, and the value of parameter π at the end of interval I as $I.end.\pi$. *Pattern intervals* are defined in a similar fashion.

10. An *abstraction* is a parameter or pattern interval $< \pi, v, \xi, I >$, where π is an abstract parameter or pattern. If I is in fact a time point (i.e., $I.start = I.end$), the abstraction can also be referred to as an *abstraction point*; otherwise, we can refer to it as an *abstraction interval*.

11. An *abstraction goal* $\psi \in \Psi$ is a proposition that denotes a particular goal or intention that is relevant to the TA task during some interval (e.g., diagnosis).

12. An *abstraction-goal interval* is a structure $< \psi, I >$, where ψ is an abstraction goal and I is a time interval. Intuitively, an abstraction-goal interval represents the fact that an *intention* holds or that a TA *goal* (e.g., the goal of monitoring AIDS patients) should be achieved during the time interval over which it is interpreted. An abstraction-goal interval is used for creating correct interpretation contexts for the interpretation of data.

13. *Induction of context intervals*: Intuitively, context intervals are inferred dynamically (at runtime) by certain event, parameter, or abstraction-goal propositions
being true over specific time intervals. The contexts interpreted over these intervals are said to be *induced* by these propositions (e.g., by the event "administration of 4 units of regular insulin"). Certain predefined temporal constraints must
hold between the inferred context interval and the time interval over which the
inducing proposition is interpreted. For instance, the effect of insulin with respect to changing the interpretation of blood-glucose values might start at least
30 minutes after the start of the insulin administration and might end up to 8
hours after the end of that administration. Two or more context-forming propositions induce a *composite interpretation context*, when the temporal spans of
their corresponding induced context intervals intersect, if the interpretation contexts that hold during these intervals belong to the SUBCONTEXT relation.
Figure 5.9 shows inducement of context intervals within, after, and even before
an inducing event.

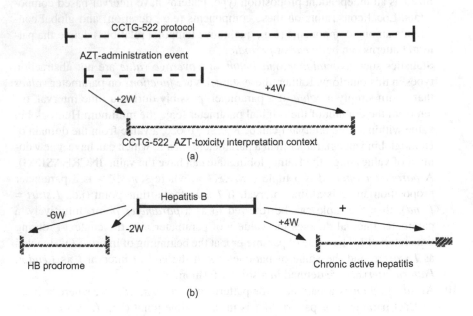

Fig. 5.9 Dynamic induction relations of context intervals (DIRCs). (a) An overlapping direct and
prospective AZT-toxicity interpretation context induced by the existence of an AZT-administration
event in the context of the CCTG-522 AIDS-treatment experimental protocol. The interpretation
context starts 2 weeks after the start of the inducing event, and ends 4 weeks after the end of the
inducing event. (b) *Prospective* (chronic active hepatitis complication) and *retrospective* (hepatitis
B prodrome) interpretation contexts, induced by the external assertion or internal conclusion of
a hepatitis B abstraction interval, a context-forming abstraction. Dashed line with bars = event
interval; striped line with bars = closed context interval; striped arrow with bars = open context
interval; full line with bars = closed abstraction interval.

Formally, a *dynamic induction relation of a context interval* (*DIRC*) is a relation on propositions and time measures, in which each member is a structure of the form $< \xi, \varphi, ss, se, es, ee >$ (see explanation below). The symbol ξ is the interpretation context that is induced. The symbol $\varphi \in P$ is the *inducing proposition*: an event, an abstraction-goal, or a parameter proposition. (An event schema is also allowed, as shorthand for the statement that the relation holds for any event proposition representing an assignment of values to the event schema's arguments.) Each of the other four symbols denotes a time measure or the wildcard symbol *. A proposition φ that is an inducing proposition in at least one DIRC is a *context-forming proposition*.

The knowledge represented by DIRCs can be used to infer new context intervals at runtime. Intuitively, the inducing proposition is assumed, at runtime, to be interpreted over some time interval I with known end points. The four time measures denote, respectively, the temporal distance ss between the *start* point of I and the *start* point of the induced context interval, the distance se between the *start* point of I and the *end* point of the induced context interval, the distance es between the *end* of I and the *start* point of the context interval, and the distance ee between the *end* point of I and the *end* point of the induced context interval. Note that, typically, only two values are necessary to define the scope of the inferred context interval (more values might create an inconsistency), so that the rest can be undefined (i.e., they can be wildcards, which match any time measure), and that sometimes only one of the values is a finite time measure (e.g., the ee distance might be $+\infty$). Note also that the resultant context intervals do not have to span the same temporal scope over which the inducing proposition is interpreted. There are multiple advantages to the DIRC representation, which separates propositions from the contexts that they induce [347].

Exactly which basic propositions and relations exist in the ontology of the KBTA theory and problem-solving method can now be clarified. They are abstraction goals, event propositions, parameter propositions, interpretation contexts, and DIRCs.

The set of all the relevant event schemata and propositions in the domain, their attributes, and their subevents forms the domain's *event ontology*. The set of all the potentially relevant contexts and subcontexts of the domain, whatever their inducing proposition, defines a *context ontology* for the domain. The set of all the relevant parameters and parameter propositions in the domain and their properties forms the domain's *parameter ontology*. The set of all patterns and their properties form the domain's *pattern ontology*. These four ontologies, together with the set of abstraction-goal propositions and the set of all DIRCs, define the domain's *TA ontology*.

To complete the definition of the TA task, the existence of a set of *temporal queries* is assumed, expressed in a predefined *TA query language* that includes constraints on parameter values and on relations among start-point and end-point values among various time intervals and context intervals. That is, a temporal query is a set of constraints over the components of a set of parameter, pattern, event and context intervals, using the domains and TA ontology. Intuitively, the

TA language is used (1) to define the relationship between a pattern-type abstraction and its defining component intervals, and (2) to ask arbitrary queries about the result of the TA inference process.

The *TA task* solved by the KBTA method is defined as follows: Given at least one abstraction-goal interval $< \psi, I >$, a set of event intervals $< e_j, I_j >$, and a set of parameter intervals $< \pi_k, v_k, \xi_k, I_k >$ (ξ_k might be the empty interpretation context in the case of primitive parameters), and the domain's temporal-abstraction ontology, produce an interpretation-that is, a set of context intervals $< \xi_n, I_n >$ and a set of (new) abstractions $< \pi_m, v_m, \xi_m, I_m >$-such that the interpretation can answer any temporal query about all the abstractions derivable from the transitive closure of the input data and the domain knowledge.

5.5.2 The RÉSUMÉ system

The KBTA method has been implemented within the RÉSUMÉ problem solver [357, 359]. In all of the domains in which the RÉSUMÉ system has been tested, the feasibility of knowledge acquisition, representation, and maintenance was evaluated, and the methodology was applied to various clinical test cases. Both the general temporal-abstraction computational knowledge and the domain-specific temporal-abstraction knowledge were found to be reusable. The RÉSUMÉ system has been evaluated for the purpose of summarizing data in multiple clinical domains, such as oncology, monitoring of children's growth [230], and management of insulin-dependent diabetes [359]. In addition, the RÉSUMÉ system and its various versions had been used to support guideline-based application [278] and guideline-based quality assessment of medical care [355], as well as for visualization and exploration of time-oriented patient data and their abstractions, a task in which the system and its various interfaces have been extensively evaluated [351, 352, 350, 354, 254, 366, 255]. The KBTA ontology and the RÉSUMÉ problem solver had even been used to successfully model and solve the task of critiquing traffic controller's actions [356]. The use of the KBTA methodology in that spatio-temporal abstraction task, has further demonstrated the generality (with respect to both domain and application) of the KBTA ontology, and the domain-independence of the computational mechanisms implemented within the RÉSUMÉ system.

The KBTA method has been implemented within multiple applications, in addition to the RÉSUMÉ system. For example, it is the basis for the ALMA temporal-abstraction system within the IDAN temporal-abstraction mediator [33]. Furthermore, The KBTA method has been extended into an incremental, data-driven version, in which raw data arrive in real-time and are continuously abstracted into meaningful interval-based abstractions. The incremental KBTA (*IKBTA*) method has been implemented within the *Momentum* system [391]. A Key operation in the IKBTA method is the incremental temporal-interpolation mechanism, whose time complexity is $O(n)$ or $O(n log(n))$, depending on whether the data are assumed to arrive in order or out of order (thus requiring a truth-maintenance system), respectively [392].

5.5.3 The Temporal-Abstraction Knowledge-Acquisition Tool

One of the principles of modern knowledge-based problem-solving methodologies is the emphasis on supporting, as much as possible, maintenance of the domain knowledge by the domain experts (as opposed to performing that task by knowledge engineers and programmers). The knowledge required by the knowledge-based temporal-abstraction method (KBTA) [347] needs to be acquired from medical domain experts. Thus, a tool for elicitation and maintenance of temporal-abstraction knowledge [347] had been designed using the *Protégé* framework. The *Protégé* project [409, 283] aims to develop a library of highly-reusable, domain-independent, problem-solving method. One advantage of the Protégé approach is the production, given the relevant problem-solving-method and domain ontologies, of automated knowledge-acquisition tools, tailored for the selected problem-solving method and domain.

Evaluation of the temporal-abstraction knowledge-acquisition tool regarding its usability, involving several medical experts and focusing on the domains of oncology (bone-marrow transplantation) and endocrinology (management of diabetes and monitoring of children's growth), has been quite encouraging [353]. The experiments proved the feasibility of semi-automated entry and maintenance of temporal-abstraction knowledge by expert care providers.

A typical view of the temporal abstraction and knowledge acquisition tool is shown in Figure 5.10.

The usability of the knowledge acquisition (KA) tool was evaluated by three expert physicians and three knowledge engineers in three domains-the monitoring of children's growth, the care of patients with diabetes, and protocol-based care in oncology and in experimental therapy for AIDS. The study evaluated the usability of the KA tool for the entry of previously elicited knowledge. The study recorded the time required to understand the methodology and the KA tool and to enter the knowledge, examined the subjects' qualitative comments, and compared the output abstractions with benchmark abstractions computed from the same data and a version of the same knowledge entered manually by KBTA experts. The results have shown that understanding the KBTA ontology required 6 to 20 hours (median, 15 to 20 hours); learning to use the KA tool required 2 to 6 hours (median, 3 to 4 hours). Entry times for physicians varied by domain-2 to 20 hours for growth monitoring (median, 3 hours), 6 and 12 hours for diabetes care, and 5 to 60 hours for protocol-based care (median, 10 hours). An increase in speed of up to 25 times (median, 3 times) was demonstrated for all participants when the KA process was repeated. On their first attempt at using the tool to enter the knowledge, the knowledge engineers

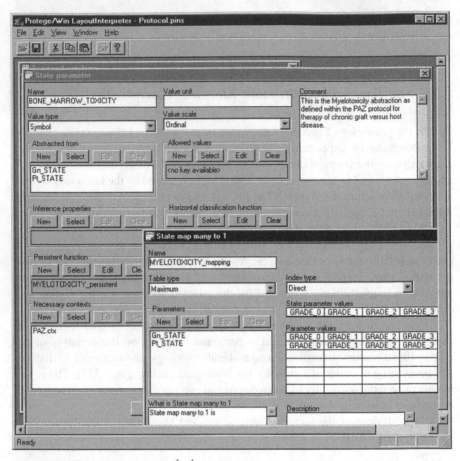

Fig. 5.10 Use of the PROTÉGÉ-based temporal abstraction knowledge-acquisition tool in the oncology domain. The user is defining a multiple-parameter abstraction, BONE_MARROW_TOXICITY, which is abstracted from they intermediate abstractions, PLATELET_STATE (Pt_STATE) and GRANULOCYTE_STATE (Gn_STATE).

recorded entry times similar to those of the expert physicians' second attempt at entering the same knowledge. In all cases the RÉSUMÉ temporal-abstraction system [359], using knowledge entered by means of the KA tool, generated abstractions that were almost identical to those generated using the same knowledge entered manually. Thus, it was demonstrated that the KA tool is usable and effective for both expert physicians and knowledge engineers to enter clinical temporal-abstraction knowledge, and that the resulting knowledge bases are as valid as those produced by manual entry.

Summary

The chapter presented different types of temporal data abstraction and then focused on various approaches to this task, relating it to the tasks of monitoring time-oriented clinical data, exploration of time-oriented clinical data, knowledge discovery from clinical databases, and summarization of on-line medical records. It then presented the concept of temporal-abstraction mediator, followed by a detailed discussion of the generic knowledge-based temporal-abstraction ontology proposed by Shahar and the method associated with this ontology.

Temporal data abstraction was singled out as the focus of this chapter because of its direct relevance to all medical tasks (diagnosis, monitoring, therapy management, guideline-based care, clinical research, etc). Any advances in temporal data abstraction will be of benefit to the higher reasoning performed by such tasks; this provided that data abstraction is modeled as a loosely coupled (sub) process. One solution that has been proposed in this chapter is the temporal-abstraction mediator. However, even granted a temporal-abstraction mediation service, further work is required to support the temporal reasoning inherent in the core medical tasks themselves. Although the key issues that need to be tackled have been identified, a number of concerns still need to be addressed in order to have a temporal-abstraction method that is both generic and easily reusable and exhibits the required computational behavior (handles data of different types and granularities, deals with uncertainty and incompleteness, can accept data in different orders, supports truth maintenance, derives and revises all possible abstractions in a time efficient manner, deals with cyclic and periodic phenomena, etc.). In addition, the technology of temporal-data abstraction should be more actively deployed in the context of medical data mining and knowledge discovery.

Bibliographic Notes

In this chapter we have tried to give a tutorial type of exposition to the relatively new but rapidly growing field of temporal-data abstraction, by overviewing the relevant concepts and illustrating them by reference to various systems developed. To the best of our knowledge there aren't any comprehensive review articles on this topic per se in the existing literature. Needless to say, however, that this chapter merely gives a flavor of the topic and in order for the reader to appreciate the relevant details and in particular in order to reproduce any of the presented concepts or methods, s/he would have to refer to the original sources. The cited references span the whole development of the field from the very early attempts (Rx, VM) to most recent ones. The structure of the chapter enables the reader to identify clusters of thematically related papers in the literature, e.g., monitoring, intensive care monitoring, etc., for a more detailed exploration.

Problems

[*To answer questions 1 to 5, refer to the chapter, to Shahar and Musen's paper about the RÉSUMÉ system [359], and to Shahar's paper about the knowledge-based temporal-abstraction (KBTA) method [347].*]

5.1. Explain in your own words the reason for having explicit temporal interpretation contexts, and list at least *four* different advantages of separating contexts from both the propositions inducing them and the parameters interpreted within their temporal span. Provide examples in each case.

5.2. Explain in your own words why having several different interpretations for each set of data at the same time is *not* contradictory, in the KBTA model.

5.3. Formation of interpretation contexts might lead to creation of new abstractions of the same data, which might in turn induce additional contexts, which might lead to inference of more abstractions. Furthermore, these contexts might be in the past, present, and/or future, and thus might affect other sets of data in the patient's record. Explain in detail (a formal proof is not necessary) why, nevertheless, the context formation process is *finite* and always ends with a *stable* database.

5.4. Consider the temporal-abstraction ontology.

1. What is the advantage of indexing parameter-specific knowledge in the temporal-abstraction parameter ontology through abstraction type (e.g., state, gradient), as opposed to indexing by the parameter name and then the abstraction type?
2. What is the advantage of specializing parameter-specific knowledge by contexts, as opposed to indexing by context (e.g., protocol CCTG522) and then parameter name and/or abstraction type?
3. Is there in fact an advantage to indexing parameter knowledge through the context and then parameter name? (Hint: consider acquisition and maintenance of the knowledge).

5.5. The RÉSUMÉ system uses a truth-maintenance system (TMS).

1. What is a TMS?
2. Why is truth maintenance relevant to the temporal-abstraction task?
3. What is the relationship between truth maintenance and temporal-semantic properties, at least within the RÉSUMÉ system? Give an example.
4. Assume that Hemoglobin (Hb) values in the range [9, 11] gr/100cc are defined (in a context not shown here) as hemoglobin state (HbS) = *moderate anemia*, and above 11 gr/100cc are defined as hemoglobin state = *normal*. Assume that the gap function for hemoglobin state = *moderate anemia* allows interpolation of up to 6 days.

Explain how exactly would the temporal semantic properties be used in the following example, by adding in the following bitemporal database all the tuples denoting abstractions with transaction times following insertion of Hb = 12.2 gr/100cc on January 18, 1996 at transaction time 96/1/20:5:00 p.m., and/or by modifying existing tuples. Show the final database view. Explain the reason for each addition or update.

Table 5.2 Bitemporal database

Para-meter	Value/unit (attribute:value)	Valid start time	Valid stop time	Transaction Time start	Transaction Time end
Hb	10.8 gr/100cc	96/1/17:9:00a.m	96/1/17:9:00a.m	96/1/17:6:00p.m.	Open
HbS	Moderate-anemia	96/1/17:9:00a.m	96/1/17:9:00a.m	96/1/17:6:00p.m.	Open
Hb	10.1 gr/100cc	96/1/19:9:00a.m	96/1/19:9:00a.m	96/1/19:5:00p.m.	Open
HbS	Moderate-anemia	96/1/19:9:00a.m	96/1/19:9:00a.m	96/1/19:5:00p.m.	Open
HbS	Moderate-anemia	96/1/17:9:00a.m	96/1/19:9:00a.m	96/1/19:5:00p.m.	Open
Hb	12.2 gr/100cc	96/1/18:9:00a.m.	96/1/18:9:00a.m.	96/1/20:5:00p.m.	Open

5.6. *To answer the following question, refer to the chapter and, in particular, to the discussion of the temporal mediator concept and to the papers of Nguyen et al. [281] and of Boaz and Shahar [33], regarding Tzolkin and IDAN temporal mediators.*

Bone-marrow toxicity in the context of the PAZ protocol is defined as a function of platelet-state in the PAZ context and hemoglobin-state within the PAZ context. These, in turn, are a function of platelet-count and hemoglobin-value, correspondingly.

Explain in detail how a query about bone-marrow toxicities in the past month would be answered by the mediator, assuming that the patient database includes only hemoglobin values, platelet counts, and PAZ protocol events. Show what the controller is doing at each phase and what is the state of the database used by the mediator. (Assume that definitions of all events, abstractions and contexts exist in the temporal-abstraction knowledge base of the oncology domain).

5.7. Consider the KNAVE-II knowledge-based visualization and exploration system's architecture which uses the IDAN temporal-abstraction mediator.

A user who uses a knowledge-based visualization and exploration distributed architecture starts a session by formulating a query about a patient. Which of the following resources does she *mostly* need to formulate that initial query?

(1) The patient database,
(2) The knowledge base,
(3) The data display module,
(4) The temporal mediator.
    Please explain.

5.8. What would "dynamic sensitivity analysis" mean in the context of the KNAVE-II system? How might a truth-maintenance system (TMS) be useful for such a task? Explain. Consider both knowledge and data aspects.

5.9. Consider Fagan's VM system for ventilation control.

1. The VM system used context-sensitive *mappings* of clinical parameters into abstractions, but context-free *rules*. What are the potential advantages and disadvantages of this approach?
2. The VM system used state-change rules to infer a new context. Why? *Are* there alternatives, in your opinion, and, if so, *What* might they be? Explain.

5.10. It has been pointed out that Russ's Temporal Control Structure (TCS) system is quite similar in certain respects to Kahn and Gorry's Time Specialist [195], an early architecture that focused on several methods of storing, organizing, and retrieving time-oriented facts in a black-box like fashion, while similar in other respects to programs such as VM, TOPAZ, M-HTP, IDEFIX and even TrenDx.

Explain in each case in detail what, in your opinion, the critics have (or might have) in mind, and what is your own opinion. Justify your comments.

5.11. Examine the Range Relation (RRel) used by Kohane's Temporal Utilities Package (TUP). Represent each of the following two temporal relations using a conjunction of one or more RRels in Kohane's TUP style:

1. A finishes B
2. A during B

5.12. Consider Haimovitz's and Kohane's TrenDx system.

1. What are the advantages of the capability for performing partial matches (matches on partial data) in the TrenDx system?
2. Why are intermediate-level abstractions useful and how might their lack create a disadvantage in the TrenDx and similar systems? Consider aspects such as knowledge acquisition and maintenance, queries by users, data mining, etc.

Part III
Time in Clinical Tasks

Chapter 6
Time in Clinical Diagnosis

Overview

This chapter aims to explain how time and temporal reasoning can feature, with critical advantage, in the task of clinical diagnosis. This is largely done through a number of representative clinical diagnostic systems. The discussion focuses on the representation of time with respect to diagnostic knowledge and patient data, and the relevant temporal reasoning. It is assumed that the reader has a reasonable understanding of diagnostic systems, in particular clinical diagnostic systems. However, for the sake of completeness, relevant fundamental notions are overviewed.

In Chapter 2 an abstract structure for representing general temporal constraints, the Abstract Temporal Graph (ATG) was presented. In this chapter a number of concrete instantiations of this structure of relevance to clinical diagnosis are discussed.

The reader is expected to gain the following from this chapter:

- An appreciation of the significance of time in clinical diagnostic reasoning.
- An understanding of the distinction between abductive and consistency-based diagnostic reasoning.
- A familiarization with aspects of time of relevance to diagnostic knowledge and patient data, namely granularity, uncertainty, incompleteness and repetition.
- An understanding of how time is represented in patient data and models of disorders, particularly causal-based models.
- More specifically, an understanding of the types of temporal constraints arising in clinical diagnostic problems.
- An appreciation of the significance of temporal data abstraction in clinical diagnosis. The importance of temporal data abstraction in general is discussed in detail in Chapter 5.
- And overall, an understanding of how the explicit representation of time influences the reasoning of clinical diagnostic systems, as well as an appreciation of the limitations of the existing technology.

C. Combi et al., *Temporal Information Systems in Medicine*,
DOI 10.1007/978-1-4419-6543-1_6, © Springer Science+Business Media, LLC 2010

Structure of the Chapter

The chapter is structured as follows. Section 6.1 gives a historical perspective on clinical diagnostic systems as a means of an introduction. Central diagnostic notions including temporal requirements are further elaborated in Section 6.2. Sections 6.3-6.8 discuss specific proposals focusing on the representation of time and the relevant temporal reasoning. The selected approaches collectively give a fairly accurate picture of the existing state of the art, although, given that this is an active area of research, this picture is changing all the time, with the appearance of new and enhanced suggestions. Through the discussion of specific approaches it will become apparent that modeling and reasoning with time in clinical diagnosis, largely means expressing and reasoning with temporal constraints. Section 6.9 presents the main forms of temporal constraints of relevance to clinical diagnosis by discussing a number of concrete instantiations of the Abstract Temporal Graph introduced in Chapter 2.

Keywords

clinical diagnosis, time representation, temporal reasoning, abductive diagnosis, consistency-based diagnosis, complex hypotheses, Causal-Temporal-Action (C-T-A) Model, temporal data abstraction, temporal uncertainty, multiple granularities, temporal consistency, Heart Disease Program (HDP), temporal Parsimonious Covering Theory (t-PCT), fuzzy t-PCT, Abstract Temporal Diagnosis (ATD), temporal abductive diagnosis, abductive diagnosis using time-objects, temporal constraints, Abstract Temporal Graph (ATG).

6.1 Introduction

6.1.1 Aim and Scope

This chapter aims to explain how time and temporal reasoning can feature, with substantial advantage, in the task of clinical diagnosis. This is done through a number of representative clinical diagnostic systems, focusing the discussion on time issues relating to diagnostic knowledge and patient data representation, and inferencing. Aspects of these systems which are not directly related to time are overlooked. It is assumed that the reader has a reasonable understanding of diagnostic systems, in particular clinical diagnostic systems. However, for the sake of completeness, relevant fundamental notions are overviewed. In order for the reader to gain a deeper understanding of the material covered, the presentation of the different diagnostic systems is unified through a classification and discussion, at a general level, of core temporal reasoning mechanisms of relevance to clinical diagnosis.

6.1.2 A Historical Perspective

Diagnostic problem solving has attracted and continues to attract considerable interest in the AI community simply because it is a difficult task to model, especially when multiple failures/disorders are involved. In a nutshell, the computer-based automation of diagnostic reasoning still presents a number of challenges. One such challenge is the modeling of time.

The majority of diagnostic systems deal with clinical domains. Pioneering diagnostic systems, such as MYCIN [378, 43], INTERNIST-1 [266], CASNET [423], and ABEL [298], were all for clinical domains. These systems, although quite successful in their respective domains, did have a number of shortcomings, some of which were attributed to their inability to model and reason with time. Time was ignored or it figured in a very implicit way in their knowledge bases and the patient data processed by them. For example, in MYCIN, whose domain was antimicrobial infections, some of the contexts (objects of reference for the system's rules) were distinguished into *past* and *current*, e.g. past cultures, current cultures, past organisms, current organisms, etc. However, the system did not have any 'understanding' of the notions of past or current and actually it did not need to have such an understanding. In fact MYCIN's shortcomings were not due to the system's ignorance of time, but rather due to its ignorance of other, more critical for the particular domain types of knowledge, the modeling of which led to the drastic reconstruction of MYCIN and the creation of the NEOMYCIN system [76].

CASNET's best known application was the domain of glaucoma. The central knowledge structure in this system was a causal network of pathophysiological states from causes to effects. Although time is inherently relevant to causality, time was again neglected. CASNET's inferencing comprised two main reasoning mechanisms; the assignment of confidence factors to simple hypotheses referring to pathophysiological states and the construction of complex hypotheses of causal chains of pathophysiological states. Both mechanisms had limitations due to the absence of time.

The creators of INTERNIST-1, another celebrated pioneering diagnostic system, for the broad domain of internal medicine, did acknowledge that the absence of time was a source of problems for the performance of the system. The *presentation time* of disorders was implicitly included in the disorder models through attribute AGE. For example, the frequency of occurrence of cardiac sarcoma primary amongst people between the ages of 16 and 25 is low and hence the hypothesis of this disorder would not be evoked for patients in this age group. In the language of INTERNIST-1 the contextual finding "AGE 16 to 25" would have an *evoking strength* of 0 and a *frequency of occurrence* of 2 with respect to disorder cardiac sarcoma primary. Such findings were treated like all other manifestations. If the earliest time that some disorder could occur were the age of 5 years, both the evoking strength and the frequency of occurrence of the finding "AGE less than 5 years" would be 0 for that disorder. If, on the basis of some other finding, the hypothesis of this disorder were to be evoked, say for a 2 year old, the scoring function used by INTERNIST-1's hypothesis evaluation mechanism would presumably render this hypothesis unattractive, but it would not necessarily refute it. However, if the system had an explicit representation of disorder presentation times, it would have the ability to directly deny hypotheses on the basis of a single item of contextual information, thus pruning more effectively its hypothesis space. This would have been especially important given the size of the system's hypothesis space and the need to keep the proliferation of active hypotheses at bay.

Causality was not a central relation in INTERNIST-1's disorder models, although a number of complementary relations between disorders, such as predisposes, causes, caused-by, etc. were included in the system's vocabulary. Such relations were used for promoting the consideration of some disorder hypothesis in light of the conclusion of one of its complementary disorders. Again such complementary relations were not associated with time, an omission prone to paradoxical conclusions. For example, let's say that condition A is a predisposing factor for condition B. By definition a predisposing factor precedes the condition for which it is a predisposing factor. Let us further assume that condition A has been concluded for some case, but there are still some significant unexplained positive findings and condition B is an active hypothesis. If the unexplained findings had actually occurred prior to the occurrence of condition A, it would be a paradox to say that the hypothesis of condition B potentially explains the given findings and that the conclusion of condition A should promote further the consideration of this hypothesis, i.e. to assume that in this case condition A has indeed acted as a predisposing factor for condition B. If condition A were in actual fact a predisposing factor for condition B, any findings caused by the occurrence of B should follow the onset of the occurrence of condition A. However, INTERNIST-1 were not in a position to carry out this sort of deeper, time-based reasoning, as neither its disorder models, nor the patient findings were associated with time. Since disorders were not modeled as dynamic processes evolving in time, the system could not distinguish between negative findings which were truly negative (non-existent) and 'negative' findings which in fact referred to past or future manifestations.

The absence of time in disorder models and patient data continued in later years as well. For example, the well known *parsimonious covering theory* [306], the development of which was principally influenced by clinical diagnostic problems, did not explicitly advocate the modeling of time, nor did the generic method of *heuristic classification* [75]. The latter method, whose applicability goes beyond diagnostic problem solving, amongst other things introduced and pointed out the significance of the process of *data abstraction*. However, the data abstractions presented did not include temporal abstractions.

Temporal data abstraction, in a rudimentary form, appeared in the PATREC system [267] that acted as an intelligent manager of patient data, in assistance to MDX [61], a diagnostic system for the syndrome of cholestasis. PATREC's 'intelligence' was primarily drawn from hierarchical structures of medical concepts, that enabled the derivation of abstractions from raw data. In addition, the system could 'understand' temporal notions such as day, week, before, during, after, same-time, etc. The diagnostic rules used by MDX could make use of such notions and the patient data could make references to time. PATREC was able to match the raw patient data against the more abstract expressions used in the diagnostic rules. This matching involved some kind of temporal reasoning.

Gradually there was an increased awareness that time was not a detail that clinical diagnostic reasoning could do without and the modeling of time started appearing in a more emphatic way. Presently, the explicit incorporation of the temporal dimension in diagnostic problem solving is receiving major attention. Representative results of these efforts, of relevance to clinical diagnosis, are the subject of this chapter. It is interesting to note that in many of the diagnostic systems discussed, time was absent originally and was subsequently included in an attempt to achieve significant performance enhancements.

A notable attempt in providing a unifying framework for temporal model-based diagnosis is found in [42]. This work represents a significant extension to previous work by some of the authors [102]. Model-based diagnosis encompasses both consistency-based and abductive diagnosis (see Section 6.2). The authors do not make any commitments regarding an ontology for time. First they distinguish between *time-varying context* (observing the behaviour of a system at different times), *temporal behavior* (the consequences of the fact that a system is in a specific (normal or faulty) mode manifest themselves after some time and for some time), and *time-varying behavior* (permitted transitions between faults; if this knowledge is missing any transition is permitted by default). In this work a categorization for diagnosis is given. The categorization starts with snapshot diagnosis (atemporal diagnosis performed at a point in time) and ends with temporal diagnosis (considers temporal behavior over a specified time window, but not time-varying behavior) and general temporal diagnosis (considers both time-varying and temporal behavior, again over a specified time window). In temporal diagnosis, the assumption that a fault persists during the considered time window is made. This assumption is waved under general temporal diagnosis.

A set-theoretic framework for diagnosis, again encompassing consistency-based and abductive diagnosis is proposed in [250]. This approach emphasizes the use of evidence functions and concentrates on interactions between faults. Time itself is abstracted away. Interactions are derived on the basis of formal properties of evidence functions rather than on the basis of any temporal considerations.

The discussion in Sections 6.3-6.8 focuses on the representation of time and the role of temporal reasoning in clinical diagnosis without going into the details of how diagnostic solutions are derived. In order to illustrate the different proposals, a simplistic pseudo-medical example is used. The proposals analysed include a Bayesian approach (HDP), a set-theoretic approach (t-PCT) and its fuzzy extensions, two logic based approaches (ATD; Console and Torasso proposal), and an object-based approach (Keravnou and Washbrook proposal). It should be said that there are many different approaches to clinical diagnosis. For example, there is a growing literature on approaches using Bayesian Belief Networks (BBNs), or simply belief networks, and their variants causal probabilistic networks, where the predominant feature is the modelling of uncertainty through probabilities [13, 118, 190, 237, 300, 390]. In many of the approaches dealing with causal probabilistic networks time is absent or implicitly represented, while in others time has been explicitly addressed (e.g. [25, 331]). A notable example is the HDP system which is included in our discussion. Dynamic Networks Models (DNMs), a temporal extension of BBNs, have forecasting/monitoring as their application scope, not diagnosis [110, 106]. Before we embark on the discussion of the selected proposals we overview central diagnostic notions from the perspective of clinical diagnosis and time.

6.2 Diagnostic Notions

A *diagnostic problem* with respect to some artificial or natural system, a person say, arises when at least one observation (of abnormality) is made suggesting that the particular individual is not functioning 'normally', i.e. as expected. A solution to the problem, referred to as *diagnosis*, is an *explanation* of the abnormal functioning.

Generally there are two paradigms of diagnostic problem solving, the *consistency-based* paradigm [116, 329] and the *abductive* paradigm [99, 312, 313, 314]. As clinical diagnosis centers around disorders the predominant paradigm used is the abductive one.

6.2.1 Consistency-Based versus Abductive Diagnosis

Consistency-based diagnosis advocates the use of models of normal behavior such as structure-and-function models describing the normal decomposition and functioning of natural organisms or artifacts. Under this paradigm, a solution to a diagnostic problem is a set of *abnormality assumptions*. An abnormality assumption states that a (replaceable) component is abnormal (thus causing some disturbance in the normal functioning). The behavior generated when the model of normal behaviour is perturbed under the given set of abnormality assumptions, must be *consistent* with the observed abnormalities. Consistency means that nothing is entailed which is in conflict with the observations to be explained, i.e. the observed abnormalities. This is the essence of consistency-based diagnostic reasoning. Under this modeling paradigm a diagnosis in terms of a set of abnormality assumptions is considered the appropriate notion of explanation. Usually the best diagnosis (explanation) is one that involves a minimal set of abnormality assumptions. This is very much in line with the well-known Occam's razor — "What can be done with fewer assumptions is done in vain with more" [312].

A malfunctioning organism is treated by reverting the abnormal components back to normal behaviour. In the case of artifacts this is easily done, simply by replacing the abnormal primitive components with new ones. However, this course of treatment is not usually viable for natural organisms. Instead the malfunctioning 'components' need to be brought back to normality, or as close to normality as possible, by other means, such as the administration of drugs, operative actions, etc. For these courses of action to be effective, it does not suffice to say that a component of the organism is malfunctioning, but also to pinpoint as accurately as possible the particular failure(s) of the component. In other words to point out, at a sufficiently detailed level, the causes of the observations of abnormality. This is the essence of abductive diagnostic problem solving. Consistency-based diagnosis can in fact be extended with abductive notions, whereby the model of normal behaviour is augmented with the expected behaviour of the different components under various failure modes [395]. A diagnosis is no longer merely a set of assumptions about ailing (replaceable) components, but a set of assumptions about operative failure modes of (replaceable) components. The explanation provided by such a diagnosis is stronger than the purely consistency-based one, in the sense that it is not just consistent with the observations of abnormality, but it actually *entails* the observations of abnormality. This is the appropriate notion of explanation when failure models are a central ingredient of the diagnostic knowledge. In this respect, a diagnosis that explains (entails) all observations that need to be explained (usually the observations of abnormality and possibly some others) is referred to as a *cover*.

As already mentioned, clinical diagnosis centers around disorders and as such the prevailing paradigm is that of abductive diagnosis. Disorder models are therefore a central knowledge structure. Models of normal behaviour are not excluded and in fact are often used as a special kind of model (background model) in conjunction with the disorder models [101, 210].

6.2.2 Diagnosing under the Assumption of Multiple Disorders

Abductive diagnostic reasoning comprises the *generation* and *evaluation* of diagnostic hypotheses as two distinct but tightly coupled steps. The reasoning is performed under the assumption of *single or multiple disorders*, thus giving rise to simple or complex hypotheses respectively. Obviously things are much simpler if the assumption of single disorder applies. The assumption of multiple disorders is broader in scope, but the generation of possible combinations of disorders is often a computationally complex process [47]. The use of appropriate heuristics can make this process tractable by pruning the search space.

If the reasoning is atemporal, multiple disorders essentially means multiple *concurrent* disorders. This is restrictive. Time-oriented reasoning broadens the coverage of real possibilities, allowing for example a multiple disorder to be the recurrence of the same disorder. Furthermore, not all disorders would need to exist in parallel, thus allowing for transient disorders with persistent manifestations. This could be so for single disorders as well.

As already mentioned, CASNET had complex hypotheses, i.e. paths in the causal network. Individual pathophysiological states on such paths constituted simple hypotheses. Given the atemporal set up under which the system reasoned, one could say that either it was assumed that once a state was brought into existence, it persisted indefinitely, thus giving rise to a nesting of state existences, or that when a state began to exist its causal predecessor(s) ceased to exist, thus giving rise to disjoint existences. The latter interpretation is in agreement with the fact that the succession of states on some path denoted a worsening situation going through various landmark states.

INTERNIST-1 also operated under the multiple disorders assumption. Above we discussed briefly how the conclusion of one disorder could promote the consideration of active hypotheses that were related with the concluded disorder through some complementary relation. This way the system tried to piece together, in a sequential fashion (see below), a complex hypothesis whose components exhibited some sort of coherence — they were not independent of each other.

When a diagnostic system deals with complex hypotheses (as it is the case when it operates under the multiple disorders assumption) a further consideration is whether a complex hypothesis is constructed in a parallel or sequential fashion. Parallel fashion means that the entire complex hypothesis is fully constructed before it may be concluded in its entirety. Sequential fashion means that the individual components comprising the (as yet unknown) complex picture are separately selected and concluded and this progressive way of construction continues, possibly in a 'blind' fashion, until the picture is considered complete.

Parallel reasoning is often based on the requirement that a complex hypothesis should be as coherent as possible, i.e. its components should be related in some way. Sequential reasoning does not necessarily guarantee coherence and has the drawback of propagating erroneous conclusions. But obviously sequential reasoning is computationally more tractable than parallel reasoning.

Example 6.1. Consider the simple causal network given in Figure 6.1. Disorder d_1 can cause manifestations m_1 and m_2, while disorder d_2 can cause manifestations m_3 and m_4. These two disorders are related in a complementary fashion since d_1 can act as a predisposing factor for d_2. Finally disorder d_3, which is unrelated to the other two, can cause manifestations m_2, m_3 and m_4.

Let us assume that the findings of the patient in question refer to the presence of all four manifestations. Applying sequential reasoning to this case would result in the conclusion of d_3 and d_1. Disorder d_3 is concluded first because it represents the 'best' single hypothesis, as it explains 3 out of the 4 findings. To complete the picture, i.e. to provide an explanation for the presence of m_1 as well, disorder d_1, that actually is the only hypothesis left, is concluded next. In this example, reasoning by considering simple, single disorder hypotheses in sequence, has resulted in an incoherent complex conclusion since its two components, disorders d_1 and d_3, are unrelated.

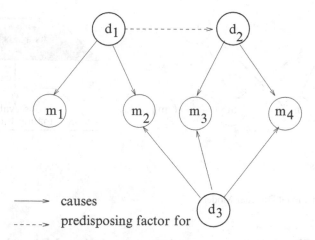

Fig. 6.1 Sequential vs Parallel Construction of Complex Hypotheses

Parallel reasoning, on the other hand, considers at once complex, multi disorder, hypotheses. In this example the competing hypotheses are $\{d_1, d_2\}$ and $\{d_1, d_3\}$. The first one is concluded because of its coherence.

Although coherence is a strong criterion for acceptance, other factors such as time could be of critical importance. For example, in the particular case, temporal considerations could lead to the conclusion that the correct diagnosis is in fact $\{d_1, d_3\}$. This is why temporal reasoning should have a key role in diagnostic reasoning, irrespective of whether the reasoning is performed in a sequential or parallel fashion. The consideration of temporal constraints gives a more accurate picture of reality and as such it provides immeasurable guidance. This is illustrated through Example 6.2, given in Section 6.2.3.

CASNET applied parallel reasoning at the level of complex hypotheses (causal chains of states). However, the system applied sequential reasoning at the level of simple hypotheses, individual states. A complex hypothesis aimed to encompass all confirmed states. Various heuristics were used to confine the generation of complex hypotheses to plausible ones. INTERNIST-1 used sequential reasoning, which was shown to be prone to erroneous conclusions. Since the presence of multiple disorders is not an uncommon phenomenon in medicine, especially in the broad domain of internal medicine, INTERNIST-1 was reconstructed so that causality featured as a central relation and complex hypotheses were formed in a parallel fashion. System CADUCEUS resulted from this reconstruction [265, 315].

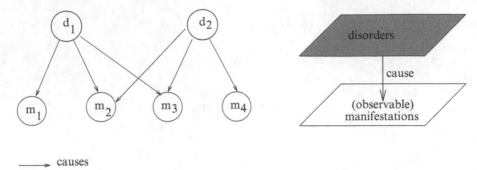

———→ causes

Fig. 6.2 Associational Disorder Models

6.2.3 The Simplest Representation: Associational Disorder Models

The simplest kind of a disorder model is an *associational model*. A disorder is modeled in terms of its external, observable manifestations (see Figure 6.2). Such a model is appropriately represented as a set of manifestations. In this set up, diagnostic knowledge forms a bipolar structure, with the disorders residing on a higher (unobservable) plane and the manifestations on a lower (observable) plane. The associations from disorders to manifestations express causality at the highest level. Hence associational models ignore the evolution of disorders, e.g. in terms of more detailed causal mechanisms, and time[1]. Moreover, a diagnostic system with a purely associational knowledge base has no knowledge of therapeutic actions, in particular no knowledge of the influences of therapeutic actions on the evolution of disorders.

Example 6.2. We continue with another simple example, that will also form the basis for the illustration of the approaches discussed in Sections 6.3-6.8. Consider the following diagnostic problem, in connection with the disorder models of Figure 6.2. The findings of the patient in question assert the presence of m_2, m_3 and m_4 and the absence of m_1. More specifically, m_2 appeared, then disappeared and then appeared again. The appearance of m_3 followed the start of the second appearance of m_2, the two overlapped and continued together. The appearance of m_4 preceded the first appearance of m_2. In addition, the contextual information that this person is currently undergoing treatment *a* is known.

The restrictions of the associational disorder models exclude the consideration of the temporal and contextual information. In fact the only information that can be considered is the presence of m_2, m_3 and m_4 and the absence of m_1, which only

---

[1] In this discussion, for the sake of simplicity the uncertainty of the associations is ignored. Such uncertainty can be modeled probabilistically or in some other fashion. For example, in a realistic system, disorders have prior probabilities of occurrence and the associations from disorders (D) to manifestations (M) have conditional probabilities, $P(M/D)$. Often, the manifestations of some disorder are classified as *necessary*, i.e., $P(M/D) = 1.0$ (or $P(\sim D/ \sim M) = 1.0$), or as *sufficient* (for the hypothesis of the disorder), i.e. $P(D/M) = 1.0$

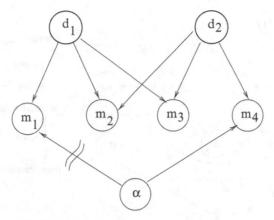

Fig. 6.3 Considering the Effects of Therapeutic Actions

gives a snapshot description of the actual patient situation and as such it is rather inaccurate.

There are three potential hypotheses, the two simple hypotheses, that either d_1 or d_2 is the cause of the patient's situation and the complex hypothesis that both disorders are operative in the patient. Disorder d_1 on its own explains the presence of m_2 and m_3 but is in conflict with the absence of m_1 and does not explain the presence of m_4. Disorder d_2 on its own explains the presence of m_2, m_3 and m_4 and is not in conflict with the absence of m_1. Given that the coverage of the hypothesis that d_1 is present is a subset of the coverage of the hypothesis that d_2 is present, there is no justification in pursuing the complex hypothesis since its explanatory power will not be any higher and what's more it will inherit the conflict between d_1 and the absence of m_1. The hypothesis that only d_2 is operative is accepted hands down as it is clearly superior to the other two. But should one rest assured that this is actually so, given that so much information about the evolution of the patient's situation has been ignored? Temporal and contextual information has been ignored, where therapeutic actions are a major part of contextual information. Ignoring the temporal information means that a dynamic, moving, picture is replaced with a static one.

Let us extend the knowledge depicted in Figure 6.2 to include the effects of therapeutic action a as well. This is shown in Figure 6.3. Action a inhibits m_1 but gives rise to m_4. When the two hypotheses are now considered in conjunction with a, they appear equally favorable. This is because the consideration of a enhances the explanatory power of the hypothesis that d_1 is present to that of the hypothesis that d_2 is present, since both the absence of m_1 and the presence of m_4 are now accounted for. Action a is immaterial to the hypothesis that d_2 is present as neither it enhances nor it reduces its explanatory power. Negative findings describing normality, such

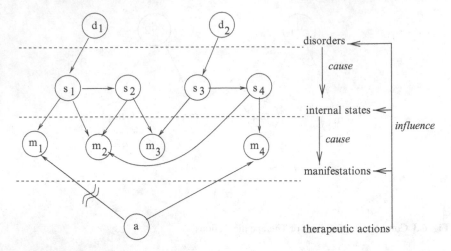

Fig. 6.4 Adding Internal States in Disorder Models and Considering the Effects of Therapeutic Actions

as the absence of m_1, that are not in conflict with expectations of a hypothesis, do not need to be explicitly accounted for. Simply they are accounted by default.

By bringing action a into the scene, a dilemma is created, since the simple associational descriptions of the two disorders and the effects of the action do not provide any means for differentiating between the two competing hypotheses. Still, the consideration of a should not be seen as a drawback. Ignoring a in the first place gave a false sense of security in believing the hypothesis that d_2 is present. Taking a into account sheds doubt in that belief. Considering the effects of actions, albeit in simple associational terms, is certainly an improvement, as the less one ignores the more informed the resulting decision would be. The next step is to take into account causality and time, thus enabling the modeling of disorders, therapeutic actions and patient states, as dynamic, evolving processes. Under this set up, the correct explanation of the given diagnostic problem is that d_1 is present, not d_2, even if this hypothesis appeared so unlikely under the initial set up that involved just the associational models of the two disorders (see below).

6.2.4 The Causal-Temporal-Action Model

An associational disorder model can be extended into a *causal model* (or causal-associational model as it is sometimes referred to[2]), by adding some detail about

---

[2] This term comes from CASNET that actually stands for Causal ASsociational NETwork. This system epitomizes the particular model.

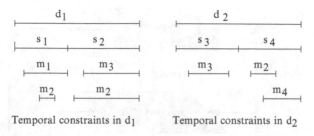

Fig. 6.5 Extending Disorder Models with Temporal Constraints

the causal mechanisms underlying the given associations. This can be done in a number of ways. A simple way often used, is to intercept another (unobservable) plane of internal pathophysiological states between the disorder and manifestation planes. The evolution of disorders is thus described in terms of intermediate states. Disorders point to their starting states, states point to states that form their direct consequents in the causal chains, as well as to external manifestations. Although the links from disorders to states, from states to states and from states to manifestations are semantically different, they could all be loosely described as causal. Figure 6.3 gives the causal models of disorders d_1 and d_2 — d_1 is refined into the sequence of states s_1 and s_2 and likewise d_2 into the sequence of states s_3 and s_4.

When the nodes and/or causal links are augmented with temporal constraints, a causal model is extended into a *causal-temporal model*. Temporal constraints can be expressed in many different ways, as will be shown in Sections 6.3-6.8. A simple way is to express them in relative terms using Allen's set of temporal relations [8]. This is the way used in Figure 6.5 to express the temporal constraints of disorders d_1 and d_2 [3]. Disorder d_1 could be some kind of a viral infection that during the first stage of its evolution (internal state s_1) causes headache (m_1) and nausea (m_2), while during the second stage (internal state s_2) causes fatigue (m_3) and more extended and severe nausea. The first bout of nausea is engulfed by strong headache, but when the headache goes, the nausea reappears, a lot more severely, followed soon afterwards with fatigue. Disorder d_2 could be a stomach disorder that causes fatigue and severe stomach upset (m_4). Action a could involve the taking of some anti-headache drug that removes the headache, but causes stomach upset as an adverse side-effect.

Temporal constraints occupy a fourth plane, orthogonal to the tri-planar hierarchical structure, since they can refer to any entity or link, within or between planes. By definition temporal constraints, specify constrains on the existence of disorders, states and manifestations.

Finally the causal-temporal disorder models could be augmented with knowledge regarding the indications and contra-indications of therapeutic actions. From

---

[3] In the evolution of d_1 there is a gap between the first occurrence of manifestation m_2 caused by internal state s_1 and the second appearance of m_2 caused by internal state s_2. If the temporal knowledge includes information on the default persistence of manifestations, i.e. their expected duration irrespective of any context of occurrence, then multiple occurrences of the same manifestation, even without gaps between them, could be inferred.

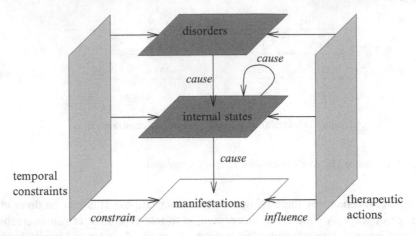

Fig. 6.6 Causal-Temporal-Action Model

this knowledge that also involves time, and part of it could in fact be expressed independently of any disorders, one could extrapolate the influences, both positive and negative, of particular actions on the evolution of particular disorders. Like the temporal knowledge, the knowledge on therapeutic actions is also orthogonal to the disorder-state-manifestation lattice as the influences could refer to any of the planes or to any inter-planar links.

The extended diagnostic knowledge model (Causal-Temporal-Action model, or C-T-A model for short) is given in Figure 6.6. This is a quintet-planar structure with a central tri-planar lattice. The existences of the entities residing on the central planes are bounded by the two orthogonal planes.

Example 6.3. Let us now go back to the diagnostic problem of Example 6.2. The patient findings that need to be explained are shown in Figure 6.7 and listed below — as can be seen the temporal information is no longer ignored.

f_1: first appearance of m_2
f_2: second appearance of m_2
f_3: appearance of m_3
f_4: appearance of m_4
f_5: absence of m_1
f_6: f_2 precedes f_3
f_7: f_4 precedes f_1

All patient data are considered, including the recurrence of m_2 and the temporal relations between manifestations. In addition the contextual finding about the ongoing execution of action a is borne in mind:

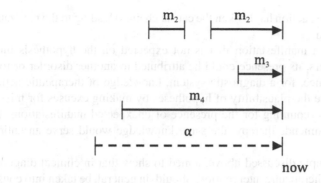

Fig. 6.7 Patient Findings

f_8: therapeutic action a is under execution, having started at a certain point in time prior to the appearance of m_4.

The following hypotheses are considered. The first hypothesis says that disorder d_1 is present. This would explain findings f_1, f_2, f_3 and f_6. However it is in conflict with f_5 and it does not cover f_4 and f_7. The consideration of the contextual finding is now called for to see whether this can resolve the conflict and provide an explanation for f_4 and f_7. This is indeed so. The absence of m_1, an expected manifestation of d_1, could be attributed to f_8 since a is known to inhibit m_1. Similarly the appearance of m_4 could be attributed to f_8 since the manifestation is a known side effect of a and the execution of the action started prior to the appearance of m_4. Notice that previously when temporal considerations were ignored, it was not possible to ensure that the necessary requirement for any causal relation, namely that an effect cannot precede its cause, was in fact satisfied. Thus the hypothesis that d_1 is present in the context of f_8, i.e. in conjunction with f_8, explains all patient findings and there is no conflict.

The second hypothesis says that disorder d_2 is present. This would explain just one appearance of m_2, f_3 and f_4. This hypothesis is not in conflict with f_5 but it is in conflict with f_6 and f_7. As before, f_8 is called into play to see whether it would enhance the plausibility of the hypothesis. In this case, the knowledge about a's execution is immaterial to the hypothesis, as already discussed. As a result, the first hypothesis wins hands down, whilst previously our ignorance of time (and the omission of contextual information) had given rise to an erroneous diagnosis.

While exploring a hypothesis, a conflict arises when an expected manifestation is missing. The conflict could be resolved either by establishing that this is a future manifestation given the state of evolution of the particular disorder, or by establishing that the manifestation is being masked or prevented by some ongoing treatment.[4] Masking means that the manifestation is actually present, but its presence is masked – its observation is being prevented – by some effect of the action. Preventing means

---

[4] This can also happen as a result of an interference by a concurrent disorder.

that some effect of the action has broken the causal chain(s) leading to the realization of that manifestation.

Likewise, when a manifestation that is not expected on the hypothesis under consideration appears, its presence could be attributed to another disorder or to an ongoing action. Hence, for a diagnostic system, knowledge of therapeutic actions functions to enhance the plausibility of hypotheses by making excuses for missing expectations or by accounting for the presence of unexpected manifestations. For a system that recommends therapy, the same knowledge would serve an entirely different purpose.

The simple example discussed above, aimed to show that in clinical diagnosis, causality, time and therapeutic interventions should, in general, be taken into consideration. This is not to say that every diagnostic domain requires all these factors; for a number of domains simple associational disorder knowledge, whereby time and therapeutic interventions are ignored and causality is viewed in the most abstract way, may well suffice. The three factors, causality, actions and time are strongly related. Changes result from causal interactions. Actions embody causal mechanisms and hence the execution of actions brings about changes, both welcome ones (positive effects) as well as unwelcome changes (negative effects or side-effects). Causal phenomena are temporal phenomena [401].

Causality and time enable the modeling of disorders as evolving dynamic processes and similarly for patient states. If the internal dysfunctional states of a disorder are modeled, it is possible to hypothesize which states, in the evolution of some operative disorder, have been reached with respect to some patient. A more accurate picture of the state of the patient can therefore be obtained that could result in the application of more effective treatment. In contrast, if disorders are not modeled as evolving dynamic processes, but as sets of temporally unrelated manifestations, as it is the case with associational models, the best one can do is to establish whether a disorder is operative or not in some patient.

The C-T-A model gives a very high view of diagnostic knowledge, that does not make any commitments as to the representation of causality, actions and time.[5] As such it can be treated as a high level specification for the knowledge of a new diagnostic system, or as a unifying framework for the comparative analysis of existing approaches to clinical diagnosis, as we do in Sections 6.3-6.8. More specifically, for each approach discussed we examine which of the five planes comprising the C-T-A model are included and in the case of the temporal constraints plane in particular (this plane is found in all approaches, otherwise they would not be included in the discussion) we see how the constraints are expressed. Temporal constraints are elaborated further in Section 6.9. Briefly, we can say that in almost every approach therapeutic interventions are ignored. A notable exception is the Heart Disease Program (HDP). Furthermore, we can identify two broad categories in general.

---

[5] In fact uncertainty is a key feature of all these types of knowledge, often represented in probabilistic terms. However, uncertainty as such is modeled at a lower level, where each top level plane of the C-T-A model can have relevant subplanes. One particular subplane could encompass the relevant uncertainty of the given type of knowledge, e.g. the probabilistic knowledge of causal interactions, or the fuzziness of temporal constraints, etc.

In one category, diagnostic knowledge is represented as a single causal network and as such the temporal constraints refer to delays between causes and effects and the persistence of states. In the other category, each disorder is modeled separately, but in associational rather than causal terms, i.e., in terms of its external manifestations. In such approaches, the temporal constraints refer to temporal relations between events marking begins and ends of manifestations. Overall, the temporal constraints plane is the part of the C-T-A model that glues everything together; as already said, time is intrinsically related to the execution of actions and to causal phenomena in general.

6.2.5 Deriving the 'Best' Diagnostic Solution

As the simple examples discussed above have demonstrated, a diagnostic problem often has multiple potential solutions. Generating such potential solutions is one challenge of diagnostic problem solving. Another challenge is selecting the 'best' potential solution. This calls for some means for evaluating competing diagnostic hypotheses (simple or complex, in partial or complete form), to see which one appears to give the best explanation of the observed malfunctioning situation. In clinical diagnosis different criteria may be used in combination for the evaluation of hypotheses. In simple domains where the *closed world assumption* may be viably applied, i.e. where it can be justifiably assumed that both knowledge and data are complete, simple criteria such as minimality based on cardinality or entailment may be sufficient for selecting the best explanation. Clinical domains are rarely simple given the inherent uncertainty and incompleteness of clinical knowledge and patient data. To start with, the requirement for a diagnostic solution to be a cover (of observations of abnormality), often is not attained. Usually diagnostic hypotheses neither entail all observations of abnormality, nor do they exhibit complete consistency with the entire relevant history (past and present) of the patient.

An evaluation mechanism would need to balance the observations of abnormality that are entailed by a diagnostic hypothesis against those that are not entailed, taking into consideration the significance of the various observations. Other criteria are also used, for example integrity and coherence are important for complex hypotheses. In [213] it is proposed that diagnostic hypotheses should be assessed as *plausible* on the basis of necessary criteria of acceptance. The best hypothesis out of the plausible ones could then be selected on the basis of additional criteria. For example, full coverage could be the criterion for plausibility and minimality the criterion for best explanation. However, as already stressed, medical diagnostic problems are rarely so well defined and clear cut for coverage to be an easily attainable requirement or for minimality to be a sufficient criterion for pinpointing the best explanation (under the assumption of single disorder, all hypotheses are minimal). In medical domains the criteria for plausible and best explanations are often complex rules composed of more primitive criteria. In [213] such primitive criteria are defined.

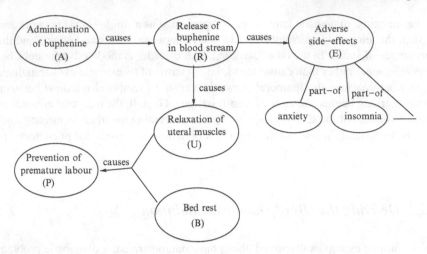

Fig. 6.8 High Level Causal Network Illustrating the Action of Buphenine

The evaluation of hypotheses involves *deductive reasoning* in order to discover what is entailed by some hypothesis. Hence, the generation and evaluation of hypotheses (comprising the abductive paradigm) is also referred to as the *hypothetico-deductive* model of reasoning. This model is evident in Bayes' Theorem that forms the basis of the reasoning for the majority of probabilistic diagnostic systems [237]. Bayes' Theorem states that $P(H/E) = P(E/H)P(H)/P(E)$ where H is a hypothesis and E the evidence. The lefthand side of the equation corresponds to abduction while the term $P(E/H)$ corresponds to deduction.

6.2.6 Illustrating some of the Temporal Requirements for Clinical Diagnosis

In Chapter 2 we discussed the time representation and temporal reasoning requirements for time-oriented systems in medicine. To a greater or lesser extent these apply to clinical diagnosis. In this subsection we illustrate, through an example, some of these requirements for clinical diagnosis.

Example 6.4. Figure 6.8 gives a causal network illustrating, at a high level, the action of buphenine. This active ingredient is administered to pregnant women to reduce the chance of premature labor. Buphenine is released rapidly into the blood stream causing the relaxation of the uteral muscles. This, in conjunction with substantial bed rest, reduces considerably the possibility of premature labour. However, the ingredient may give rise to various adverse side-effects, such as palpitations, anxiety, insomnia, etc.

Figure 6.9 analyses the relevant processes at a more detailed level. At this level we see that the administration of buphenine (process A) is a periodic process, involving a sequence of 'instantaneous' events, "oral administration of a 15mg capsule of buphenine". These events are considered instantaneous because it would make no sense to try and analyse them at a more detailed level (pop capsule in the mouth and then swallow it with a small quantity of water). The interval between successive occurrences of this event is 4 to 8 hours and the duration of the entire process A is between 3 to 22 weeks. This is because the earliest time that such a course of treatment makes sense to be applied is the 15th week of gestation and the latest the 34th week of gestation. There is no point to administer this drug after the 37th week of gestation. These temporal considerations give rise to the minimum and maximum duration of process A.

The release of the active ingredient into the blood stream (process R) is a direct effect of process A. More specifically, process R is also a periodic process with a direct pairwise correspondence between its components and those of process A. The components of process R are interval events, starting 10 minutes after the happening of the corresponding point event of process A and lasting for 8 hours. Thus each component of process R overlaps with its preceding component. The overlap is minimal (10 minutes duration) when the distance between the components of process A is maximal (8 hours), and it is maximal (4 hours) when the distance between the components of process A is minimal (4 hours). As a result, process R is a continuous one since there are no gaps (periods of inactivity) between the components comprising it — the time interval spanning it is *convex*. In contrast process A is an intermittent process since there are gaps between its discrete components — the time interval spanning it is *nonconvex*. Process R starts 10 minutes after the initiation of process A and terminates 8 hours and 10 minutes after the termination of process A.

The relaxation of the uteral muscles (process U) is treated as a direct effect of process R, although in reality there is a whole chain of reactions linking the release of buphenine in the blood stream and the relaxation of the uteral muscles. There would be no point in explicating this causal chain if the reasoning performed does not make use of it (although it may be required for the purpose of more detailed explanations). Strictly speaking, process U is also a periodic process with a pairwise (causal) correspondence between its components and the components of process R. Again this level of detail is uncalled for. We could simply consider process U as a continuous, atomic (non-decomposable) process running concurrently with process R.

The analysis of process B, 'Bed rest' is more qualitative. One extreme scenario is that it is also a continuous process, running concurrently with process U. A more likely scenario, though, is that it is an intermittent process spanning process U, with the constraint that the gaps do not exceed, say 20%, of the overall span of the process.

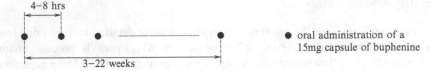

A: Administration of buphenine

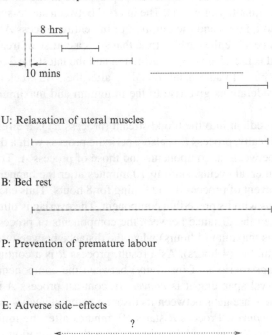

R: Release of buphenine in blood stream

U: Relaxation of uteral muscles

B: Bed rest

P: Prevention of premature labour

E: Adverse side-effects

Fig. 6.9 Temporal Analysis of Processes of Figure 6.8

Processes U and B jointly cause process P, 'prevention of premature labour'. Process P is a continuous process that hopefully lasts the duration of process R.

Finally process E, 'adverse side-effects', can be analysed into the individual adverse side-effects. These represent its components. However, none of these components is a necessary one, i.e. it is not necessary that at least one of these components should materialize, in which case process E may not materialize at all. Thus it's not possible to pinpoint accurately the extent of this process. The maximal extent is the duration of process R (at any point throughout the activity of the drug, at least one adverse side-effect is materializing) and the minimal extent is 0 (the process does not materialize at all). If a component of a process is not a necessary one it may be associated with conditions (involving temporal and other factors) underlying its materialization. In addition, its potential existence is delimited by the existence of the overall compound process. For example, palpitations have a high incidence of occurrence. However if they do occur, they will do so intermittently during the first 48 to 72 hours from the start of the administration of the drug. Anxiety has a lower incidence of occurrence whose severity varies with the concentration of the drug in the blood, i.e. the duration of the overlap of the components of process R. Hence the analysis of process E reveals a more complex temporal pattern with higher uncertainty and without the regularity characterising periodic processes.

The above description of the particular therapeutic process could be of relevance to a diagnostic system whose scope includes diagnostic problems relating to pregnancy. If a pregnant woman presents with anxiety and insomnia, and her record shows administration of buphenine, the incidents of anxiety and insomnia could be attributed to the administration of this drug provided the temporal constraints associated with the particular administration of the drug and the particular incidents of anxiety and insomnia match the generic picture. Otherwise they could be attributed to other causes such as hormonal changes that occur normally during pregnancy. Hence a diagnostic system that does not know about the evolution of a normal pregnancy and the effects (both correcting and adverse effects) of potential courses of actions relating to problems associated with pregnancy, but only knows about disorders associated with pregnancy, would have a deficient knowledge base.

Example 6.4 has illustrated a number of temporal requirements: uncertainty, expressed in a metric (ranges of intervals) or relative way (possible temporal relations), convexity or non-convexity of processes (continuous or non-continuous processes), periodicity, temporal constraints of causal relations, compound processes with conditional (variant) components and multiple granularities (minutes, hours, weeks). These requirements apply both to diagnostic knowledge and patient data. In addition, the need for *temporal data abstraction* with respect to patient data arises. Raw data are low level, non decomposable entities. Since diagnostic knowledge is expressed at a much higher level that involves more complex entities (trends, periodic occurrences, temporal patterns, etc.) temporal data abstraction acquires critical significance.

6.3 The Heart Disease Program

The Heart Disease Program (HDP) [247] diagnoses disorders of the cardiovascular system. As the author puts it (p.196) "Most of the cardiovascular disorders of concern to HDP are chronic, progressive, and many can not be corrected short of a cardiac transplant. As a result, patients typically arrive with existing diseases and existing therapies. The problem is to determine what new diseases or complications are now present and their relationship to the known diseases. Thus, the therapies with both their beneficial effects and side effects are an important part of the domain.". The diagnostic knowledge is modeled as a Bayesian probabilistic network. The nodes of the network represent pathological states, manifestations, or therapies.

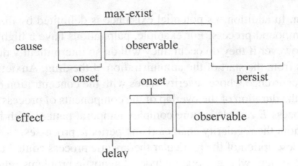

Fig. 6.10 Temporal Model of a Causal Relation in HDP (adapted from (Long, 1996), p.202)

The arcs represent causal relations, although an arc from a therapy to a pathological state represents a 'correcting' action (the therapy causes the ceasing of the pathological state). A strictly probabilistic approach which was used in the early version of the system had the limitation of generating hypotheses which were impossible given the temporal constraints involved. As a result temporal constraints on the causal relations were then explicitly represented and the patient data became time-stamped entities. A difficulty of the particular medical domain is that the available observations are limited.

The temporal semantics of the causal relation is quite rich (see Figure 6.10). *Onset* is the range of time that can be assumed for the effect when it is observed while *delay* is the range of time the cause must be true before the effect can start (this includes the onset time). *Persist* is the range of time that the effect will remain if the cause ceases to be true and *max-exist* is the maximum time the cause will remain, even though the effect continues. Onset, delay, etc., are time intervals. The possible durations of these intervals are expressed probabilistically. For example, the probability that myocardial infarction (MI) persists for less than one hour is 0, while it is 0.5 for a persistence between one hour and one day. The probability that MI is over after a day has gone by is 1.0. Another example is that MI causes low left-ventricular-systolic-function (LVF) with a probability of 0.6. The probability that low LVF ceases immediately or it persists up to 6 hours after the end of MI is 0.3. The probability that low LVF returns to normal after 6 hours, but less than infinity, is 0.8. Hence the probability for an infinite persistence of low LVF is 0.2. The means that low LVF can not be reverted.

Furthermore a causal relation is classified as *self-limiting* if the abnormality ceases by max-exist without any rectification action while a state is classified as *intermittent* if it is absent over subintervals of the interval in which it is true. The temporal constraints associated with causal relations are interpreted on the basis of the following rules:

1. When a node is observed, it is assumed to already be producing effects for onset time.

2. Effects are observable at a time after the cause given by the delay, if it exists, otherwise by the onset.
3. Effects are observable after the cause is observable and overlap the cause.
4. Effects continue until the cause ceases, unless the max-exist is exceeded or the effect is intermittent.
5. Effects continue after the cause ceases, in accordance with the persist.

The above temporal model of a causal relation enables the representation of different patterns of causality such as *immediate* (the effect happens immediately), *progressive* (the effect, once it takes place, continues and often worsens), *accumulative* (when a cause is required to exist over a period of time), or *corrective* (when a state causes another state to return to normality; here the causes are often therapies but they can also be pathophysiological states such as dehydration "correcting" high blood volume).

Patient findings comprise observations, symptoms, history and tests. Observations are point-based, referring to *now*; they report results of physical examination. Tests are like observations except that they refer to the past. Symptoms are reported by the patient and have duration; their start and end are expressed relative to now, e.g. six hours ago. History is like symptoms.

During a consultation with HDP, nodes in the Bayesian network are instantiated. This essentially means deriving their observable extents (earliest and latest, begin and end), again relative to the current point in time. The temporal constraints and the rules interpreting them (see above) enable the derivation of such temporal extents both for a cause given its effect and vice versa. For example, the earliest begin of a cause is the earliest begin of the effect plus the maximum delay of the effect, minus the maximum onset of the cause (see Figure 6.10). The times are all temporal distances from the current time, so 1 day + 1 day is two days prior to the reference time. Since diagnostic reasoning is abductive reasoning the most common reasoning step is to infer a cause from an effect, starting from the effects matching patient findings. The aim is to find the relationship between internal states and observed findings. A diagnostic hypothesis consists of a complete causal explanation detailing how the diseases and mechanisms in the hypothesis provide a consistent accounting of the patient findings (consistency means that all the temporal constraints of the given causal relations are satisfied).

The essential functions of the temporal reasoning performed by HDP are therefore to deduce and maintain the causal temporal constraints in hypotheses and to support the possibility of nodes having different values over different time intervals. For example, to allow left atrial pressure to have a value of high for some period after the myocardial infarction but also to be normal by the time of the examination, say due to the administration of nitroglycerin 4 hours ago. In addition, the temporal characteristics permit the adjustment of probabilities when there are different causal situations with different probabilities. The probability along a pathway from a known node or primary node is recomputed using the more specific temporal information from the case.

The issue of multiple granularities is very relevant to the domain of HDP since some processes take place over minutes while others take place over months or years. However, the multiplicity of granularities is not handled in a systematic way. Furthermore, the Bayesian network representing the diagnostic knowledge does not allow for cyclic phenomena, although such phenomena do happen in the particular medical domain. Cyclic (periodic) phenomena in the patient data are treated as atomic findings in the same way as all other findings. The author justifies this by saying that the system does not need to reason about cyclic phenomena, which we interpret to mean that the system does not need to look inside cyclic phenomena. In fact it appears that the derivation of temporal data abstractions does not form part of the reasoning of HDP. The system does know about relevant therapies though, as already mentioned. Therapies and their effects are modeled similarly to causes. This is a strength of the representation. For example, a therapy may take a period of time (onset) before it produces the desired effect. Persistence or the other temporal properties are not used in the model of the effects of therapies. Most drugs have some persistence (the pharmacokinetic half-life), which can be considered part of the treatment time.

The author stresses that the incorporation of time enhanced the performance of HDP not only because the evaluation of hypotheses was more accurate but also because this resulted in constraining the generation of hypotheses.

Let us now analyze this system with respect to the C-T-A model. All five planes are included. The Bayesian probabilistic network is a causal network with probabilities, since the central relation is that of causality. Temporal constraints are expressed with respect to causal links (see Figure 6.10). Onset, delay, max-exist, etc. are intervals of time whose durations are expressed as sets of possible values with corresponding probabilities. Hence temporal uncertainty is expressed through ranges of absolute durations and probabilities. During a consultation, nodes in the causal network are assigned earliest and latest begins and ends with respect to the reference time point, i.e. now.

There is a single causal network encompassing all relevant internal states, external manifestations and actions. In some of the other approaches we shall examine, each disorder is modeled separately on its own. Hence the causal network of Figure 6.4 can be converted into this representation by adding to each causal link the relevant temporal constraints as illustrated in Figure 6.10. The network is stored with respect to its nodes, where each node points to its possible causes. For example, manifestation m_2 would point to both internal states, s_1 and s_2, of d_1 as well as to s_4. This way the recurrence of m_2 in the context of d_1 is expressed. If the separate appearances of m_2 were caused by the same state, say s_1, or more generally if m_2 were repeatedly caused by s_1, this could not be expressed.

6.4 Temporal Parsimonious Covering Theory

The parsimonious covering theory (PCT) of Peng and Reggia (1990) [306] is a well known general theory of abductive diagnosis. The basic version of the theory models diagnostic knowledge in terms of a set of causes (disorders), D, a set of manifestations, M, and a causal relation, $C \subseteq D \times M$, relating each cause to its effects. This is the simple associational model discussed above (see Figure 6.2). An efficient algorithm, BIPARTITE, is defined, which incrementally constructs, in generator-set format, all the explanations of a set of occurring manifestations, $M^+ \subseteq M$, for some case (patient). An explanation is a subset of D which covers completely the set of occurring manifestations M^+ (i.e., every element of M^+ is an effect of at least one element of the explanation) and satisfies a specified *parsimony criterion*. There are three parsimony criteria which in ascending order of strictness are *relevancy* (every disorder in the explanation is a cause of at least one element of M^+), *irredundancy* (no subset of the explanation is also a cover of M^+) and *minimality* (the explanation has the minimum cardinality among all the covers of M^+). The preferred parsimony criterion is that of irredundancy and this is the one embedded in the BIPARTITE algorithm.

Wainer and Rezende (1997) [418] have proposed a temporal extension to the basic version of PCT, referred to as t-PCT, which they have applied to the domain of food-borne diseases. Their argument is that irredundancy as the parsimony criterion is too weak to significantly reduce the number of alternative explanations.

In PCT the diagnostic knowledge is a collection of causal associations between disorders and manifestations. In t-PCT each disorder is modeled, separately, as a *temporal graph*. This is a directed, acyclic, transitive, not necessarily connected, graph the nodes of which represent manifestations and the directed arcs represent *temporal precedence* — a directed arc from manifestation m_i to manifestation m_j means that the begin of m_i must precede the begin of m_j. A temporal graph is not necessarily a causal graph since the precedence relation does not necessarily imply causality. If there is quantitative information about the duration of a manifestation, it is associated with the corresponding node; similarly if there is quantitative information about the elapsed time between the start of two manifestations, it is associated with the corresponding arc. Durations and delays are time intervals expressed as pairs of numbers denoting minimum and maximum extents. HDP uses the same, metric or absolute, way for expressing temporal uncertainty at the level of diagnostic knowledge. This representation also allows for temporal incompleteness as not every delay or duration needs to be specified. Moreover, the precedence relation does not need to be completely specified. Hence if no temporal information is known, the "temporal" graph of a disorder is nothing else than the (disconnected) set of the effects of the disorder, just like in PCT. Such a temporally unconstrained model matches against any temporal constraints in the patient information, but of course if the diagnostic knowledge is temporally unconstrained it would not make sense for the patient information to be temporally constrained.

Figure 6.11 gives the temporal graphs of disorders d_1 and d_2. Here we assume that no quantitative information about the duration of manifestations or about the elapsed time between the starts of two manifestations is given. In the temporal graph of d_1, the first and second appearances of m_2 are denoted as $m_2^{(1)}$ and $m_2^{(2)}$ respectively.

A similar representation to the disorder model is used for patient information. The patient information consists of a set of actual manifestations. Again, where the

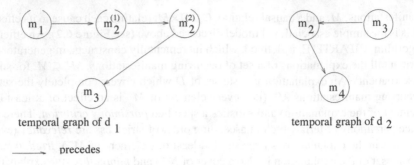

temporal graph of d $_1$ temporal graph of d $_2$

$\longrightarrow$ precedes

Fig. 6.11 Temporal Graphs of Disorders d_1 and d_2 in t-PCT

duration of a manifestation is known, a range for this is specified. In addition a range for the beginning of a manifestation, if known, is specified relative to some arbitrary origin. For example, it can be said that manifestation m_1 started between 2 and 3 weeks ago and it lasted for 1-2 days, and that manifestation m_2 is present but there is no information as to when it started. Here the arbitrary origin is taken to be the present point in time. Different granularities are implicated both in the representation of the diagnostic knowledge and the patient information but as with HDP, the multiplicity of granularities is not handled in a systematic, semantically-based, fashion.

An interval, $I = [I^-, I^+]$, is defined to be a non-empty, convex, set of time points. The time period modeled by the interval I is represented as a range of values between a min extent (I^-) and a max extent (I^+)[6]. When I^- equals I^+ there is no uncertainty, i.e., the duration of some manifestation (at a given granularity) is known exactly, or the distance between two times (say 'now' and the begin of some manifestation) is known exactly. Finally, it is inconsistent to have $I^- > I^+$ as this is saying that the min extent exceeds the max extent (inconsistencies are used for constraining the formation of hypotheses — see below).

Two binary, arithmetic, operations are defined on intervals, intersection ($\cap$) and sum ($+$):

$$I \cap J = [max(I^-, J^-), min(I^+, J^+)]$$
$$I + J = [I^- + J^-, I^+ + J^+]$$

Intersection gives the common subrange, and hence the two intervals must denote the same temporal measure. For example, let the expected duration of manifestation m_1 under disorder d_1 be between 2 and 4 days, and let the actual duration of that manifestation in some patient be between 1 and 2 days. Formally these durations are expressed as the intervals,

$$dur_{d_1}(m_1) = [2, 4]$$
$$dur^+(m_1) = [1, 2]$$

---

[6] The notation used can be confusing as one would expect I^- to denote the begin time-point and I^+ the end time-point of the interval I; normally an interval is equated with its absolute extent rather than a range for its duration.

The intersection of these two intervals is the interval $[2,2]$. So the two intervals agree on a duration of 2 days. If the actual duration of m_1 were 1 day, the intersection between $[2,4]$ and $[1,1]$ would give the inconsistent interval $[2,1]$ that denotes a disagreement between the actual and the expected.

Sum propagates delays (between the begins of manifestations) along a chain of manifestations, or positions a manifestation on the time line (the sum of the delay of its begin, from some origin, and its duration). The result of a sum operation is always a consistent interval. For example, let's say that in disorder d_1 the expected delay between the start of manifestation m_1 and the start of the first appearance of m_2 is between 2 and 3 days and the expected delay between the starts of the two appearances of m_2 is between 3 and 4 days. The sum of the intervals $[2,3]$ and $[3,4]$ gives the interval $[5,7]$ which is the expected delay between the start of m_1 and the start of the second appearance of m_2.

Let us further say that in the given patient, m_1 and (the first appearance of) m_2 started 5 and 2 days ago respectively. The actual begins of these manifestations are expressed as the intervals,

$begin^+(m_1) = [-5,-5]$

$begin^+(m_2) = [-2,-2]$

Since the reference point is now, a time before it is denoted with a negative sign. By applying a sum operation between the delay ($[2, 3]$) associated with the arc from m_1 to m_2 (in the temporal graph of d_1) and the actual begin of m_1 ($[-5,-5]$), it is inferred that the begin of m_2, should d_1 hold, ought to be between 3 and 2 days ago ($[-3,-2]$). This inference is consistent with the actual begin of m_2, as can be shown through an intersection operation between the actual ($[-2,-2]$) and the inferred ($[-3,-2]$) that yields the consistent interval $[-2,-2]$. If m_2 had started 4 days ago, this would have generated an inconsistency with respect to the particular arc on the basis of the actual begin of m_1 since $[-3,-2] \cap [-4,-4]$ would give the inconsistent interval $[-3,-4]$.

The operations of sum and intersection are therefore applied for deciding whether the model of some disorder (temporal graph) is temporally inconsistent with the patient information. *Temporal inconsistency* arises if

1. there is at least one arc inconsistency, or
2. the actual duration of a manifestation is different from its expected duration under the given disorder.

An *explanation* that contains a disorder which is temporally inconsistent with the patient information is removed. More specifically a set E of disorders is a temporally consistent explanation of M^+, if E covers M^+ (each element of M^+ appears in the temporal graph of at least one element of E), E satisfies a parsimony criterion (preferably irredundancy) and every disorder in E is not temporally inconsistent with the patient information. Algorithm BIPARTITE has been extended to t-BIPARTITE implementing the notion of temporal inconsistency. It is important to note that temporal inconsistency is not implemented as a filter but is used to direct the formation of explanations.

In summary, PCT is a well known theory of abductive diagnosis, although its basic version is only applicable to very simple medical diagnostic problems. The temporal extensions of PCT discussed above, yielding t-PCT, certainly expand the application scope to more realistic medical diagnostic problems. The principal objective is to use time for constraining the generation of diagnostic explanations beyond what is possible from the irredundancy parsimony criterion.

If we are to examine this approach with respect to our analytical framework, the C-T-A model, we see that the internal states and actions planes are missing. These omissions restrict the range of applicability of this approach, not least because causality in terms of internal mechanisms is left out. The temporal constraints plane affects the external manifestations plane. Temporal constraints are expressed through relation precedes as well as through quantitative ranges for delays between begins of manifestations and for durations of manifestations. Moreover, it is not necessary to specify every delay or duration, with respect to the temporal graphs of disorders, or to give the actual begin and duration of every manifestation with respect to the patient in question. Hence temporal incompleteness and uncertainty is supported. The authors note, however, that the omission of cycles is a major restriction on the expressive power of their representation since it is not possible to represent general recurring events. This restriction is imposed in order to reduce the complexity of reasoning. One way of controlling this complexity is to hide the repetition (i.e. the cycles) behind compound (periodic) occurrences which at a higher level of abstraction (coarser granularity) can be treated as atomic, indivisible, entities. However, compound occurrences and the systematic handling of multiple granularities are not addressed in this proposal, and neither is temporal data abstraction.

6.5 Fuzzy Temporal/Categorical Diagnosis

The work on t-PCT overviewed above has also been extended on a number of directions using fuzzy sets [419], namely:

- The crisp representation of a time interval as a range of values has been replaced with a fuzzy set giving ranges for typical as well as possible extents.
- The manifestations of a disorder are distinguished into necessary and possible. In addition, the intensity of manifestations, both in the disorder models and the patient information, is expressed in terms of fuzzy sets.
- The speed of evolution of the 'concluded' disorder is computed on the basis of the patient data. This is used to make predictions about past and future events.

A number of *consistency measures* between the patient's data and a disorder model are defined and used to define diagnostic explanation. The diagnostic explanation is based on the single disorder assumption but apart from giving its definition the authors do not specify the means for deriving it. However it is interesting to examine the various temporal extensions proposed. The test domain, as in t-PCT, is infectious diseases.

The diagnostic knowledge consists of models for the various disorders. Each disorder is modeled separately as a *temporal graph*. A single time scale, Θ, is assumed. D and M are the sets of disorders and manifestations as before. Each disorder has a set of effects from M (the causal relation used before). The effects of a disorder are distinguished into *necessary* and *possible*. In addition, a disorder is associated with a set of (instantaneous) events. This set must definitely include events corresponding to the beginning of each of the disorder's effects. Events corresponding to the end of some of these effects may also be included, as well as non-observable events, e.g. the event denoting the onset of the disorder itself, i.e. the onset of the infection. Some pairs of these events, say (e_i, e_j), are associated with a *fuzzy temporal interval* which states the elapsed time between event e_i and event e_j. If e_i occurs before e_j this is a positive fuzzy set. The temporal graph modeling the disorder is a graph of its associated events. The events are the nodes of the graph and where the elapsed time between a pair of events can be specified there is a directed arc joining the pair of events whose label is the particular fuzzy temporal interval. As with t-PCT the temporal graph of a disorder is not a causal graph. It simply gives the temporal distances between events and such distances could be the duration of a manifestation, the delay between the start of the disorder and the begin of a manifestation, the delay between the begins of two manifestations, the elapsed time between the end of one manifestation and the begin of another, etc.

Figure 6.12 gives the temporal graph of disorder d_1[7]. The graph includes nodes for the begins and ends of all manifestations. In addition there is special node m_0 that represents the event signalling the start of the disorder. The included arcs are annotated with fuzzy intervals (see below) giving the elapsed time, at the granularity of days, between the given pairs of events. Contrary to the arcs of the temporal graphs of t-PCT, an arc here does not necessarily imply precedence. The elapsed time between a pair of events, if not explicitly given, may be possible to infer. This can be done through the computation of the minimal graph of the disorder (see below). For example, the duration of the second appearance of manifestation m_2 which is not explicitly given can be inferred as the sum of the elapsed time between the start of this manifestation and the start of m_3, the duration of m_3, and the elapsed time between the end of m_3 and the end of $m_2^{(2)}$. The latter is 0 meaning that the two manifestations terminate together (they persist until the end of the disorder).

The patient information consists of a set of manifestations known to be or to have been present in the patient, M^+, and a set of manifestations known to be absent, M^-. The begin of each manifestation in M^+ is denoted by an (instantaneous) event. The end of some manifestations in M^+ may also be denoted by an event. Each event is associated with a fuzzy temporal interval representing the possible moments on which that event has happened, relative to the present moment of time, θ_0, representing the moment of diagnosis.

A *fuzzy temporal interval*, which is the central notion in the proposed temporal representation of the diagnostic knowledge and patient information, is a *fuzzy interval*. A fuzzy interval is a fuzzy set [434] with a convex membership function. A

---

[7] Note that the temporal constraints of d_1 depicted in Figure 6.12 are somewhat extended from those shown in Figure 6.5 in order to illustrate disjunctive constraints as well.

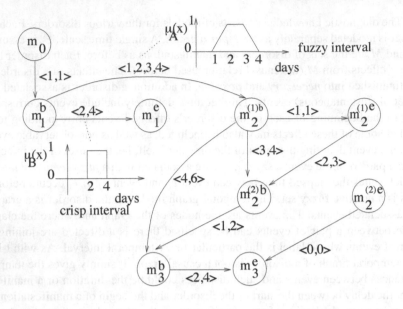

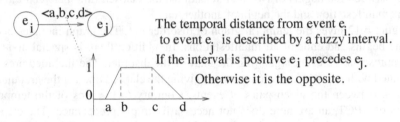

m_i^b event denoting the begin of manifestation m_i.

m_i^e event denoting the end of manifestation m_i.

Fig. 6.12 Temporal Graph of Disorder d_1 in Fuzzy t-PCT

normalized fuzzy set is one for which at least one element belongs to it with certainty. A *positive interval* is a fuzzy interval whose domain is the real line and every $x < 0$ does not belong to the interval, i.e. $\mu(x) = 0$ where $\mu(x)$ is the membership function. An interval A is *tighter* than an interval B, i.e. more constrained, if for every element from the relevant domain, its degree of membership in A does not exceed its degree of membership in B.

The fuzzy intervals used in the definition of the disorder models and the expression of the patient information are assumed to be *trapezoidal intervals*. A trapezoidal interval, A, is a normalized interval that can be expressed as a quadruple $\langle a_1, a_2, a_3, a_4 \rangle$. The range $[a_2, a_3]$, where $\mu_A(x) = 1$, is the *core* of A and the range

$[a_1, a_4]$, where $\mu_A(x) > 0$, is the *support* of A. If $a_1 = a_2$ and $a_3 = a_4$ the interval is crisp, denoted as the range $\langle a_1, a_4 \rangle$. The core of a trapezoidal interval gives the typical range of values, nested within the possible range of values given by the support of the interval. In a disorder model typical and possible ranges for some temporal distance are based on a large number of actual cases. The authors argue that a fuzzy interval is easily obtained by asking the expert to provide two nested intervals to account for the lapse of time between two events: one containing the interval of time between which the events typically occur, and another interval comprising all possible cases. Figure 6.12 illustrates the trapezoidal interval denoting the elapsed time between the begin of m_1 and the begin of the first appearance of m_2. The typical range for this temporal distance is 2 to 3 days. However it is possible to be from about 1 day up to about 4 days. All other distances in the temporal graph of Figure 6.12 are expressed in a crisp way as ranges of min and max extents, just like in t-PCT. The crisp interval denoting the duration of m_1 (between 2 and 4 days) is also shown in Figure 6.12.

To perform the relevant temporal reasoning, a *minimal network* is computed from the temporal graph modeling a disorder. A minimal network gives the most tight interval possible between any pair of events (from t_{ij}, the interval from event e_i to event e_j, and t_{jk}, the interval from event e_j to event e_k, an interval from event e_i to event e_k, t_{ik}, is obtained through a sum operation; the interval from e_i to e_k is the intersection of all intervals – given or derived – between the two events). If nothing is known or can be derived for some pair of events, the default interval, 'anytime', is assumed.

Regarding patient information, the temporal distances, as fuzzy intervals, between all pairs of the relevant events are computed from the temporal distances between each event and the present moment of time. All these events refer to the past (or the present). The temporal distance from event e_i to event e_j is obtained by subtracting the time of occurrence of e_i from the time of occurrence of e_j.

A number of consistency measures between a disorder model and the patient information are defined. These are:

- *Temporal consistency*, computed on the basis of the expected and actual temporal distances between pairs of events. More specifically, for each pair of events in the patient information that is also included in the minimal network of the disorder model, the *height* of the intersection of the given pair of intervals is computed (the height is the maximum value of the membership function). The temporal consistency is the minimum between the obtained heights. Hence a temporal consistency of 1 means that the actual and expected temporal distances of every pair of relevant events match on a typical value, while a temporal consistency of 0 means that for at least one pair of relevant events the actual and expected temporal distances do not match even on a possible value. Hence a temporal distance of less that 1 but greater than 0 means that for at least one pair of relevant events the match was for a possible value. It is interesting to note that if there are no common events, and hence no common manifestations, between a disorder model and the patient information, temporal consistency is either undefined or taken to be 1 by default. As already discussed, the notion of consistency is weaker than the notion of entailment (causality) used in abductive diagnosis; anything which is not in direct conflict with an expectation is taken to be consistent with. Of course temporal consistency is only one requirement for a disorder to constitute a diagnostic explanation. Another requirement is for every patient manifestation to be an effect of the disorder and this is where entailment comes into the explanation.

- *Categorical consistency* reflecting the fact that a necessary manifestation of a disorder must happen, assuming that there was enough time for it to happen. Enough time means that either an event following the begin of the necessary manifestation has happened or an event preceding the begin of the manifestation has happened and the expected delay time between it and the begin of the necessary manifestation has expired. Categorical consistency is calculated as temporal consistency by assuming that all necessary manifestations that have not yet occurred will start sometime after the present moment of time. For example, let's say that m_1 is a necessary manifestation for d_1. If event m_2^b, signalling the begin of some appearance of m_2, has happened it may be deduced that there was enough time for m_1 to develop. Similarly, let's say that m_3 is necessary for d_1. If event m_1^b, signalling the begin of m_1 has happened 10 days ago, which is the maximum elapsed time between the begin of m_1 and the begin of m_3, again it may be deduced that there was enough time for m_3 to develop.
- *Intensity consistency* reflecting the match between the expected and actual intensities of the relevant manifestations in the disorder model and patient information. Intensity consistency is not a temporal measure.

A single disorder which is temporally, categorically and intensity consistent with the patient information, i.e. none of these measures equals 0, and includes all patient manifestations in its effects is a plausible diagnostic explanation. As already noted, the means for deriving such plausible diagnostic explanations, in an effective and expert-like manner, have not yet been specified. The authors also point out that further work is required in defining the criteria for selecting the best explanation. In addition, they want to relax the assumption of a single disorder to allow for the possibility of multiple disorders.

Once a diagnostic explanation is selected and assuming it has 100% temporal consistency, the temporal information associated with the model of the particular disorder can be used to reduce the temporal uncertainties in the patient information, i.e. to tighten up the relevant intervals as much as possible. Furthermore, the *speed of development* of the disorder in the given patient can be conjectured from the revised patient information. Thus, predictions about past manifestations, that have not yet been tested for, or about future manifestations, can be more accurate. For example, if the disorder is progressing at a slower speed than average, a necessary manifestation may need more time to develop. Hence the categorical consistency can be revised. Similarly if the disorder appears to be progressing much faster than average, quicker therapeutic action may be necessary. Computing the speed of development of the selected disorder is a noteworthy aspect of this proposal and the interested reader is referred to the literature for the relevant details.

The analysis of t-PCT with respect to the C-T-A model given in the previous section applies to this approach as well. The internal states and actions planes are missing. The expression of temporal constraints has of course been usefully extended. A constraint is no longer confined to delays between begins of manifestations and durations of manifestations and expressed as a range of typical values. A constraint can now refer to any temporal distance between any pair of events (begins and/or ends of manifestations) and can be expressed as a fuzzy interval encompassing all typical as well as possible values and not just as a crisp interval. This is a more appropriate way of handling temporal uncertainty. Temporal incompleteness also continues to be supported as with t-PCT since not every distance needs to be specified, but it could simply be unknown. General recurrence of events is not addressed. In our example, the recurrence of m_2 in the context of d_1 is a very simple case of recurrence that can be handled by the proposed approach. If m_2 has occurred just once in a particular patient, the reasoning could go as follows regarding the evaluation of the hypothesis that d_1 is present. Does the occurrence of m_2 match the temporal characteristics of the first or the second expected appearance of m_2? If a match is obtained, could the missing occurrence, first or second, be excused, e.g. is it only a possible expectation, has it been masked in some way, etc?

In summary, this proposal, like its predecessor, t-PCT, and in turn its predecessor, PCT, is characterized by a rigorous set-theoretic exposition that makes the proposed notions clear and easy to reproduce. Although it improves t-PCT on a number of points, there are still a number of critical issues to be tackled, e.g. multiplicity of time scales, recurring manifestations, temporal data abstraction, etc.

6.6 Abstract Temporal Diagnosis

Gamper and Nejdl (1997) [150] have developed a logic-based framework for Abstract Temporal Diagnosis (ATD) that has been applied to the domain of hepatitis B with promising results. This framework, in contrast with the previous approaches discussed, emphasizes the need to automatically derive abstract observations over time intervals from (direct) observations at time points. This is an important aspect of the framework despite the fact that the actual temporal data abstraction mechanism used is very simplistic; the importance is in signifying the essence of temporal data abstraction in diagnostic contexts, in addition to its predominant application context which is patient monitoring.

ATD is based on Allen's (1984) [8] time-interval logic. More specifically the temporal primitive used is the convex time-interval. A linear and dense time structure, called time line, is assumed which is unbound in both directions, past and future. The time line is represented by real numbers. So intervals have explicit Begins and Ends, $[B, E]$. Although the time structure is dense, direct observations are made at discrete points in time. At this level a specific, single, granularity, say month, is used. The overall time of relevance to some diagnostic problem is denoted by i_{max} which includes all other time intervals. Time invariant properties hold throughout

i_{max}. Temporal propositions are represented by introducing time as an additional argument to the relevant predicates, e.g., fever(high, i), meaning that fever is high during interval i. Furthermore, this framework adopts Allen's algebra of temporal relations between intervals.

The designers of ATD observe that a process-oriented ontology is more suitable for representing dynamic systems than a component-based ontology. A process-oriented ontology gives the internal pathophysiological states and the causal mechanisms between them that result in the manifestation of the particular observations of abnormality. A component-based ontology gives the static decomposition of a system into components and subcomponents. A component-based ontology is used in the consistency-based paradigm of diagnosis while a process-oriented ontology is akin to the abductive paradigm. In ATD diagnostic knowledge is represented as a set of process descriptions, one of which corresponds to normal behaviour while the others represent different abnormal behaviours. The temporal evolution of a (normal or abnormal) process is expressed through a chaining sequence of states, described in terms of *process state assumptions*, $s(p, i)$. This says that process p assumes state s throughout interval i. Interval i is constrained through temporal relations with other intervals such as i {starts, during, finishes, equal} [1, 10] which says that i either starts, is during, finishes or is equal to the absolute interval [1, 10], i.e., i is constrained to be within that period.

A process is non observable. There are, however, observable parameters (manifestations). Manifestation propositions (or simply manifestations) are expressed as $m(v, i)$ which says that manifestation m assumes value v throughout interval i. A manifestation can assume different values over different time intervals but only one value at a time (hence recurrence at the level of manifestations is allowed). Such mutual exclusivity constraints for manifestations as well as process states are expressed, again in first order logic, as a kind of background knowledge.

The *State Description Model* (SDM) of a process p is a logical formula $\alpha \wedge \tau_\alpha \supset \beta \wedge \tau_\beta$ where α and β are conjunctions of temporal propositions (process state assumptions and/or manifestations) while τ_α and τ_β are conjunctions of temporal relations between the implicated time intervals. These temporal constraints denote the temporal behaviour of the given process. The SDM formula says that if process p assumes the states as specified in α such that the temporal relations in τ_α are satisfied, the manifestations in β and the temporal relations in τ_β are predicted. So the entailment in the formula captures causality (the process states in α are possible causes for the manifestations in β) and hence it can be said that an SDM integrates the causal and temporal knowledge for the given process. For illustration, the SDM of disorder d_1 is given below:

$\forall I_{s_1} \forall I_{s_2} \exists I_{m_1} \exists I_{m_2^{(1)}} \exists I_{m_2^{(2)}} \exists I_{m_3}$

$s_1(d_1, I_{s_1}) \wedge s_2(d_1, I_{s_2}) \wedge I_{s_1} \{meets\} I_{s_2} \supset$

$m_1(present, I_{m_1}) \wedge m_2(present, I_{m_2^{(1)}}) \wedge m_2(present, I_{m_2^{(2)}}) \wedge$

$m_3(present, I_{m_3}) \wedge$

$I_{m_1} \{finishes, during\} I_{s_1} \wedge$

$I_{m_2^{(1)}} \{during, finishes, is\_met, after\} I_{m_1} \wedge$

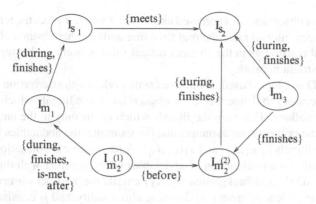

Fig. 6.13 Constraint Network from the SDM of d_1

$$I_{m_2^{(1)}}\{before\}I_{m_2^{(2)}} \wedge I_{m_2^{(2)}}\{during, finishes\}I_{s_2} \wedge$$
$$I_{m_3}\{finishes\}I_{m_2^{(2)}} \wedge I_{m_3}\{during, finishes\}I_{s_2}$$

The temporal constraints included in τ_α and τ_β form a binary constraint network, where nodes are time intervals and arcs are labeled with (disjunctions) of temporal relations. Inverse relations or self-referencing are not permitted. As the temporal relations are mutually exclusive, under a consistent scenario every arc should be labeled with a single relation. Figure 6.13 gives the constraint network corresponding to the SDM of disorder d_1.

Raw patient data consist of a set of direct observations, where an observation is a measurement (v) of an observable parameter (m) at a particular time point t (with respect to the real time line). This is expressed as $m(v,i) \wedge i\{equal\}[t,t]$. A temporal data abstraction mechanism is applied to the direct, point-based, observations in order to derive more useful abstract, interval-based, observations. As already indicated the abstraction mechanism of ATD is very simplistic; namely all parameters are considered concatenable and thus all "meeting" intervals denoting the same parameter and value are joined in order to obtain maximal validity periods for the given parameter-value. Since direct observations are discrete the sense of "meeting" or consecutiveness is with respect to the relevant granularity. Apart from simple state abstractions, no other form of temporal abstraction is performed, e.g. trends. The temporal abstraction mechanism is based on the needs of the particular medical domain, that appear to be rather simple. Thus the distinction between concrete and abstract observations is that the former refer to time points while the latter to intervals. If at every measurement, a parameter changes value, no abstraction is possible but in reality this does not happen since once a change takes place it persists over a period of time. The sampling frequency of a parameter (temporal granularity) is chosen so as not to miss any significant changes. Often, however, it is not possible to have "complete" information on a parameter even with respect to the given sampling rate and hence one of the challenges of temporal data abstraction is how to fill such gaps.

An abstract observation is expressed as $m(v, i) \wedge \tau$ where τ gives the temporal constraints between interval i and the real time line and/or other abstract observations. The temporal constraints in the abstract patient information are also represented as a binary constraint network.

Since ATD is a logic-based, abductive framework, for the derivation of explanations it is necessary to define the set of *abducibles*, those literals which are acceptable as explanations. These are the literals which occur only in the antecedents of SDMs, e.g. the process state assumptions. For example, the abducibles in the SDM of d_1 are the literals $s_1(d_1, I_{s_1})$ and $s_2(d_1, I_{s_2})$. A (diagnostic) explanation of the patient information is a conjunction of abducibles which together with the diagnostic knowledge (SDMs and background theory) entails the abstract observations (derived from the patient information) denoting abnormality, and is consistent with all abstract observations (such explanations are obtained by deploying a logic-based, abductive reasoning engine). This is a typical logic-based definition of diagnosis. It should be reemphasized, though, that the criterion of complete coverage of observations of abnormality that appears in this framework, in t-PCT and its fuzzy extensions as well as in the following approach is a very strict one for realistic medical diagnostic problems. However, this issue is of relevance to general diagnosis and not specifically to temporal diagnosis. Both this and the previous two approaches use time for constraining the set of plausible explanations, where a necessary criterion for plausibility is complete coverage of the observations of abnormality. Temporal constraints should be explicitly modeled with the aim of directing the formation and evaluation of hypotheses. Where complete coverage is attainable, temporal constraints do provide the means for differentiating between the covers.

The ATD framework does not include the actions plane of the C-T-A model. The constraints plane refers to the time intervals designating the period of occurrence, on the time line, of internal states and observable manifestations. The constraints are expressed as relative temporal relations (using Allen's set of relations) between intervals or between intervals and absolute periods of time. Temporal uncertainty and incompleteness is supported since the relation between a pair of intervals can be expressed as a disjunction of possible relations and it is not necessary to specify the relation between any pair of intervals. Relations can be inferred by computing the transitive closure of the network of constraints. In this framework causality is loosely expressed through the implications in the SDMs. The implication in an SDM bundles together all the individual causal links between internal states and external manifestations. Individual causal links between internal states are not given either, although the evolution of a disorder, in the particular domain examined, is modeled as a chaining sequence of internal states. The pairwise causality links between successive states are implicitly expressed through temporal relation *meets*, between the intervals corresponding to these states. Hence temporal constraints with respect to causal links are not explicitly given, as it is the case with the previous two proposals discussed.

6.7 Console and Torasso's Temporal Abductive Diagnostic Architecture

Console and Torasso [103] propose a logic-based, temporal abductive architecture for medical diagnosis. This architecture is an extension of the causal component of CHECK [100, 404], an atemporal diagnostic system which was applied to the domains of cirrhosis and leprosis.

In this proposal diagnostic knowledge is represented as a single, acyclic, causal network where the arcs are associated with temporal information denoting a range (minimum and maximum) for the delay between the start of the cause and the start of the effect. These delays can be expressed with respect to various granularities and in fact some delay may involve mixed granularities such as a minimum of 1 hour and a maximum of 1 day. As with the previous approaches discussed the multiplicity of granularities is mentioned only in passing.

The temporal primitive of the proposal is the time point. A time interval is a convex set of time points. In order to allow uncertainty with respect to temporal existences, *variable intervals* are used which are expressed as quadruples of time points where the first pair of time points gives the range for the begin and the second pair the range for the end of the interval. A variable interval encompasses a set of precise intervals. This is a fairly standard representation for absolute temporal uncertainty. We have already seen it in HDP.

The nodes of the causal network define findings (manifestations), pathophysio-logical states (initial and intermediary), and contextual information (c-nodes). Manifestations and contextual information are directly obtainable (observable entities) while pathophysiological states (p-states) are non observable, possibly with the exception of some initial states, and therefore need to be inferred. Some of the p-states define diagnoses. All types of nodes are associated with a set of single-valued attributes and the value set of each attribute is expressed in linguistic terms. Instantiating such a node means assigning values to its attributes and determining its temporal extent as a (variable) time interval.

A causal arc in the network relates an antecedent to a consequent where a consequent is either a p-state or a manifestation. In the simplest form, an antecedent is a single p-state, but it can also be a conjunction of p-states and c-nodes involving at least one p-state (although the authors note that the majority of causal relations encountered in the medical domains investigated were of the simple form). Figure 6.14 shows the types of causal links. For example, in Figure 6.4 all the causal links are of a simple type. Furthermore no c-nodes are involved. The link from s_1 to s_2 is a causal link from a p-state to a p-state and the link from s_1 to m_1 is a causal link from a p-state to a manifestation. Where complex causal antecedents are involved the delay information can be interpreted either in a weak way as the delay between the first point in time where all the elements in the antecedent hold, and the start of the consequent, or in a strong way as the separate pairwise delays between the start of each conjunct of the antecedent and the consequent. The weak interpretation is more intuitive, but is computationally intractable, while the strong interpretation

Simple causal links

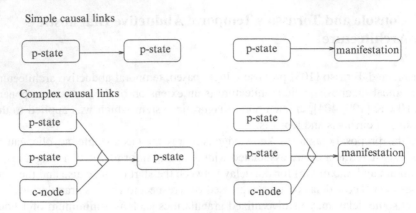

Complex causal links

Fig. 6.14 Types of Causal Links in Console and Torasso Architecture

is restrictive (incomplete relative to the weak interpretation) but computationally tractable. This is because, as already said, the principal step in medical diagnostic reasoning, in an abductive sense, is to infer causes from effects. Under the strong interpretation the temporal extent for each conjunct node in the antecedent can be deterministically computed (albeit as variable intervals) from the temporal extent of the consequent node and the delay information. Actually the delay information is used to compute the bounds for the start of an antecedent node from the bounds of the start of a consequent node (this also ensures that an effect cannot precede its cause). Bounds for the end of an antecedent node are obtained on the basis of two general temporal constraints, namely that a cause cannot 'outlive' its effect and that there cannot be a gap between the two. Under the weak interpretation the only thing that can be deterministically obtained is the temporal extent of a so called virtual state that defines the co-existence of the various nodes involved.

A causal arc can be associated with an arbitrary logical expression the validity of which in a given diagnostic situation is a necessary condition for assuming the presence of the causal arc. The authors do not explain whether such a condition can involve further temporal constraints. The impression given is that such conditions express atemporal contextual factors for the materialization of the causality relation.

As mentioned above the nodes in the causal network are characterized by a set of attributes. A causal arc in fact encompasses a set of associations relating combinations of node-attribute-value triplets (for nodes of the antecedent) to combinations of attribute-value pairs for the consequent. If none of these associations is satisfied, the causal arc is again revoked. Although not explicitly discussed, it appears that the delay information of a causal arc applies to all potential combinations of attribute-values specified.

Finally there can be multiple causal arcs (either with simple or complex antecedents) sharing the same consequent.

The representation of the causal relation in this proposal is rich, but from the atemporal rather than the temporal sense. In fact the authors themselves admit that

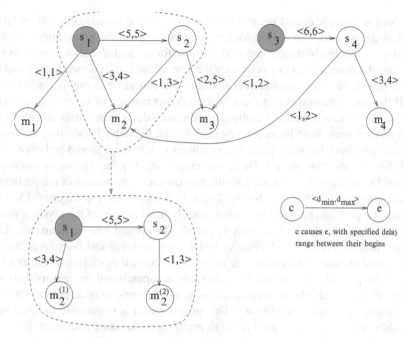

Fig. 6.15 Causal Network of Figure 6.4 Annotated with Delay Information

the expressive power allotted to the causal relation through complex antecedents would be uncalled for in most medical domains. Furthermore, such complex causality reverts to a collection of simple causal relations, just sharing the same consequent (and underpinned by the same condition, if any) if computational reasons enforce the strong interpretation of the delay information. From the temporal perspective, this representation is not rich; the temporal semantics of a causal arc are both simple (just delay information) and restrictive (a cause cannot outlive an effect, e.g. in d_1, s_1 cannot outlive m_2). Many causality patterns in medicine, e.g. those identified in HDP, cannot be expressed in this formalism from the temporal perspective.

A difference between this approach and the previous approaches discussed is that in this architecture no general duration information for the nodes in the causal network is given. The authors emphasize that in medicine the available temporal knowledge tends to be just delay information, while temporal extents can only be dynamically inferred and constitute factual information for the particular diagnostic cases. Actual temporal extents of unobservable states may indeed need to be inferred but general knowledge about the persistence of the relevant conditions can be of great assistance in such reasoning. Such general temporal knowledge can be available in medical domains, as the HDP, the t-PCT, and the next approach's domains illustrate.

Figure 6.15 gives the causal network of Figure 6.4, annotated with delay information (at the granularity of days), as proposed by this architecture. As can be seen there is a single causal network encompassing all disorders, just like in HDP, not separate networks for each disorder, as in the other approaches. Internal states s_1 and s_3 respectively denote the initial states for d_1 and d_2. Such states do not have causal antecedents. Let us consider the part of the causal network enclosed by dashed line. How does one interpret the multiple causal links impinging on node m_2? If the interpretation is that there is a single occurrence of m_2 with potentially multiple concurrent causes, whereby the individual delay constraints are mutually satisfied, then there is an inconsistency namely that the delay between s_1 and m_2 is required to be both between 3 and 4 days (direct link from s_1 to m_2) and between 6 and 8 days (connection via s_2). These two ranges are disjoint. The inconsistency is resolved by splitting the m_2 node into the two distinct occurrences of the particular manifestation, as it is the case for d_1. The split is also shown in Figure 6.15. This resolves one problem but creates another, namely to which one of the two new nodes referring to m_2 should the link from s_4 point? The correct answer is no one. This problem wouldn't arise if each disorder had its own causal model. Moreover, the explicit representation of the recurrence of m_2 in the context of d_1 is done for nothing since in the patient information, for reasons of computational tractability, it is not allowed to give multiple distinct occurrences of the same manifestation, but only one occurrence at most (see below). This means that it is impossible to instantiate more than one node in the causal network referring to the same manifestation.

Overall, the C-T-A model is not completely supported by this proposal since the actions plane is missing. The constraints are expressed as ranges of delays between the begins of causes and effects; the periods of time designating the existences of states and manifestations are dynamically obtained as variable intervals. Hence temporal uncertainty is supported.

The authors view temporal reasoning in medical diagnosis as a temporal constraint satisfaction problem. The patient information consists of a collection of manifestation instances (findings), and their (possibly variable) temporal extents. The objective is to determine the path in the network, whose delays are temporally consistent with the extents of the findings, and which accounts for the findings. This is done by propagating backwards the temporal extents of the findings. More specifically, the integration of an abductive engine with a temporal constraint satisfaction engine is advocated. As with HDP, t-PCT and ATD, the constraint satisfaction engine participates actively in the formation of diagnostic explanations. However, the authors of this architecture believe that a cleaner setup would be to use the abductive engine to generate potential hypotheses and then to use the temporal constraint engine as a filter to throw away those hypotheses that are not viable from a temporal perspective. This pipeline approach was not used, though, because the abductive engine was throwing away viable explanations due to its ignorance of the temporal evolution of a pathophysiological situation. The authors believe that the use of the temporal engine in a tightly integral fashion with the abductive engine increases the overall complexity of the system. The view of abduction employed in this proposal is again the strict, logic-based, view, where all observations of abnormality need to be covered and consistency with all observations is required, a view that is certainly challenged by many real-life medical domains, as already said. The authors of HDP, t-PCT and ATD see the tight integration of temporal reasoning with their respective diagnostic engines as a strength of their approaches because it makes the formation of diagnostic explanations more efficient by preventing the proliferation of temporally inconsistent possibilities.

The representation of patient information in this approach, in the authors' own admission, is rather restrictive; more specifically an observable parameter can attain an abnormality value during a single interval only and throughout the rest of the period of relevance to the diagnostic activity, this parameter is assumed to be normal. This is very restrictive because it prevents the expression of repetitive happenings, as illustrated above; of course periodicity is absent at the level of diagnostic knowledge as well since the causal network is acyclic and each disorder is not modeled separately through its own causal model that would have allowed multiple instantiations giving distinct occurrences of the same disorder. The authors indicate that in order to allow recurring happenings they would need to use nonconvex time intervals. This would mean a drastic reconstruction of their proposal. Apart from nonconvexity, compound occurrences and temporal data abstraction (e.g. temporal trends) would need to be supported.

6.8 Abductive Diagnosis Using Time-Objects

From the discussion so far on the various proposals, two broad categories can be identified. In one category diagnostic knowledge is represented as a single causal network and as such the temporal constraints refer to delays between causes and effects and the persistence of states. The HDP system and Console & Torasso's architecture belong to this category. In the other category each disorder is modeled separately, but in associational rather than causal terms, i.e., in terms of external manifestations. In these approaches, the temporal constraints refer to temporal relations between events marking begins and ends of manifestations. The ATD system, and the t-PCT and its fuzzy extensions, belong to this category. In every proposal, temporal uncertainty and incompleteness are recognized as necessary representation features, expressed either in an absolute or metric way (ranges for delays, durations, begins and ends of actual existences), or a relative way (disjunctions of temporal relations between validity intervals of occurrences). Regarding therapeutic interventions, these are ignored in all proposals with the exception of the HDP system.

In all these approaches, occurrences (internal states and manifestations) are treated as indivisible entities. Moreover, strictly speaking, occurrences are not treated as dynamic entities, embodying time as an integral aspect, and actively interacting with each other; time is loosely associated with them by pairing an atemporal entity with some time interval, e.g., by including the interval as yet another argument of the relevant predicates. Recurring phenomena and periodicity in general, as well as temporal trends are not addressed. These require compound occurrences. Temporal data abstraction is addressed only in ATD, but what is provided is not a fully fledged temporal data abstraction engine. However, [53] reports very interesting

results in the diagnosis of arrhythmias from electrocardiograms, using temporal abstraction and inductive logic programming. Temporal abstraction identifies the sequences of significant events (in symbolic form) from the ECG signals, where recurrence and periodicity is allowed. Inductive logic programming is used for learning
the models of the particular disorders from sets of examples.

Another issue which is generally identified as significant, but only mentioned
in passing is that of multiple temporal granularities. Bettini et al [28] have given a
formal exposition to this topic with respect to databases, data mining and temporal
reasoning.

The approach to be discussed in this section is based on a time ontology
[204, 205, 209] the central primitives of which are the time-axis, enabling a multi-
dimensional and multigranular model of time [208], and the time-object, a dynamic
entity embodying time as an integral aspect, and bringing together temporal, structural and causal knowledge.

A *time-axis*, α, represents a period of valid time from a given conceptual perspective. It is expressed discretely as a sequence of time-values, $Times(\alpha) = \{t_1, t_2, ..., t_n\}$,
relative to some origin. All time-axes used in some application could share the same
origin. Time-axes are of two types, atomic and spanning. An atomic axis has a single
granularity (time-unit), that defines the distance between successive pairs of time-
values on the axis. Its time-values are expressed as integers. A spanning axis spans a
chain of other time-axes. It has a hybrid granularity formed from the granularities of
its components, and its time-values, also inherited from its components, are tuples.
An application can involve a single atomic axis and a single granularity, or a collection of time-axes and multiple granularities where the same period of time can be
modeled from different conceptual perspectives. Example time-axes could be fetal-
period, infancy, childhood, puberty and maturity, the latter four collectively forming
a spanning axis, lifetime. These are general, or abstract, time-axes whose origin,
say birth, is a generic time-point. Such time-axes can be instantiated for specific
cases by binding their abstract origin to an actual time-point (a particular individual's birth), thus obtaining concrete time-axes. The distinction between abstract and
concrete, that applies to time-objects as well, is important; diagnostic knowledge is
abstract, a patient history is concrete.

A *time-object* is a dynamic entity that has time as an integral aspect. It is an association between a *property* and an *existence*. The existence of abstract/concrete
time-objects is given with respect to abstract/concrete time-axes. Formally a time-
object, τ, is defined as a pair $\langle \pi_\tau, \varepsilon_\tau \rangle$ where π_τ is its property and ε_τ its existence
function. The time-axis that provides the most appropriate conceptual context for
expressing the existence of τ is referred to as the *main time-axis* for τ and the expression of τ's existence with respect to its main time-axis, as its *base existence*.
The existence function, ε_τ, maps the base existence of τ to other conceptual contexts
(time-axes). A time-object has a *valid* existence on some time-axis iff the granularity of the time-axis is meaningful to the property of the time-object and the span of
time modeled by the time-axis covers (possibly partially) the base existence of the
time-object. A (known) valid existence is expressed in an absolute way in terms of
earliest and latest begins and ends, given as time-values on the relevant time-axes.

Time-objects can be compound, involving component time-objects. Compound time-objects can be viewed from different temporal perspectives. At gross perspectives some components may not be visible, if their presence would be meaningful only at finer granularities. The existence of a component of a compound time-object is bound by the existence of the compound time-object. More specifically, there should be at least one component starting at the same time as the compound time-object and there should be at least one component finishing at the same time as the compound time-object. Periods of inactivity within the existence of a compound time-object could be denoted as special kinds of components; this way there would be some operative component at any time.

Causality is a central relation in this ontology. More specifically, the ontology includes relations *causes*, *causality-link* and *cause-spec*, that are defined at the level of abstract time-objects, concrete time-objects and abstract properties, respectively. Relation $causes(\tau_i, \tau_j, cs, cf)$, where τ_i and τ_j are abstract time-objects, cs is a set of temporal and other constraints, and cf is a certainty factor, is used in the following axiom for deriving a *causality-link* between a pair of concrete instances of τ_i and τ_j. A general constraint, that always needs to be satisfied is that a potential effect cannot precede its potential cause:

Axiom 1: $causality\text{-}link(\tau_i, \tau_j, cf)$

$\Leftarrow causes(\tau_i, \tau_j, cs, cf) \wedge conds\text{-}hold(cs) \wedge \sim starts\text{-}before(\tau_j, \tau_i).$

Predicate $starts\text{-}before(\tau_j, \tau_i)$ expresses that τ_j starts before τ_i. Even if all the specified conditions are satisfied, by some case, still it may not be definite that the *causality-link* actually exists owing to knowledge incompleteness. This is modeled by the certainty factor.

Relation *cause-spec* between properties has six arguments, where the first two are properties, the third is a granularity, the fourth and fifth are sets of relative (*TRel*) and absolute (*TAbs*) temporal constraints respectively and the last one is a certainty factor. This relation also enables the derivation of a *causality-link* between a pair of (concrete) time-objects through the following axiom:

Axiom 2: $causality\text{-}link(\tau_i, \tau_j, cf)$

$\Leftarrow cause\text{-}spec(\rho_i, \rho_j, \mu, TRel, TAbs, cf) \wedge$

$\pi(\tau_i) = \rho_i \wedge \pi(\tau_j) = \rho_j \wedge$

$r\text{-}satisfied(\tau_i, \tau_j, \mu, TRel) \wedge a\text{-}satisfied(\tau_i, \tau_j, \mu, TAbs) \wedge$

$\sim starts\text{-}before(\tau_j, \tau_i).$

Predicates *r-satisfied* and *a-satisfied* express the satisfiability of the relative and absolute temporal constraints respectively. The granularity defines the base granularity, i.e. the finest granularity under which it is meaningful to discuss a causal relation between the given pair of properties. Where a granularity is not explicitly given in the expression of some relative or absolute temporal constraint, the base granularity is assumed. Hence the base granularity is the default granularity, unless a coarser granularity is called for. Relations *causes* and *cause-spec* are illustrated below.

A concrete time-object can be viewed as an *agent* which is extinguished, alive and active, or asleep and waiting, if its existence refers to the past, present, or future respectively. It is a dynamic entity that exhibits a certain behaviour in terms of the

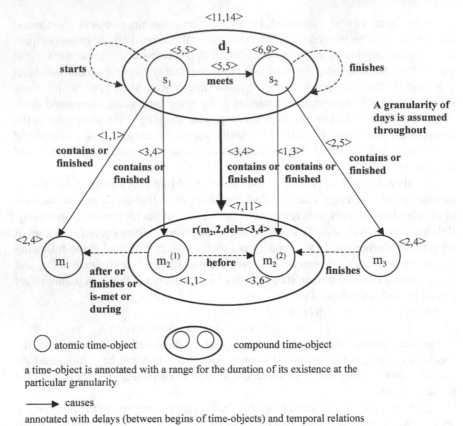

Fig. 6.16 Modeling d_1 as a Causal-Temporal Structure of Atomic and Compound Time-Objects

changes that it can bring about either to itself or to other time-objects. A change is *constructive* if it brings about the creation of another time-object or it is *destructive* if it brings about the destruction of some time-object, possibly its own self. So a time-object interacts with other time-objects in a positive (constructive) or negative (destructive) manner. Such interactions are modeled through causality. Furthermore, part of the behaviour of a time-object may deal with its defense mechanisms against other time-objects with negative intentions towards it. Disorder processes, the behaviour of therapeutic actions and patient histories can be modeled in terms of time-objects [204]. For example, a disorder process can be modeled as a compound time-object, the components of which are the disorder's internal states. The behaviour of a disorder time-object is to materialize its components in accordance with temporal and other constraints and to 'defend' itself against therapeutic action time-objects. A therapeutic action time-object behaves with negative intentions towards disorder time-objects.

In this proposal each disorder is modeled separately. Figure 6.16 gives the model of disorder d_1 as a causal-temporal structure of (abstract) time-objects. In this model the expression of temporal constraints, both absolute and relative (see below), assumes a granularity of days. The model involves two compound time-objects, the disorder per se, the evolution of which is analyzed in terms of a chaining sequence of the states s_1 and s_2 (its components) and the recurrence of manifestation m_2. The other time-objects (s_1, s_2, m_1, $m_2^{(1)}$, $m_2^{(2)}$ and m_3) are atomic, although some, say s_1 and s_2, could also be decomposed into finer components. The compound time-object representing the recurrence of m_2 has the complex property $r(m_2, 2, del = \langle 3, 4 \rangle)$. Three-place predicate r denotes a recurrence where its arguments, in order, give the property that recurs, the number of repetitions and the pattern of repetition. In the particular instance the pattern of repetition is that there is a delay of 3 to 4 days between the begins of the (two) occurrences of the given property. Recurrence time-objects are by definition compound [207]. The components of the particular recurrence time-object are also explicitly given since it is necessary to represent their separate causal antecedents. If the recurrence denoted an indefinite repetition, the enumeration of the individual components, at an abstract level, would not be possible. A recurrence time-object, by default, has a nonconvex existence, since any gaps, at a given granularity, between the existences of the individual components represent periods of inactivity, as it is the case with the particular recurrence time-object. Other time-objects, both atomic and compound, are assumed to have convex existences.

Decomposition (and its inverse aggregation) is one of three central types of relationships between time-objects. It enables the definition of compound time-objects and their break down at finer levels. The other two central relationships are causal and temporal relationships. Both of these relationships are necessary in the definition of the model of d_1. In Figure 6.16, each continuous (directed) arc represents an instance of relation *causes*, annotated with absolute and relative temporal constraints. The absolute constraints could refer to ranges for the duration of a cause or an effect (in the figure a range for a duration annotates the relevant node of the model as it applies to any causal arc having the particular node as cause or effect), or the delay between the begins of a cause and an effect (the ontology allows for other types of absolute temporal constraints but a full listing is not necessary for the purposes of our discussion). Relative temporal constraints for relation *causes* refer to relative temporal relations (from Allen's set plus some other complex relations) between the cause and effect, e.g. the *causes* arc from s_1 to m_1 states that s_1 either contains m_1 or it is finished by m_1.

It is important to note that temporal relations between time-objects are granularity-sensitive, meaning that the relation between a given pair of time-objects can be viewed differently under a finer granularity than under a grosser granularity. Of course the picture revealed under a finer granularity is more accurate and hence closer to reality than than revealed under a grosser granularity. However, modeling and reasoning at a grosser level means higher abstraction and by definition abstraction hides detail (even disguises reality), that is deemed unnecessary or even obstructive under the particular frame of reasoning. Hence at a grosser granularity the temporal relation between s_1 and m_1 might be that of equality and at a finer granularity the relation could be just contains.

Formally the instance of relation *causes* from s_1 to m_1 can be expressed as the following ground atom:

$causes(s_1, m_1,$
$\{dur(s_1, days) = \langle 5, 5 \rangle, dur(m_1, days) = \langle 2, 4 \rangle, delay(days) = \langle 1, 1 \rangle,$
$temp\text{-}rels(days) = \{finished, contains\}\}, 1.0)$

A certainty factor of 1.0 implies that if all the constraints given in the third argument are satisfied by two concrete instances of s_1 and m_1 and provided that m_1 does not precede s_1 (actually this is impossible given the delay constraint) a *causality-link* from the concrete instance of s_1 to the concrete instance of m_1 definitely holds (this is Axiom 1). As another example, let us give the formal expression of relation *causes* from s_1 to s_2:

$causes(s_1, s_2,$
$\{dur(s_1, days) = \langle 5, 5 \rangle, dur(s_2, days) = \langle 6, 9 \rangle, delay(days) = \langle 5, 5 \rangle,$
$temp\text{-}rels(days) = \{meets\}\}, 1.0)$

The relative temporal constraint that the cause meets the effect in the above instance of relation *causes* is superfluous as it can be inferred from the absolute constraints. In general, constraints (absolute and relative) could be inferrable from other constraints.

Where compound time-objects participate in causal relations, causality can be expressed at different levels of abstraction. This is also illustrated in the model of d_1 in the case of the causal relation from the compound time-object denoting the disorder per se and the compound time-object denoting the recurrence of m_2; hence both cause and effect are compound time-objects. Formally the relation is expressed as:

$causes(d_1, e, \{\pi(e) = r(m_2, 2, delay = \langle 3, 4, days \rangle),$
$dur(d_1, days) = \langle 11, 14 \rangle, dur(e, days) = \langle 7, 11 \rangle, delay(days) = \langle 3, 4 \rangle,$
$temp\text{-}rels(days) = \{contains, finished\}\}, 1.0)$

A couple of points are worth noting. Firstly the recurrence time-object is denoted by a dummy name, e, for effect, and its property is included in the list of constraints, since it specifies a complex (absolute) constraint. Secondly the disorder is given a finite and explicit duration, namely between 11 and 14 days. Hence, coining HDP's terminology, d_1 is a self-limiting disorder that can come to an end on its own without treatment action, e.g. d_1 could be a kind of viral infection, as already mentioned, that runs its course from between a week and a half, to a fortnight.

At a more detailed level the gross causal relation between d_1 and the recurrence of m_2 can be refined into two causal relations between the internal states of d_1 and the respective occurrences of m_2. If a compound time-object can be accounted for at a more abstract level, reasoning at a lower level, dealing with more detailed states and causal relations, may not be necessary. Hence the ability to have compound causes and effects and to reason causally at different levels of abstraction provides more flexibility in reasoning.

Although the precedence relation, *before*, between the two distinct occurrences of m_2 can be inferred from the constraints associated with relation *causes* from s_1 to

s_2, and the pairwise *causes* relations between the two internal states and the occurrences of the manifestation, still this is explicitly given. Some redundancy is useful, especially if complicated inferencing is avoided this way. Other relative temporal relations between time-objects in the model of d_1 are explicitly given. In Figure 6.16 these are given as annotations of dashed, directed arcs. All these involve entities that are not causally related, namely external manifestations. Alternatively, or in conjunction, such arcs can also be annotated with absolute temporal constraints. The temporal constraints should be mutually consistent. More specifically, the metric constraints alone should be consistent, the relative constraints alone should be consistent, and there should be consistency across the two types of constraints. Initially, a minimal set of relations can be given that can be easily shown to be consistent, and other relations can be subsequently derived.

As already explained, in a diagnostic context the most frequent, and critical, reasoning step is to hypothesize a cause from an established effect. This is an abductive step. While exploring a hypothesized cause, a frequent reasoning step is to hypothesize further observable, but as yet unknown, effects of the cause. This is a deductive step. Deducing the presence of a causality-link between a pair of concrete (either established or hypothesized) entities is a less frequent step. Relation *causes* can support all three reasoning steps. Its use in the last of these steps is illustrated through Axiom 1. The application of Axiom 1 refers to a pair of concrete time-objects, whose absolute existences under a concrete time-axis are defined. In this respect the list of constraints (third argument of relation *causes*) is treated as a conjunction.

The other two reasoning steps (hypothesizing a cause or an effect) entail the construction of a time-object, i.e. the derivation of its existence under the relevant concrete time-axis, so that all the constraints that underlie the presence of a *causality-link* between the given entity and the constructed entity are satisfied. In this respect the list of constraints given in the relevant instance of relation *causes* provides the means for inferring the existence of a cause/effect on the basis of the existence of a potential effect/cause.

Relation *causes* is defined at the level of abstract time-objects and as such it is used in the definition of disorder models. Relation *cause-spec* is similar to *causes* but as it is defined at the level of properties its use is independent of disorder models. Instances of this relation represent a kind of background knowledge, e.g. for expressing effects of therapeutic actions. In the case of multi-disorder diagnostic hypotheses, i.e. hypotheses consisting of multiple disorder model instantiations, relation *cause-spec* provides a means for investigating possible links, of a causal nature, between time-objects belonging to different disorder model instantiations, thus enabling the construction of a more coherent complex hypothesis. This is done through Axiom 2 (see above). Where *cause-specs* define effects of therapeutic-actions, their usage could be in establishing links (positive or negative) between components of a diagnostic hypothesis and contextual information about ongoing (therapeutic) actions.

An example of relation *cause-spec* is the following:
cause-spec(bacteria(present), throat(sore), *days*,
{*finished*}, {*dur*(bacteria(present)) = ⟨7, 10⟩,

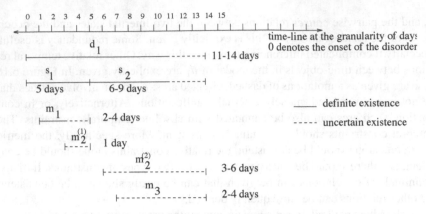

Fig. 6.17 Existences of Time-Objects Comprising the Model of d_1 (relative to the start of the disorder and at the granularity of days)

$delay = \langle 1, 3 \rangle \}, 0.9)$

This instance of the relation specifies a general causal association between bacteria and soreness of throat, i.e. properties *bacteria (present)* and *throat (sore)*. More specifically, for an actual incidence of bacteria to be the cause of an actual incidence of soreness of throat, it is necessary for the lifecycle of the particular bacteria to be between 7 to 10 days and for the soreness to have started within 1 to 3 days from the acquisition of the bacteria. Finally once the soreness starts it should last until the demise of the bacteria. Still, in 10% of cases of an actual incidence of bacteria acquisition and an actual incidence of soreness of throat, where all the above conditions hold, the cause of the particular soreness of throat is not the particular bacteria (this is expressed as a certainty factor of 0.9). [8]

Similarly to *causes, cause-spec* can be used both for deriving (hypothesizing) causal links or for hypothesizing causes from effects or vice versa.

The existences, at the granularity of days, of the time-objects comprising the model of d_1, relative to the start of the disorder, are given in Figure 6.17. A continuous line indicates a definite existence while a dashed line an indefinite existence.[9] Only time-object s_1 has a completely definite existence (coinciding with the first 5 days of the existence of the disorder). Some time-objects, e.g. time-object m_3, have a completely indefinite existence. In the case of m_3 any continuous subperiod of duration 2-4 days within the demarcated period is a possible existence for it. In the case of $m_2^{(2)}$ the possible subperiods must include the definite part of its existence. The set of possible existences of some time-object can be further constrained through relative temporal relations with other time-objects.

---

[8] If some condition is unknown, or possibly true due to the inherent temporal uncertainty, a causal link may be hypothesized on the assumption that the given condition holds. If the condition is subsequently refuted, the causal link should also be erased.

[9] Recall that an absolute existence is expressed as a range for its begin and a range for its end. If the two ranges overlap the existence has no definite part.

Disorder d_1, say a viral infection of finite duration, belongs to the category of finitely persistent disorders which can recur (see below). In the models of such disorders it would be meaningless to give absolute existences (i.e. in terms of time-values on a time-axis) to the time-objects concerned. Instead existences are expressed relative to the start of the disorder as shown in Figure 6.17 for d_1. When the model of such a disorder is instantiated under a concrete temporal context, its constituents can be assigned absolute existences [214, 213]. In fact, in the case of recurring disorders, multiple instantiations of the same disorder with distinct absolute existences may be called for in a particular situation.[10]

The models of disorders that belong to the category of infinitely persistent disorders (see below) can be given absolute existences with respect to the relevant abstract time-axes. This is illustrated by the model of disorder Spondyloepiphyseal Dysplasia Congenita (SEDC), a skeletal dysplasia, given in Figure 6.18. The (spanning) abstract time-axis, *lifetime* that begins at birth and terminates at death, and involves mixed granularities, is assumed in this model.

The early roots of the time-object based framework for abductive diagnosis are found in the temporal reasoning framework of the Skeletal Dysplasias Diagnostician (SDD) system [212]. SDD is an expert diagnostic system for the domain of skeletal dysplasias and malformation syndromes [211]. These are developmental abnormalities that affect the skeletal system to varying degrees. Time is intrinsically relevant to this domain and temporal reasoning was included in the design aims of SDD from the beginning. The simplified description of disorder SEDC given in Chapter 2 is repeated below.

SEDC *presents from birth* and may be lethal. It *persists throughout life.* Symptoms can include: short stature, owing to short limbs, *from birth*; mild platyspondyly *from birth*; absent ossification of knee epiphyses *at birth*; bilateral severe coxa-vara *from birth, worsening with age*; scoliosis, *worsening with age*; wide tri-radiate cartilage *up to about the age of 11 years*; pear-shaped vertebral-bodies *under the age of 15 years*; variable-sized vertebral-bodies *up to the age of 1 year*; and *retarded ossification* of the cervical spine, epiphyses, and pubic bones.

In Figure 6.18 non-observable entities are represented by ovals while 'observable', often through x-raying, are represented by hexagons. Since the evolution of SEDC is not analyzed in terms of internal states, and only its external manifestations are depicted, this is really a (temporal) associational model. However the model illustrates the need for temporal data abstraction as some manifestations express trends, namely the worsening of scoliosis or coxa-vara.

In this ontology, disorders and their manifestations are classified as follows from the temporal perspective:

- Infinitely persistent, either with a fixed or a variable initiation margin (e.g. SEDC);
- Finitely persistent, but not recurring, again either with a fixed or a variable initiation margin (e.g. chicken pox) ; and

---

[10] If instead of separate disorder models there were a single causal network, this would not be possible.

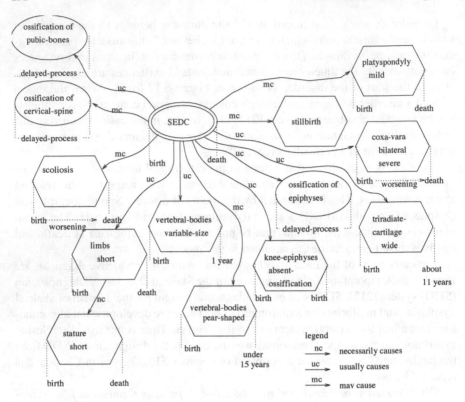

Fig. 6.18 Model of Skeletal Dysplasia SEDC

- Finitely persistent which can recur (here the initiation margin is variable), e.g. flu.

The time-object proposal conforms to the C-T-A model. All five planes are included. The actions plane is not discussed here, but the interested reader could refer to [214]. Temporal constraints apply both to causal relations and to structural relations in the case of compound time-objects. Furthermore, the very existence of a time-object is a temporal constraint, and the existences of separate time-objects can be temporally constrained, even if the time-objects are not causally or structurally related. Temporal constraints can be expressed both in metric and relative terms. However, handling together metric and relative constraints can increase the complexity of the reasoning of the system. As with other approaches, temporal uncertainty and incompleteness are addressed, but unlike the other approaches discussed, multiple granularities are an inherent aspect of the underlying temporal ontology.

$$c_{ik}$$

c_{ij}	<	=	>
<	<	<	<,=,>
=	<	=	>
>	<,=,>	>	>

c_{kj}

Fig. 6.19 Transitivity Table for Relative Temporal Relations between Instantaneous Events

6.9 Temporal Constraints

In this section we examine further the temporal constraints encountered in clinical diagnosis. More specifically, we consider a number of instantiations of the Abstract Temporal Graph (ATG), discussed in Chapter 2, emanating from the proposals discussed above. The particular cases of the ATG structure considered are: (a) the temporal entities are instantaneous events and the constraints are relative (case I) or absolute (case II); and (b) the temporal entities have duration and the constraints are relative (case III). The matter of mixed constraints is also discussed.

6.9.1 Case I: Instantaneous Events and Relative Constraints

Here the temporal entities are instantaneous events and $C = \{<,=,>\}$ or $\{before, equal, after\}$, i.e. the domain of relative temporal relations. Special constant $self\_ref$ is set to $=$, while relations $\not<, \neq$, and $\not>$ are respectively denoted by the disjunctive constraints $\{=,>\}, \{<,>\}$ and $\{<,=\}$. The *inverses* of $<, =$, and $>$ are respectively $>, =$, and $<$. Finally function *transit* is given by the transitivity table of Figure 6.19. As can be seen everything is deterministic, except when n_i is $after\ n_k$ but n_k is $before\ n_j$, and vice versa.

Case I is the simplest instantiation of the ATG. The domain of constraints is a three element set and the corresponding transitivity table is largely deterministic. In spite of its simplicity this structure is adequately expressive for the needs of various problems, not only in clinical diagnosis, and as such it has been investigated by many researchers, e.g. [152].

As an illustration, Figure 6.20 gives the case I ATG corresponding to disorder d_1. The nodes represent instantaneous events marking the begin or end of the disorder per se, its internal states and its external manifestations.

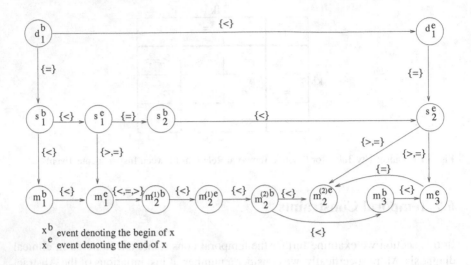

Fig. 6.20 Case I ATG for disorder d_1

6.9.2 Case II: Instantaneous Events and Absolute Constraints

The temporal entities are again instantaneous events, but the domain of constraints is expressed in a metric way, in terms of temporal distances. Such distances could be expressed discretely with respect to a particular granularity, say *days*, i.e.

$C = \{-\infty, ..., -4, -3, -2, -1, 0, 1, 2, 3, 4, ..., +\infty\}$,

where a negative number means "days after", a positive number means "days before" and 0 means "equal". Hence C is an infinite set. If finite bounds are set on the maximum days before and after, C becomes a finite domain.

Special constant *self_ref* is set to 0,

inverse$(c) = -c$ and

transit$(c_{ik}, c_{kj}) = \{c_{ik} + c_{kj}\}$.

For example *transit*$(-3, 2) = \{-1\}$. This says that if some event e_i is 3 days after some other event e_k, which is 2 days before a third event e_j, then event e_i is 1 day after event e_3. Thus with metric constraints, *transit* is implemented through integer addition and no uncertainty arises. With relative constraints (case I above and case III below) *transit* is implemented through a transitivity table and uncertainty can arise.

The disjunctive constraint $\{-3, 2\}$ means either 3 days after, or 2 days before, i.e. just these two possibilities excluding the in between. Hence it is different from the disjunctive constraint $\{-3, -2, -1, 0, 1, 2\}$ which denotes any (discrete) distance from 3 days after to 2 days before. This constraint is equivalent to the (discrete) range $\langle -3, 2 \rangle$. If constraints are expressed as ranges, the definitions of functions *match* and *propagate* given in Chapter 2, would need to be changed and in fact they would respectively correspond to functions *intersection* and *sum* of t-PCT. Similarly the minimization algorithms given in Chapter 2 should be appropriately modified.

Metric or absolute constraints can be expressed in various ways. What is described above is just one way. If instead of integers, reals are used for C, we no longer have a discrete domain, but any fractional distance (with respect to a unit granularity) can be expressed. In a more complicated representation, the elements of C could be tuples of reals allowing the expression of temporal distances with reference to a hierarchy of granularities, etc.

An arc in an ATG whose label is a disjunctive constraint, entails uncertainty since the temporal relationship between the existences of the particular pair of temporal entities cannot be given "precisely", where the precision is always relative to the domain of constraints C. Further uncertainty can be captured if the elements of C are extended from simple atomic values (either giving relative or metric constraints) to pairs of values, $\langle c, u \rangle$, where c is an atomic constraint as before and u is an uncertainty measure, e.g. a degree of belief, a certainty factor, a probabilistic value, a fuzzy membership function value, etc. The elements of such a domain of constraints are still mutually exclusive. The uncertainty measure, u, has its own domain of values ranging from a value denoting absolute uncertainty to a value denoting absolute certainty. This domain could be qualitative or quantitative. In the latter case even if the atomic constraints are relative and finite in number, C would be infinite.

The semantics of a pair, $\langle c, u \rangle$, would depend on the semantics of the particular uncertainty measure used, which in turn would depend on whether the ATG in reference models diagnostic knowledge or patient data. For example, at the level of diagnostic knowledge, u could denote the frequency of occurrence of the particular temporal relationship, c, between the relevant pair of temporal entities, while at the level of patient data, it could denote the observer's degree of belief that the temporal relation between the relevant pair of concrete temporal entities is c. Function *transit* and the other access functions should be extended to deal with the uncertainty measures.

If the constraints are associated with uncertainty, complete temporal knowledge is denoted by a single pair $\langle c, u \rangle$ where u represents absolute certainty and complete temporal ignorance is denoted by the disjunctive constraint encompassing all possible temporal relations where each is associated with absolute uncertainty.

Let us have a specific example, taking as atomic temporal constraints the discrete domain of days before or after, and interpreting the uncertainty measures as values of (normalized) fuzzy membership functions. Hence the domain of u is [0..1]. The disjunctive constraint $e_1 \{\langle -3, 1.0 \rangle, \langle 1, 0.05 \rangle\} e_2$ from event e_1 to event e_2, says that typically (hence a membership value of 1.0), where 'typically' could be interpreted as over 95% of the cases, e_1 occurs 3 days after e_2 but in less than 5% of the cases it could occur 1 day before e_2. In contrast, the simple constraint $\{-3, 1\}$ gives equal weight to 3 days after, or 1 day before. Fuzzy intervals, as used in fuzzy t-PCT, are captured by the representation of an ATG. However, it would be more compact to use fuzzy set definitions as arc labels, e.g. tuples of 4 values for normalized trapezoidal fuzzy intervals, instead of sets of constraint-uncertainty pairs.

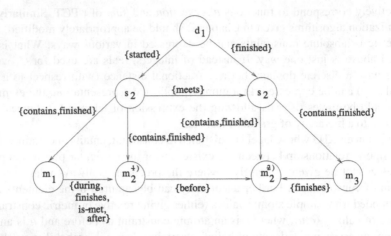

Fig. 6.21 Case III ATG for disorder d_1

The precedence graphs of disorders in t-PCT and the temporal graphs in fuzzy t-PCT belong to case II. The nodes of a precedence graph are occurrences with duration, not instantaneous events. However the constraints refer to temporal distances between the starts of such occurrences (delays) or between the start and end of some occurrence (duration). Hence it would be more appropriate to replace each occurrence with a pair of instantaneous events denoting its start and end, just like in the temporal graphs of fuzzy t-PCT. This is a more general representation that completely subsumes the precedence graph representation. In a precedence graph a node points only to nodes that follow it in time. Absolute constraints are more intuitively expressed with respect to instantaneous events while relative constraints are more appropriately applied to occurrences with duration. This brings us to the third instantiation of the ATG.

6.9.3 Case III: Occurrences with Duration and Relative Constraints

This is the generalization of case I. The temporal entities represent any occurrence and not just instantaneous events. The domain of constraints is Allen's set of relative relations, namely $C = \{before, after, meets, is\_met, overlaps, overlapped, starts, started, during, contains, finishes, finished, equal\}$, where *after* is the inverse of *before* and vice versa, etc. Special constant *self_ref* is set to *equal*. Function *transit* is defined in terms of Allen's transitivity table [10]. The constraint networks of ATD are covered by this case. As an illustration, Figure 6.21 gives the case III ATG corresponding to disorder d_1.

6.9.4 Temporal Constraints in Causal Networks

The temporal constraints associated with causal relations can be represented in terms of an ATG. Causal links are not explicitly represented in an ATG; hence at ATG is not a causal graph.

In Console and Torasso's proposal the temporal constraints of causal relations are expressed in a metric way as ranges for the delay between a cause and an effect.[11] In addition there are two general constraints, namely that a cause cannot 'outlive' its effect and that there cannot be a gap between the two. The first general constraint says that the end of a cause is before or equal to the end of its effect. The second general constraint says that the end of a cause is equal to or after the start of its effect.

Each metric constraint refers to the start of a specific cause and the start of a specific effect. These constraints are directly represented as a case II ATG. The nodes denote the starts of causes and effects and there is an arc from the start of some cause to the start of each of its direct effects, labeled with the relevant range. The fully connected version of the ATG would include arcs from the start of every state (cause or effect) to the start of every other state. Hence a state would be linked to all its effects, direct and remote, the starts of independent causes would also be linked, as well as the starts of independent effects. An arc joining a cause to a remote effect would be initially labeled $\{0, ..., +\infty\}$. An arc joining independent causes or effects would be initially labeled $\{-\infty, ..., +\infty\}$. If the labels are then minimized through propagation and matching operations, it may be possible to derive finite ranges for the temporal distances between the starts of independent causes sharing an effect, or the starts of independent effects sharing a cause. In addition, where there are multiple causal paths between the same cause and effect, a conflict will be raised if the separate delays are not mutually consistent, i.e. if there is no match between the delay derived via one route and the delay derived via another route.

The two general constraints refer to the end of a cause and the end of its effect or the end of a cause and the start of its effect. These constraints can be instantiated with respect to every specific cause and its (direct) effect. Such constraints are more intuitively expressed in a relative way (see above). However they can also be expressed in a metric way, again as a case II ATG. Hence the simplest thing to do is to extend the ATG discussed above by adding nodes denoting the ends of causes and effects. Arcs are then included from the end of a cause to the end of a direct effect of it, labeled $\{0, ..., +\infty\}$, and from the end of a cause to the start of a direct effect of it, labeled $\{-\infty, ..., 0\}$.

Hence the specific constraints, constrain the start of states while the general ones, their ends. When used together, they enable the derivation of ranges for the duration of states. Hence in the ATG we can include arcs from the start to the end of a state, initially labeled $\{0, ..., +\infty\}$. A minimization process may be able to reduce

---

[11] Although different granularities are implicated, for the sake of simplicity one granularity is assumed; after all the multiplicity of granularities appears to be handled in an ad hoc fashion in this proposal.

the infinite ranges for the state durations and the other temporal distances to finite ranges.

Here we have mixed constraints; the specific constraints are expressed in a metric way while the instantiation of the two general constraints is more intuitively expressed in a relative way. However, both types of constraints, explicitly or implicitly, refer to the same type of temporal entities, namely instantaneous events denoting starts or ends of states. Hence for the sake of simplicity all constraints are given in a metric way. We also note that each arc has a single label. More generally, the same arc could be associated with multiple, mixed labels and the question that arises is how to handle mixed constraints with respect to the same set of temporal entities. These temporal entities are in fact instantaneous events (cases I & II discussed above), since temporal entities with duration arise only in case III ATGs with respect to relative constraints. But we could have relative constraints with respect to temporal entities with duration and metric constraints with respect to their start and/or end events.

Thus how do we handle metric and relative constraints with respect to some collection of instantaneous events? One option, the one used above, is to translate constraints of the one type to the other type. Moreover, if the same arc is annotated with mixed constraints these could be interpreted either as a conjunction or as a disjunction. The interpretation of a conjunction is more appropriate since the additional constraints of the one or the other type for the same arc should reduce further rather than relax the possibilities. Hence the relevant operation, once the translation takes place, is that of a match.

Metric constraints are more precise than relative constraints, even if, in the case of discrete metric constraints, the abstraction of a single granularity is imposed. For example, in relative terms, the (metric) delay between any cause and its effect (as the number of days between the start of the cause and the start of its effect), would become {*before, equal*}. This translation results in information loss. On the other hand translating a relative constraint to a metric expression could introduce much uncertainty, e.g. the relative constraint {*before, equal*} would be translated to the infinite range $\{0, ..., +\infty\}$. If the translation option is viable, in the sense that it does not result in undue loss of information or unnecessary uncertainty, the benefit of it is that the minimization algorithms can be directly applied.

The other option is to handle each type of constraint separately and then to perform the relevant matches across the same arcs. Let C_m and C_r be respectively the domains of metric and relative constraints. There is one ATG, but each arc has both a metric and a relative label. If some arc does not initially have a metric/relative label, it is given the label C_m/C_r. The minimization is performed separately for each type of label. Assuming that no conflict is raised in connection with the one or the other type of label during the relevant minimization process, the minimized pair of labels for each arc is finally compared for consistency (assuming the two labels form a conjunction – see above), e.g. the labels {*after*} and $\{0, 1\}$ are inconsistent but the labels {*equal, after*} and $\{-3, -2, -1\}$ are consistent. If everything is consistent the corresponding labels are appropriately reduced, e.g. {*equal, after*} in the above example is reduced to {*after*}, the pair of labels {*equal, after*} and $\{0\}$

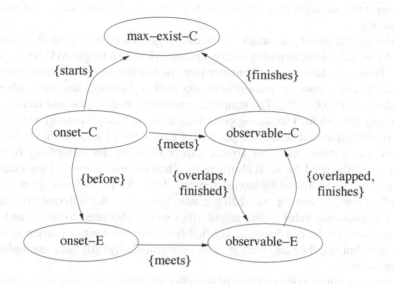

Fig. 6.22 A Case III ATG for the Relative Temporal Constraints of HDP

becomes {*equal*} and {0} and the pair of labels {*equal*, *after*} and {−3, −2, −1, 0, 1, 2}
becomes {*equal*, *after*} and {−3, −2, −1, 0}.

As a case III ATG (relative constraints on occurrences with duration) can be converted to a case I ATG (relative constraints on instantaneous events),[12] a single ATG whose nodes are instantaneous events, suffices for the representation of a mixed set of relative constraints between occurrences with duration and metric constraints between the starts and ends of such occurrences.

In HDP the temporal constraints of causal relations are more complicated than simple delays between the starts of causes and effects. Here the existence of some occurrence is refined into interval *onset* and interval *observable*; possible durations for these intervals are expressed probabilistically. *Delay* is the period from the start of the onset of a cause to the start of the observability of its effect. *Persist* is the period from the end of the observability of a cause to the end of the observability of its effect. As before there are a number of general constraints or interpretation rules that can be instantiated with respect to specific pairs of causes and effects. The metric constraints can be represented as a case II ATG while the relative constraints as a case III ATG. In the latter case occurrences are decomposed in accordance with their onset and observability periods (see Figure 6.22). The case III ATG could be translated into a case I ATG as explained above. The case I and II ATGs can then be merged into a single ATG with mixed constraints. This can be handled as discussed above. Alternatively the case II and III ATGs can be handled separately. This means

---

[12] Any relative relation from Allen's set can be translated into a set of relative relations between the starts and ends of the given occurrences.

checking each one separately for consistency and then checking across the two for consistency.

The temporal constraints arising in the time-object based models of disorders are also of a mixed nature, requiring either separate ATGs or a single ATG with hybrid labels. However, there is another matter here, that of mixed granularities (the general problem of temporal constraint networks with granularities has been addressed by Bettini et al (2000)[28]. The temporal constraints, both metric and relative, are granularity dependent. Translating everything to some base granularity is probably not a viable option as it defeats the purpose of having mixed granularities in the first instance, i.e. it throws away the abstractions provided by the multiplicity of granularities and the need for such abstractions. Thus the multiplicity of granularities should be better addressed by having separate ATGs for the different granularities as well as mechanisms for translating constraints across the different granularity spaces. Consistency rules across granularities would also need to be defined. For example would it be consistent to say that two temporal entities overlap at some granularity but that they are equal at some finer granularity? All these are topics for further research.

Let us now return to the problem of deciding whether some constraint is satisfied by a given set of mutually consistent constraints. The solution to this problem given in Chapter 2 is based on the assumption that all constraints, queried and given, are of the same form. Here we relax this assumption and say that the queried constraint qc could be of a different form to the ATG representing the given constraints, e.g. the qc could be $n_i\{before, meets\}n_j$ while the given constraints form a case II ATG (metric constraints at the granularity of days on instantaneous events). Such a heterogeneous situation could be handled by converting qc to the form of the ATG and proceeding in the same manner as before. In this example the translation of the qc proceeds as follows:

$$n_i\{before, meets\}n_j \longrightarrow$$
$$n_i\{before\}n_j \vee n_j\{meets\}n_j \longrightarrow$$
$$n_i^e\{+\infty, ..., 1\}n_j^s \vee n_i^e\{0\}n_j^s \longrightarrow$$
$$n_i^e\{+\infty, ..., 1, 0\}n_j^s$$

where n^s/n^e denotes the start/end of n.

Finally, a minimal ATG representing the model of some disorder encompasses all feasible temporal scenarios about the evolution of the disorder. The specific scenarios can be obtained by a topological sorting of the ATG.

Summary

Computer-based support for the task of diagnostic problem solving has attracted considerable research work in the field of Artificial Intelligence, in general, and expert systems in particular, for a number of years now. Diagnostic problem solving with respect to medical problems is by and large more challenging than the troubleshooting of artificial devices due to the general complexity of medical problems.

The topic of this chapter is temporal clinical diagnosis. In the early efforts, time was ignored but in recent years there is increased effort in modeling time as a necessary and critical component of diagnostic reasoning. Temporal considerations participate actively in the formation and evaluation of diagnostic hypotheses, thus representing an integral aspect of the relevant abductive reasoning processes.

In this chapter we have used a high level analytical framework, the Causal-Temporal-Action Model (C-T-A Model), for comparing a number of approaches to temporal clinical diagnosis. Causality is necessary if we want to move away for the simplest, associational representation of disorders, and be able to capture the evolution of disorders in terms of internal states. In addition, knowledge of the effects (both positive and negative) of therapeutic actions (or interventions in general) is necessary in order to be able to interpret situations (patient states, hypothesized diagnoses, etc) in a context-sensitive fashion. Hence the C-T-A model advocates both the explicit modeling of causality in some form or another, as well as the modeling of actions. Time is the part of our analytical model that glues everything together, since actions embody causal mechanisms that are set in operation in a time-related fashion, changes and dynamic situations are inherently causal in nature and time is intrinsically related to causality. Uncertainty is a feature that characterizes all five planes of the C-T-A Model. Since the focus of our attention in this chapter is time and temporal reasoning, we have only touched upon the matter of uncertainty in relation to time.

The sample of proposed approaches discussed in this chapter is by no means exhaustive but it is adequately illustrative of the various issues under consideration. Through this discussion it transpires that in systems where causality has a central position, the diagnostic knowledge is often represented in terms of a single (acyclic) causal network, while in systems that use associational representations, each disorder is modeled separately. In the first case, the temporal constraints refer mainly to delays between causes and effects, but also to the persistence of states. In the second case, the temporal constraints relate, temporally, pairs of (instantaneous) events marking begins or ends of happenings, etc. The proposed time-object based framework for abductive diagnostic reasoning aims to model occurrences as dynamic entities embodying time as an integral aspect and allows the viewing of such occurrences from different conceptual temporal contexts referring to different granularities. This way the multiplicity of granularities is naturally interwoven into the framework and compound time-objects can be decomposed at finer levels or aggregated at grosser levels thus providing the necessary detail or relevant abstractions from the appropriate temporal contexts. Overall, this framework emphasises the need for higher power of expression, allowing the expression of mixed temporal constraints, i.e., both metric and relative constraints.

In order to bring out more clearly the differences between the different approaches regarding the representation of temporal constraints, the chapter rounds off the discussion by including a section on temporal constraints, where a number of specific instantiations (of relevance to clinical diagnosis) of the Abstract Temporal Graph (ATG) structure introduced in Chapter 2 are presented and discussed.

In concluding this chapter it is fair to say that the work done so far in temporal clinical diagnosis is substantial. However, more effort is called for, in order to fully address the temporal needs of realistic medical diagnostic problems and to use time in the derivation of solutions in a computationally efficient and effective manner. A number of important topics such as multiplicity of granularities, repetition, compound occurrences, and the efficient use of temporal constraints in the formation of hypotheses warrant further research effort. It is satisfying to note that such research effort is indeed being expended.

Bibliographic Notes

The literature on diagnostic problem solving is extensive given that the computer-based modeling of diagnostic reasoning attracted substantial interest in the AI community from the early days and continues to be an area of interest. Clinical diagnosis represents a major subset of these efforts. This chapter focused on the temporal aspects of clinical diagnostic systems by overviewing a representative sample of such systems. In order to glean out all the relevant details, the interested reader should refer to the original sources cited. In addition, a recent comprehensive review on temporal reasoning for decision support in medicine that covers the tasks of diagnosis, monitoring and therapy planning is given in [17]. Finally, the use of fuzzy theory in temporal model-based diagnosis for medical domains is attracting further interest (see for example [293] in conjunction with a historical account on the foundations of fuzzy reasoning in medical diagnosis [344]).

Problems

6.1. In the early diagnostic systems, medical and other, temporal reasoning was not included in any explicit way. Why do you think this was so? As a project you can investigate the reasoning of pioneering medical diagnostic systems and more recent systems based on them, trying to see how temporal considerations have gradually filtered in.

6.2. Implement the minimization algorithms of Chapter 2 and apply them to a number of example ATGs.

6.3. Try to define some disorder process (either a common disorder you are familiar with or some pseudo disorder) emphasizing its temporal aspects. If you cannot think of a disorder process, you may try to define some process that occurs naturally in human beings. The next step is to try and model the identified temporal constraints in terms of the various types of ATGs discussed in Section 6.9 of this chapter.

6.4. Write a small essay identifying the kind of diagnostic situations for which temporal reasoning would not be called for. This is the so called *snapshot diagnosis* that features as the simplest type of diagnosis in Brusoni et.al.'s categorization of temporal diagnosis [42]. Read about the other types of temporal diagnosis, as well as the distinction drawn by these authors between *temporal consistency* and *temporal entailment*, and try to relate these to simple medical diagnostic problems.

6.5. Think of various ways in which time can participate in the (abductive) generation of diagnostic hypotheses. Consider both simple and complex hypotheses. In the case of complex hypotheses argue in favour and against, sequential and parallel reasoning.

6.6. Argue in favour and against the following representations for diagnostic knowledge, taking into consideration time and the potential need for recurring situations. (a) a single causal network, and (b) separate models for each disorder.

6.7. The research work for the fuzzy extensions of t-PCT has considered the problem of dynamically determining the speed of evolution of a hypothesized/confirmed disorder in some patient, so that more accurate predictions/conclusions may be drawn for the particular patient. Think about this problem and of ways of dealing with it, in particular for domains where causal temporal knowledge is available. You may also read about the proposal of the fuzzy t-PCT authors, the application of which does not require causal knowledge at the level of internal states.

6.8. Assuming we have detailed causal-temporal models of disorder processes and (therapeutic) interventions, define a theoretical framework for the dynamic derivation of potential interactions between a set of instantiations of disorders and/or external interventions. Such a framework would be very useful for deriving the potential behaviour of a complex situation arising from multiple disorders and external contextual factors.

6.9. The C-T-A Model is a high level analytical framework for temporal diagnosis that takes into consideration contextual factors about the execution of actions. This model can be refined in different ways, both by focusing within a particular plane, or the relations between planes, e.g. by defining relations between disorders, or modeling the uncertainty within or between planes. Think of different ways of refining this model, with respect to diagnostic domains you may be familiar with. In particular, think of ways of refining the model by taking into consideration general issues of uncertainty both within and between the various planes.

6.10. Finally, try to formulate your own framework for temporal clinical diagnosis that includes both the generation and evaluation of temporal diagnostic hypotheses.

Chapter 7
Automated Support to Clinical Guidelines and Care Plans

Overview

Clinical guidelines are a powerful method for standardization and uniform improvement of the quality of medical care. Clinical guidelines are often best viewed as a set of schematic plans, at varying levels of abstraction and detail, for screening, diagnosis, or management of patients who have a particular clinical problem (e.g., fever of unknown origin) or condition (e.g., insulin-dependent diabetes). Clinical guidelines typically represent the consensus of an expert panel or a professional clinical organization, and, as much as possible, are based on the best evidence available. However, unless automated support is provided, guidelines are not easily accessible, nor applicable, at the point of care. As we shall see, reasoning about time-oriented data and actions is essential for guideline-based care.

In this chapter, we will discuss in detail the tasks involved in providing automated support to guideline-based care, survey several of the major current approaches, and exemplify them by presenting one of them in detail. We will also discuss the issue of how to convert text-based guidelines into a formal executable format, and conclude with several insights into the future.

Structure of the Chapter

We shall start by explaining what are the main tasks involved in providing support for guideline-based care, and continue by briefly reviewing several of the major approaches currently available for the representation of clinical guidelines, in particular, those that need to be executed over significant time periods. We will then exemplify these notions by focusing on a particular approach, the *Asgaard* project and the intention-based *Asbru* guideline-representation language. We shall also discuss in detail one particular but rather generic approach, the *DeGeL* framework, to the pressing problem of how to transform masses of clinical guidelines from free

C. Combi et al., *Temporal Information Systems in Medicine*,
DOI 10.1007/978-1-4419-6543-1_7, © Springer Science+Business Media, LLC 2010

text to a formal machine-comprehensible representation, and store, retrieve, browse, maintain, and apply them conveniently, using a distributed digital library architecture.

We will conclude with a brief discussion and several insights regarding future trends in providing automated support to guideline-based care.

Keywords

Clinical guidelines, protocols, care plans, automated therapy, planning.

7.1 Clinical Guidelines: an Introduction

Clinical guidelines (sometimes referred to as *Care Plans*) are a powerful method for standardization and uniform improvement of the quality of medical care. According the *Institute of Medicine's (IOM)* definition, *clinical guidelines* are "systematically developed statements to assist practitioner and patient decisions about appropriate health care for specific clinical circumstances" [141]. Clinical guidelines typically represent a medical expert consensus regarding the screening, diagnosis or management, over either limited or extended periods of time, of patients who have a particular clinical problem, need, or condition (e.g., fever of unknown origin; or therapy of insulin-dependent diabetes). In this chapter, we will be focusing mainly on guidelines for management of patients over extended periods, namely, on management of chronic patients. (*Clinical protocols* are highly detailed guidelines, often used in areas such as oncology and experimental clinical trials.)

The application of clinical guidelines by care providers typically involves collecting and interpreting considerable amounts of data over time, applying standard therapeutic or diagnostic plans in an episodic fashion, and revising those plans when necessary. Thus, reasoning about time-oriented data and actions is essential for guideline-based care.

Clinical guidelines can be viewed as reusable *skeletal plans* [146], namely a set of plans at varying levels of abstraction and detail, that, when applied to a particular patient, need to be refined by a care provider over significant time periods, while often leaving considerable room for flexibility in the achievement of particular goals. Another view, which we shall dwell upon in more length in this chapter, is that clinical guidelines are a set of *constraints* regarding the *process* of applying the guideline (i.e., care-provider actions) and its desired *outcomes* (i.e., patient states), to which we refer to as process and outcome *intentions*. These constraints are mostly *temporal*, or at least have a significant temporal dimension, since most clinical guidelines concern the care of chronic patients, or at least specify a care plan to be applied over a significant period.

It is now universally agreed that conforming to state-of-the-art guidelines is the best way to improve the quality of medical care, a fact that had been rigorously demonstrated [165], while reducing the escalating costs of medical care. Clinical guidelines are most useful at the point of care (typically, when the care provider has access to the patient's record), such as at the time of order entry by the care provider. In such a context, even simple reminders and alerts have powerful effects, especially in outpatient contexts. In deed, significant enhancement of compliance with preventive-care guidelines, such as the rate of using pneumococcal and influenza vaccinations when appropriate, has been demonstrated, even within an inpatient (hospital) context, by integrating several simple alerts within the hospital's order-entry system [121].

In this chapter, we shall briefly review several of the major approaches currently available for the representation of clinical guidelines that need to be executed over significant time periods, and then focus on a particular one, the *Asgaard* project and the intention-based *Asbru* guideline-representation language, and the knowledge structure underlying it, also known as a guideline *ontology*. Additional information regarding automation of clinical guidelines can be found in a useful primer on computer-based clinical guidelines [400].

Since automated analysis and application of unstructured text-based guidelines is not feasible, due to limitations of current technologies, such automation requires formal representations of clinical guidelines that can be parsed and executed by machines. We call such representations *machine comprehensible*. (The term "comprehension" is used here in a strictly formal sense, not a cognitive one.) We shall therefore discuss in detail one approach, the *DeGeL* framework, to the pressing problem of how to transform the masses of existing clinical guidelines from free text to a formal machine-comprehensible representation. These guidelines are intended to handle multiple different clinical situations; sometimes there exists more than one guideline per situation (e.g., guidelines differ among different professional or national organizations, and there are often slight variations, depending on intended clinical context). Even when the guidelines are transformed to machine-comprehensible formats, they might be represented in multiple different guideline-specification languages. We shall examine the fashion in which the DeGeL architecture handles the tasks of storage, retrieval, browsing, maintenance, and application of these guidelines, using a multiple-ontology, distributed digital library architecture.

We will conclude with a brief discussion and several insights regarding future trends in guideline use.

7.2 The Need for Automated-Support to Clinical Guidelines

Most clinical guidelines are text-based and inaccessible to the physicians who most need them. Even when guidelines exist in electronic format, and even when that format is accessible online, physicians rarely have the time and means to decide which of the multiple guidelines best pertains to their patient, and, if so, exactly what does applying that guideline to the particular patient entail. Furthermore, recent health-care organizational and professional developments often reduce guideline accessibility, by creating a significant information overload on health care professionals. These professionals need to process more data then ever, in continuously shortening periods of time. Similar considerations apply to the task of assessing the quality of clinical-guideline application.

To support the needs of health-care providers as well as administrators, and ensure continuous quality of care, more sophisticated information processing tools are needed. Due to limitations of state-of-the-art technologies, analyzing unstructured text-based guidelines is not feasible. Thus, there is an urgent need to facilitate guideline dissemination and application using machine-readable representations and automated computational methods.

Several of the major tasks involved in guideline-based care, which would benefit from automated support, include specification (authoring) and maintenance of clinical guidelines, retrieval of guidelines appropriate to each patient, runtime application of guidelines, and retrospective assessment of the quality of the application of the guidelines.

Table 7.1 describes these and other tasks and the knowledge required to perform them.

Supporting guideline-based care implies creation of a *dialog* between a care provider and an automated support system, each of which has its relative strengths. For example, physicians have better access to certain types of patient-specific clinical information (such as their odor, skin appearance, and mental state) and to general medical and commonsense knowledge. Automated systems have better and more accurate access to guideline specifications and detect more easily pre-specified complex temporal patterns in the patient's data. Thus, the key word in supporting guideline-based care is *synergy*.

7.3 Automation of Clinical Guidelines

Several approaches to the support of guideline-based care permit hypertext browsing of guidelines via the World Wide Web [20] but do not directly use the patient's electronic medical record.

Several simplified approaches to the task of supporting guideline-based care that do use the patient's data, encode guidelines as elementary state-transition tables or as situation-action rules dependent on the electronic medical record [370]. The *Arden Syntax* [181, 394], which will be described later in this chapter in more detail, encodes medical knowledge as a set of individual rules. It is quite useful for representing alerts and reminders and for encoding various conditions (e.g., eligibility) as part of more extended frameworks for guideline-based care. *Augmented decision tables* [371] extend the rule-based framework with additional knowledge, such as probabilities and utilities.

Table 7.1 The tasks involved in guideline-based care and the knowledge required for performing them.

Task	Questions to be answered	Required Knowledge
Verification of a guideline	Are the intended plans, or processes (potentially) achievable by following the prescribed *actions*? (*a syntactic check*)	Prescribed actions; intended overall action pattern (i.e., the intended plan, or process)
Validation of a guideline	Are the intended *outcomes* (potentially) achievable by the prescribed *actions* and intended *plan*? (*a semantic check*)	Prescribed actions, intended overall action pattern (process); intended outcomes; action/plan effects
Applicability of guidelines	What guidelines or protocols are applicable at this time to this patient?	Filter and setup preconditions; overall intended outcomes; the patient's state
Eligibility of patients	Which patients are currently eligible for the given guideline?	Filter and setup preconditions; overall intended outcomes; the patients states
Application of a guideline	What should be done at this time according to the guideline's prescribed actions?	Prescribed actions and their filter and setup preconditions; suspension, restart, completion, and abort conditions; the patient's state
Recognition of the care-provider's intentions	Why is the care provider executing a particular set of actions, especially if those deviate from the guideline's prescribed actions?	Executed actions and their abstraction to executed plans; process and outcome intentions; the patient's state; action/plan effects; revision strategies; preferences
Quality assessment	Is the care provider deviating from the prescribed actions or intended plan? Are the deviating actions compatible with the author's plan and state intentions?	Executed actions and their abstraction to plans; action and state intentions of the original plan; the patient's state; action/plan effects; revision strategies; preferences
Evaluation of a guideline	Is the guideline working?	Intermediate/overall outcome intentions; the patient's state; intermediate/overall process intentions; executed actions and plans
Modification of a potential (candidate) or currently running (being applied) guideline or action	What alternative guidelines or actions are relevant at this time for achieving a given outcome or process intention?	Intermediate/overall outcome intentions; action/plan effects; filter and setup preconditions; revision strategies; preferences; the patient's state

However, representing complex, long-term care plans that should be applied over significant time periods, and which include multiple components (in particular, other plans) requires more expressive approaches, which will be discussed in detail in this chapter.

7.3.1 Prescriptive versus Critiquing Approaches

Most of the approaches described in this chapter can be described as being *prescriptive* in nature, specifying *what* actions need to be performed and *how*. However, several systems, such as Miller's *Attending* system [264], have used a *critiquing* approach, in which the physician suggests a specific therapy plan and gets feedback from the program. The advantage of a critiquing approach is that the system's output is focused on the user's input plan and directly relevant to it. It is particularly relevant when the user has significant domain expertise, but might still need additional comments (e.g., a resident in anesthesiology suggesting her plan for inducing anesthesia in a particular surgery patient). However, the critiquing approach does require significant user input and, from the psychological point of view, needs to be carefully worded.

As we shall see, The *Asgaard* project [355], an example framework described in detail later in this chapter, uses both a prescriptive methodology for specification and application of prescribed interventions, and a critiquing methodology for retrospective quality assessment. The *Asbru* language developed within the Asgaard project includes both prescriptive and critiquing knowledge roles.

7.3.2 The Arden Syntax

The *Arden Syntax* [181, 394] is named after the Arden Homestead in NY, in which representatives from ten universities discussed sharing of medical knowledge. It represents medical knowledge as independent units called *Medical Logical Modules (MLMs)*, which use a Pascal-like programming language to encode highly specific rules, grounded in the local institution's database schema (Table 7.2). The general medical logic (encoded in the Arden syntax) is separated from the institution-specific component (encoded in the local query language and terms).

Table 7.2 Example of a Medical Logical Module (MLM) in the Arden syntax.

- *Maintenance*:
    - title: Agranulocytosis and trimethoprim/sulfamethoxazole
    - author: Dr. Bonzo
- *Library*:
    - keywords: granulocytopenia; agranulocytosis; trimethoprim; sulfamethoxazole
    - citations: 1. Anti-infective drug use ... Archives of Internal Medicine 1989; 149(5): 1036-40
- *Knowledge*:
    - type: data driven;
    - data:
        · anc:= read last 2 from (query for ANC where it occurred within the past 1 week);
        · pt_taking_tms := read exist query for TMS order;
        · evoke: on storage of ANC;
    - logic:
        · if pt_taking_tms and last anc < 1000 and decrease of anc > 0 then conclude true else conclude false;
    - action:
        · store "Caution: The patient's relative granulocytopenia may be exacerbated by trimethoprim/sulfamethoxazole.";

The actual grounding of the generic terms in MLMs is done by enclosing in curly braces terms in the general medical logic that need to be replaced by local terms, then substituting them at implementation time (at the actual site) by local database queries and expressions. The problem of linking an MLM to a local site (e.g., a local electronic medical record database) has thus become to be known as "the curly brackets problem" [324], although it occurs, of course, in any knowledge-representation format that attempts to be general, but that needs to be applied within a particular site or a new environment.

The Arden syntax is an established (by the *American Society of Testing and Materials*, or *ASTM*) medical-knowledge representation standard. Since 1999, it is maintained by the HL7 organization, which includes academic center and various vendors, and focuses on standards for sharing clinical information. Several MLM libraries exist.

There are several issues, however, arising when using the Arden syntax:

- Difficulty in reuse of general clinical knowledge within different contexts, even within a single system (e.g., what is "mild anemia?") leading to difficulties in maintenance of the knowledge [380].
- Sharing problems, as were encountered when MLMs were transported between the Columbia Presbyterian medical center, New York City, New York and the LDS medical center, Salt Lake City, Utah. Most of the difficulties were due to local query and vocabulary differences (the "curly braces" problem) as well as local practices [324].
- Difficulty in representation of continuous therapy plans: Each MLM represents a well-defined, independent rule, typically intended to fire in a data-driven manner when relevant, and is thus not suitable for representing a long-term therapy plan.
- Lack of ability to represent and reuse higher, meta-level problem-solving knowledge (e.g., temporal-abstraction knowledge of the type discussed in earlier chapters).
- Besides the "curly braces" problem (which actually is not unique to MLMs, but is more emphasized in the Arden Syntax low-level representation format), another potential limitation is the lack of explicit notification mechanisms for alerts and reminders; like database queries, notification is contained in curly braces in an MLM and left to local implementers. Explicit notification mechanisms in the Arden Syntax itself may be a part of a future edition.

In general, rule-based approaches to the representation of clinical guidelines, such as the Arden Syntax, typically do not include an intuitive representation of the guideline's clinical logic, have no semantics for the different types of clinical knowledge represented, lack the ability to easily represent and reuse guidelines and guideline components as well as higher, meta-level problem-solving knowledge, do not allow for inherent, intended, ambiguities in the therapy algorithm (such as when considerations exist for and against different therapy options, which need to be evaluated by the attending physician, or when the patient's preferences need to be explicitly considered), and do not support application of guidelines over extended periods of time, as is necessary to support the care of chronic patients guideline-based care over extended periods in automated fashion.

Nevertheless, MLMs are an excellent option when simple, one-time, reminders and alerts need to be written and used, without the heavier machinery of more complex guideline representations, and in that sense they complement more expressive guideline-representation formats [302, 304].

7.4 Automation of Complex, Longitudinal Guideline-Based Care

During the past 20 years, there have been several efforts to support complex guideline-based care over time in automated fashion. Before we delve into a particular approach (the Asgaard project and its Asbru ontology) in the next section, we present a brief summary of several of the influential architectures and representation languages. A good review comparing six of the main approaches along multiple semantic and syntactic dimensions is by [305].

7.4.1 The EON Project

Stanford University's *EON* project [278, 407, 408] is a general, client-server architecture that developers can use to build systems that support automated reasoning about guideline-directed care. The EON framework has gradually evolved from Stanford University's domain-specific *ONCOCIN* [406] project, which applied oncology protocols at the Stanford lymphoma clinic, and *T-HELPER* project, which encoded and applied therapeutic and prophylactic guidelines in the AIDS domain in a large silicone-valley clinic [277]. The result is a sophisticated type of "middleware, " that is, a set of software components designed for incorporation within other software systems (e.g., hospital information systems) for various applications of guideline-based care.

The EON framework includes reusable problem-solving components that have specific functions (e.g., planning, classification of time-oriented clinical data), such as:

- A *therapy planner*, based on a version of the *episodic skeletal-plan-refinement* problem-solving method [406], which incrementally refines a predefined therapy plan over long time periods, given the patient's data;
- A *temporal mediator* [281], which can answer queries to a patient database about either primitive time-oriented patient data or their abstractions; the mediator includes two components:

  - The *RASTA* temporal-abstraction module [268], a simplified version of the *RÉSUMÉ* temporal-abstraction system [359];

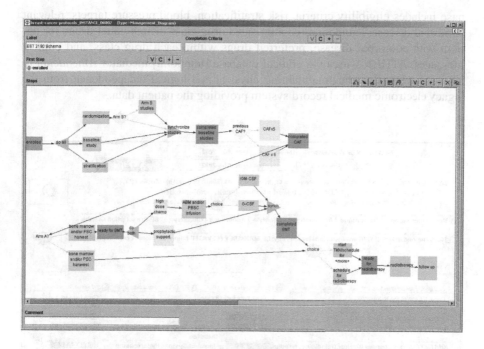

Fig. 7.1 A graphical interface generated by the Protégé-2000 system for specification of clinical guidelines in the EON framework. In this case, an experimental protocol for bone-marrow transplantation is being defined by a medical expert or a knowledge engineer. Green boxes, such as "high-dose chemo," represent actions; pink boxes, such as "ready for radiotherapy," represent patient states; yellow circles, such as "arm A?" represent decision (choice) points. (Figure supplied courtesy of S. Tu.)

- The *Chronus-II* temporal-maintenance system [284], an extension of the *Chronus* System [111].
- An *eligibility-determination* module, *Yenta*, which matches patients and their characteristics to eligibility criteria of clinical guidelines.
- A *knowledge-acquisition* module to acquire and maintain clinical guidelines (Figure 7.1). The module was designed by using tools from the *Protégé* project's knowledge-acquisition framework [166]. The Protégé project focuses on automated generation of graphical knowledge-acquisition tools, given the ontology of the problem-solving method (in this case, the EON guideline ontology).
- A shared knowledge base, maintained by the Protégé-based knowledge-acquisition module, of clinical guidelines and general medical concepts.

A specific successful instance of the EON architecture is the *ATHENA* system for management of hypertension, originally installed at the Palo Alto Veterans Administration Health Care System [155, 156] (Figure 7.2). ATHENA operationalizes guidelines for hypertension. Its main user interface is an intuitive one (see Figure 7.2), and its knowledge base is maintained using the Protégé-2000 system. The knowledge base includes eligibility criteria, risk stratification, blood pressure targets, relevant comorbid diseases, guideline-recommended drug class or individual drugs for patient with comorbid disease, preferred drugs within each drug class, and clinical messages. ATHENA uses the clinical database [temporal] mediator *Athenaeum* to access patient data, thus maintaining physical and logical independence from the legacy electronic medical record system providing the patient data.

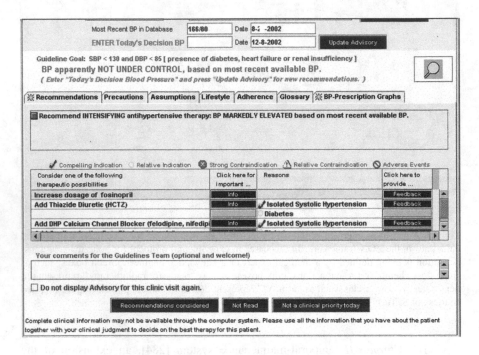

Fig. 7.2 An example of the *ATHENA* system interface. Having entered the patient's most recent blood pressure, the care provider is offered advice based on a relevant guideline for management of hypertension. (Figure supplied courtesy of M. Goldstein.)

7.4.2 The PROforma Methodology

The *Proforma* framework was developed by Fox et al. [145] at the *Advanced Computation Laboratory* (ACL) of *Cancer Research UK* (formerly the *Cancer Research Campaign* and *Imperial Cancer Research Fund*, London, UK). The underlying language, technology and illustrative applications are comprehensively described in a book dedicated to safety-critical applications in general, and in medicine in particular, focusing on the *Proforma* framework, by Fox and Das [144].

The basic modeling approach used in *Proforma* integrates object-oriented (currently, implemented in the Java language) and logic programming (currently, implemented in the Prolog language) techniques. The *Proforma* approach explores the expressivity of an intentionally minimal set of constructs, all subtypes of the Task entity: *Plans, Actions, Decisions*, and *Enquiries*. All tasks inherit the generic attributes *name, caption, description, goals, preconditions, trigger conditions*, and *postconditions*. Attributes of the four task subclasses, such as the *components* and *scheduling constraints* of a Plan, or the *candidates* of a Decision, are specific to that subclass, and are only inherited by tasks of that subclass type. Any specific clinical task is seen as an instance, or a specialization, of a more general task class. For example, Plans may be specialized into *research protocols* and *routine guidelines*, while Decisions can be specialized into *diagnosis* and *treatment* decisions. Actions are typically clinical procedures (such as the administration of an injection) which need to be carried out. Enquiries are actions returning required information; typically requests for information or data from the user, such as laboratory-test results. The attributes of each task subtype are summarized in a template.

PRO*forma* supports the definition of clinical guidelines and protocols in terms of a set of tasks that can be composed into networks representing plans or procedures carried out over time, representing the high level structure of a guideline, and logical constructs (such as pre- and post-conditions) which allow the details of each task and inter-relationships between tasks to be defined using the task-specific templates.

A graphical editor for PROforma guidelines has been developed (Figure 7.3), and PROforma applications demonstrating guideline enactment over the World Wide Web are currently under development.

7.4.3 The Guideline Interchange Format (GLIF)

The *InterMed* multiple academic-center (Columbia, Harvard, Stanford) collaboration effort led to the development of a guideline-specification language named the *Guideline Interchange Format*, or *GLIF* [286, 303, 302, 304, 38].

The GLIF ontology attempted to integrate key lessons from the Arden syntax and MLMs, the EON project, the Asgaard project and the Asbru language, and other guideline-representation projects. It was intended to enable representation of complex multi-step plans with branching logic, as well as simpler alerts, and therapeutic as well as diagnostic guidelines. The current core language, GLIF-3 [303, 38] includes several core ontological entities in the context of a flowchart, such as a *patient-state* (determination) step, an *action* (intervention) step, a *branch* (decision) step, a *case* (multiple branching) step, and a *synchronization* step. Figure 7.4 shows a typical set of screen shots while editing a cough-management guideline [183] using the GLIF-3 ontology and tools generated by the Protégé-2000 framework.

The GLIF default medical data model is based on the HL7 *reference information model (RIM)* [342]. GLIF includes a formal syntax for querying about patient states. Originally, queries included only free-text strings, and were later extended

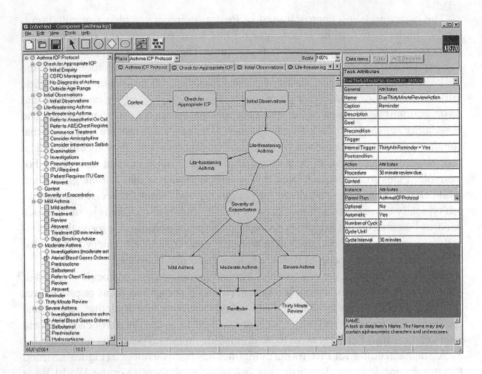

Fig. 7.3 An editing environment for Proforma guidelines. The user can select several types of ontological entities, such as rectangles (plans) and diamonds (enquiries). (Figure supplied courtesy of J.Fox.)

to use the Arden syntax; that representation was further extended to the *Guideline Expression Language* (*GEL*) [304], and eventually to the object-oriented *GELLO* language [285].

GLIF can represent temporal operators from the Arden syntax, but not complex temporal expressions or temporal abstractions, and does not assume the ability to automatically query patient records for such abstractions. The handling of uncertainty regarding patient data is not incorporated into the core language either.

An execution engine that simulates guidelines represented in the GLIF language using an internal patient data source was developed at Columbia University [420].

Thus, a somewhat *hybrid* structure (see the section on the DeGeL project) exists in the GLIF3 language and tools: A top-level graphical representation of the guideline created by the physicians with the help of knowledge engineers; a computable representation created by informaticians; and a bottom-level representation, customized to the local needs of different medical centers, with respect to both guideline logic, local preferences, and local database schemas and data vocabularies [303, 302, 304].

An example of using the earlier GLIF-2 ontology for representation of complex clinical guidelines was reported by the Boston-based Partner's Healthcare System

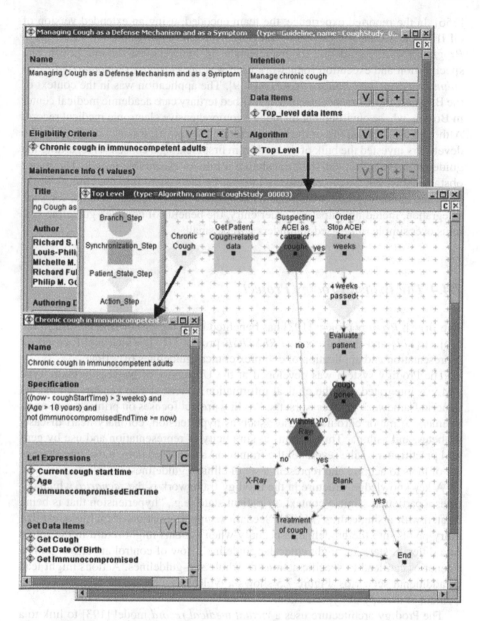

Fig. 7.4 Part of a cough guideline authored in the GLIF-3 ontology. Shown the top of the screen (in a Protégé-2000 generated interface) is a guideline instance. The details of its Algorithm slot are shown on the screenshot on the bottom right. The details of one of the steps of the algorithm are shown in the screen shot on the bottom left. Note the various types of ontological entities, such as a patient-state step (detection/determination), action step and branch step (decision). (Courtesy of M. Peleg).

[256]. In the reported experience, the team encoded, using an extended version of GLIF-2, the secondary prevention portion of the *National Cholesterol Education Program* (*NCEP*) guideline for management of hypercholesterolemia [280]. The specification and execution framework used was the *Partner's Computerized Algorithm Processor and Editor* (*PCAPE*) [439]. The application was in the context of the Brigham and Women's hospital, a 700-bed tertiary care academic medical center in Boston, whose environment includes a comprehensive electronic medical record. Although the expressivity of the representation framework was not an obstacle, the developers invested the bulk of their effort in first creating a consensus regarding the guideline's semantics, and then translating it into an executable format. The major obstacle, however, was in effectively integrating the system into the clinical workflow; the authors' opinion was that without more sophisticated methods for such integration, including outpatient order entry, the benefits of complex guideline systems over simple rule-based reminders will be small.

7.4.4 The British Prodigy Project

The British *Prodigy* project [192] has benefited from the earlier British *DILEMMA* project [177] and the European *PRESTIGE* project [161], which had focused on representation and application of complex clinical guidelines over time.

The Prodigy system was developed at the University of Newcastle upon Tyne and was integrated into the clinical information system of two vendors. Unlike its generic-framework predecessors, the Prodigy project focuses on primary-care management of major chronic diseases, such as hypertension, coronary heart disease, diabetes, and asthma. Thus, it aims at simplicity of representation and use by general practitioners. Like several of the frameworks discussed here, it uses the Protégé set of tools to acquire and represent a set of clinical guidelines.

A key knowledge structure in the Prodigy framework is the *scenario*, which defines a particular clinical context, or patient state (e.g., "hypertension that is being treated with a single drug") [192]. Scenarios support creation of multiple, explicit entry points into the guideline, especially when patients might return in the future in a different state, or might enter the guideline's flow of control at various points. Scenarios can lead to specific actions or whole sub-guidelines. Actions might lead to additional scenarios. Figure 7.5 shows a typical Prodigy set of scenarios in the domain of therapy of hypertension.

The Prodigy architecture uses a *virtual medical record* model [193] to link to a specific electronic patient record structure.

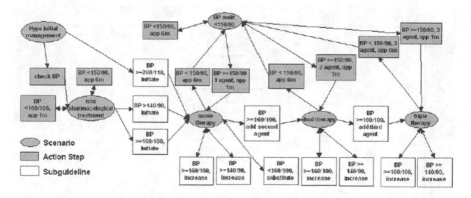

Fig. 7.5 Clinical scenarios in the Prodigy project, representing in this case the treatment of hypertension. (Modified from [192].)

7.4.5 The Italian GUIDE Project

The *GUIDE* project [326] is part of a more general framework, *Careflow*, developed at the University of Pavia, Italy, for modeling and applying clinical guidelines in the broader context of general medical care.

Rather uniquely, relative to the other approaches described in this section, the GUIDE framework focuses on two important aspects:

1. The integration of guidelines into an *organizational workflow*, including and emphasis on the local organizational model and on modeling the available resources, and
2. Supporting the use of decision-analytical models, such as decision trees and influence diagrams, in addition to standard procedural models.

The underlying Guideline representation in the GUIDE framework uses the *Petri net* formalism for modeling concurrent processes, to model clinical processes [325]. The GUIDE project has extended Petri Nets to support improved modeling of time, data and plan hierarchies. The GUIDE model uses the underlying Petri Net model to be able to support the representation of sequential, parallel and iterative control structures.

As part of the overall project, a commercial system was used to develop the Careflow management system; business-process management tools were used to support extensive simulation studies to test the careflow models embedded in that system.

The graphical GUIDE authoring tool (Figure 7.6) enables designers to interactively author a guideline flowchart as a Petri net. Computational tools enable simulation of the resulting guideline using the Petri net semantics.

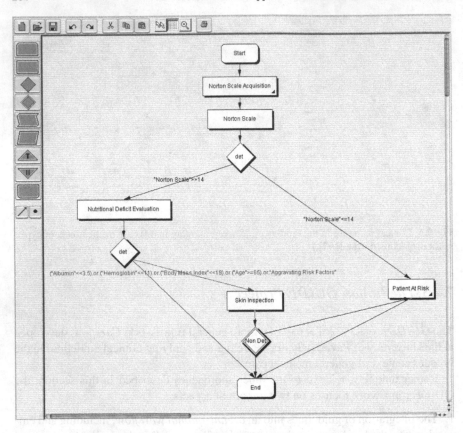

Fig. 7.6 The highest-level page of a guideline for pressure-ulcer prevention in the GUIDE system. The guideline's target audience is nurses. First, the nurse must evaluate for each patient a clinical measure called the *Norton scale*. The small triangle in the *Norton Scale Acquisition box* indicates that this box may be expanded to see the Norton scale details. The *Norton Scale box* represents the calculation of the Norton Scale index, based on the previously acquired scale values. Then there is a deterministic decision box: if the Norton index <= 14, the patient is defined as "at risk", and clicking on the box *Patient at risk* shows the detailed treatment. If the Norton index > 14, other risk factors must be evaluated, starting with the nutritional status. If low albumin, low hemoglobin, low body mass index, old age, or other particular diseases are present (*Aggravating risk factors* is an abstraction based on presence of diabetes, cancer, etc.), a skin inspection is suggested. After the skin inspection, the nurse herself decides if the patient has to be considered at risk or not, indicated by the *non det*, or *non- deterministic* (one of several non predetermined options) diamond. (Courtesy of S. Quaglini).

7.4.6 The GEM Ontology

A detailed representation format, *GEM* [372], and an associated mark-up tool, *GEM CUTTER* [202], both developed at Yale University, New Haven, CT, USA, enable structuring of a text document containing a clinical guideline as an extensible markup language (XML) document, using a well-defined XML schema. GEM is mainly a guideline-documentation ontology. The GEM markup tool is an application running on a stand-alone computer, and not intended for distributed use. The GEM framework was not intended to support computational tools that can interpret the resulting semi-structured text, since it does not include a formal computational model.

The feasibility of creating an XML-based implementation framework for GEM-encoded guidelines has been demonstrated [153], although it does not seem to support extended care over significant time periods, due to the lack of persistent-memory mechanisms, interaction with an electronic patient record, and complex control structures.

However, the carefully designed XML schema underlying GEM has been found quite useful for documenting clinical guidelines and as a semi-structured format for sharing guidelines.

7.4.7 Commercial Approaches

Other approaches to support guideline use at the point of care enable a Web-based connection from an electronic patient record to an HTML-based set of rules, such as is done in the *Active Guidelines* model [397], which is embedded in a commercial electronic medical record and order-entry system developed by the Epic Co. In that model, access to a specialized set of hidden tags, available only when an *active guideline interpreter* (*AGI*) module is part of the environment, enables the user to exploit the tags so as to automatically add them at the point of care to the electronic order.

However, such approaches have no standardized, sharable, machine-readable representation of guidelines that can support multiple tasks such as automated application and quality assurance, and are not intended for representation of complex multi-step care plans over time. They can be viewed as special cases of alerts and reminders that are incorporated into an order-entry system and, from that point of view, certainly have a value in well-defined applications.

7.5 The Asgaard Project and the Asbru Language

In this section, we will take a detailed look at one particular, highly expressive approach for guideline-based-care support, which demonstrates many of the issues discussed in the previous sections. In the next section we will see how that approach was used within a larger-scale framework for specification, retrieval, and application of clinical guidelines.

The research over the past two decades has demonstrated that automating complex guideline-based care requires the use of an underlying richly expressive, machine-readable formal language, specific to that task, which enables specification of (1) multiple types of clinical actions over time (e.g., sequential, parallel, periodic) and associated temporal and other constraints (e.g., administration between 8:30 to 10:00 a.m.), and (2) the intermediate and overall clinical-processes and patient-outcome goals of the therapy plan, that is, the *process* and *outcome intentions* of the guideline, or the (physician) *action* and (patient) *state* intentions of the guideline [355]. These intentions are in fact *temporal-pattern constraints* (e.g., a process intention to administer regular insulin twice a day; an outcome intention to maintain fasting blood glucose within a certain range over at least 5 days a week) that have individual *weights* signifying their relative importance. Such knowledge is necessary to determine, for instance, whether a care provider is still following most of the guideline, or, at least, its spirit. Such a provider might be applying the guideline in modified fashion, as is, in fact, the case in 50% of inspected guideline applications.

Indeed, precisely such an expressive, intention-oriented language *Asbru*, had been designed within the *Asgaard*[1] project [263, 355, 345]. Figure 7.7 presents the overall Asgaard architecture for automated guideline support.

7.5.1 The Design-Time Versus Execution-Time Intention-Based Model

Underlying the Asgaard project is a set of basic assumptions regarding the typical scenario involved in specification and application of clinical guidelines. During *design time* of a clinical guideline, an *author* (or a committee) designs a guideline (Figure 7.9). The author prescribes (1) *actions* (e.g., administer a certain drug in the morning and in the evening), (2) an *intended plan*-the intended intermediate and overall pattern of actions, that is, the intended process, which might not be obvious from the description of the prescribed actions and is often more flexible than prescription of specific actions (e.g., use some drug from a certain class of drugs, twice a day), and (3) the intended intermediate and overall pattern of *patient states*, or *outcomes* (e.g., morning blood glucose should stay within a certain range). Intentions are temporal patterns of provider actions or patient states, to be achieved, maintained, or avoided.

During execution (application) time, a care provider applies the guideline by performing actions, which are recorded, observed, and abstracted over time into an

---

[1] In Norse mythology, Asgaard was the home and citadel of the gods, corresponding to Mount Olympus in Greek mythology. It was located in the heavens and was accessible only over the rainbow bridge, called Asbru (or Bifrost).

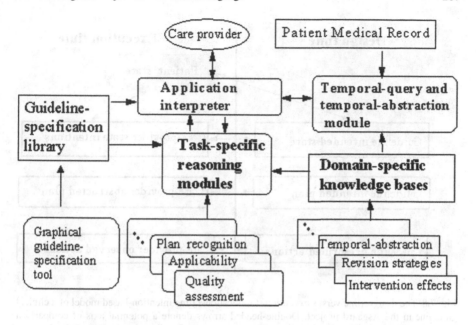

Fig. 7.7 The Asgaard guideline-support architecture. Arrows denote data or knowledge flow.

abstracted plan (see Figure 7.8). The state of the patient also is recorded, observed, and abstracted over time. Finally, the intentions of the care provider might be recorded too-inferred from her actions or explicitly stated by the provider.

7.5.2 Asbru: A Global Ontology for Guideline-Application Tasks

Shahar, Miksch, and Johnson have developed for the purposes of the Asgaard project a language specific to the set of guideline-support tasks and the problem-solving methods performing these tasks, called Asbru [355, 263]. *Asbru* enables a designer to represent a clinical guideline, including all of the knowledge roles useful to one or more of the problem-solving methods performing the various tasks supporting the application of clinical guidelines.

The major features of Asbru are that prescribed actions can be continuous; plans might be executed in parallel, in sequence, in a particular order, or every time measure; temporal scopes and parameters of guideline plans can be flexible, and explicit intentions and preferences can underlie the plan. These features are in contrast to most traditional plan-execution representations, which have significant limitations and are not applicable to dynamic environments such as clinical domains. Medical domains have certain characteristic features: (1) actions and effects are not necessarily instantaneous: actions are often continuous (have duration) and might have

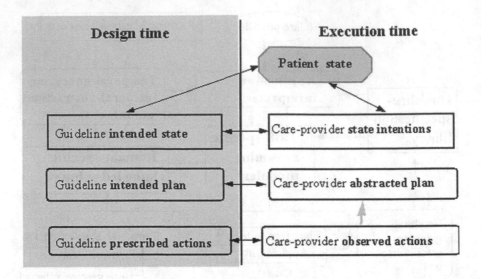

Fig. 7.8 The design-time versus execution- (application-) time intention-based model of a clinical guideline in the Asgaard project. Double-headed arrows denote a potential axis of comparison (e.g., for critiquing purposes) during runtime execution of the clinical guideline. Striped arrows denote an abstracted-into relationship.

delayed effects; (2) goals often have temporal extensions; (3) there is uncertainty regarding the effect of available actions; (4) unobservable, underlying processes determine the observable state of the world; (5) a goal may not be achievable; (6) parallel and periodic execution of plans is common.

The requirements of plan specifications in clinical domains (i.e., guidelines) are a superset of the requirements of typical toy domains used in planning research. The Asgaard team had defined a formal syntax for the Asbru language in Backus-Naur form [263]. The Asbru language combines the flexibility and expressivity of procedural languages (e.g., rules in the Arden syntax) with the semantic clarity of declaratively expressed knowledge roles. These roles (e.g., preferences and intentions) are specific to the ontology of the methods performing the guideline-support tasks.

7.5.2.1 The Asbru Time Annotation

The time annotation used within the Asbru language allows a representation of uncertainty in starting time, ending time, and duration of a time interval. The time annotation supports multiple time lines by providing different reference annotations. The *reference annotation* can be an absolute reference point, a reference point with uncertainty (defined by an uncertainty region), a function (e.g., completion time) of a previously executed plan instance, or a domain-dependent time point variable (e.g., CONCEPTION). Temporal shifts from the reference annotation represent uncertainty in the starting time, the ending time, and the overall duration (Figure 7.9). Thus, the *temporal annotation* represents for each interval the *earliest starting shift* (ESS), the *latest starting shift* (LSS), the *earliest finishing shift* (EFS), the *latest finishing shift* (LFS), the *minimal duration* (MinDu) and the *maximal duration* (MaxDu). Temporal shifts are measured in time units. Thus, a temporal annotation is written as ([ESS, LSS], [EFS, LFS], [MinDu, MaxDu], REFERENCE). All temporal-shift constraints can be unknown (unbound, denoted by an underscore, "_") to allow incomplete time annotations.

To allow temporal repetitions, the Asbru designers have defined the notion of cyclical time points (e.g., MIDNIGHTS, which represents the set of midnights, where each midnight occurs exactly at 0:00 a.m., every 24 hours) and cyclical time annotations (e.g., MORNINGS, which represents a set of mornings, where each morning starts at the earliest at 8:00 a.m., ends at the latest at 11:00 a.m., and lasts at least 30 minutes). In addition, certain short-cuts are allowed, such as for the current time, whatever that time is (using the symbol *NOW*), or the duration of the plan (using the symbol *). Thus, the Asbru notation enables the expression of interval-based intentions, states, and prescribed actions with uncertainty regarding starting, finishing, duration, and the use of absolute, relative, and even cyclical (with a pre-determined granularity) reference annotations.

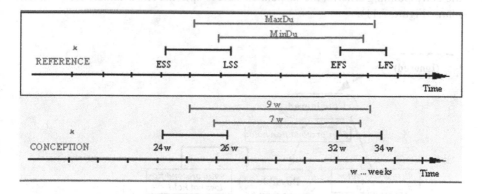

Fig. 7.9 A schematic illustration of the Asbru time annotations. The upper part of the Fig. presents the generic annotation. The lower part shows a particular example representing the time annotation [[24 WEEKS, 26 WEEKS], [32 WEEKS, 34 WEEKS], [7 WEEKS, 9 WEEKS], CONCEPTION]), which means "starts 24 to 26 weeks after conception, ends 32 to 34 weeks after the conception, and lasts 7 to 9 weeks." REFERENCE = reference annotation, ESS = earliest starting shift, LSS = latest starting shift, EFS = earliest finishing shift, LFS = latest finishing shift, MinDu = minimal duration, MaxDu = maximal duration. The annotation is thus ([ESS, LSS], [EFS, LFS], [MinDu, MaxDu], REFERENCE).

7.5.2.2 The Semantics of the Asbru Task-Specific Knowledge Roles

A guideline, or *plan* in the guideline-specification library is composed of a set of plans with arguments and time annotations. A decomposition of a plan into its subplans is always attempted by the runtime application module, unless the plan is not found in the guideline library, thus representing a non-decomposable plan. These runtime semantics can be viewed as a "semantic" (as opposed to a purely syntactic) halting condition, which increases runtime flexibility, since the same plan might imply an atomic action for one clinical site, but might be decomposable into more primitive actions at another clinical site. A non-decomposable plan is executed by the clinical user or by an external call to a computer program. The library includes a set of primitive plans to perform interaction with the user or to retrieve information from the medical patient record (e.g., OBSERVE, GET-PARAMETER, ASK-PARAMETER, DISPLAY, WAIT)). All plans have return values.

Generic library plans that are mentioned as part of an executing guideline have *states* (*considered, possible, rejected,* and *ready*), that determine whether the plan is applicable and whether a plan instance can be created (Figure 7.10). At execution time, a *ready* plan is instantiated. A set of mutually exclusive plan states describes the actual status of the plan instance during execution. Particular *state-transition criteria* (conditions) specify transition between neighboring plan-instance states. Thus, if a plan is *activated*, in can only be *completed, suspended,* or *aborted* depending on the corresponding criteria; the suspended state is optional and available for complex plans (Figure 7.11).

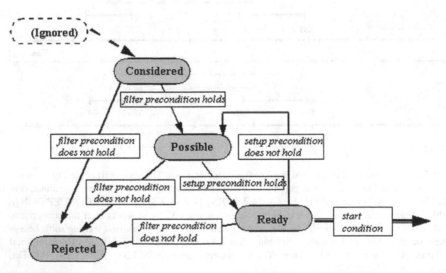

Fig. 7.10 Plan-selection states and their state-transition conditions in Asbru. These states are relevant to guideline-library plans that are considered for execution.

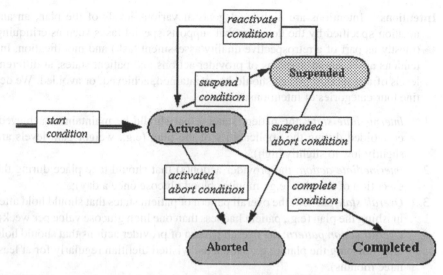

Fig. 7.11 Plan-execution states and their state-transition conditions in Asbru. These states are relevant to instances of guideline-library plans that are being executed.

The semantics of different state-transition conditions are explained below. A plan consists of a name, a set of arguments, including a time annotation (representing the temporal scope of the plan), and five (optional) components: *preferences*, *intentions*, *conditions*, *effects*, and a *plan body*, which describes the actions to be executed. All components are optional. Every subplan has the same structure. Thus, a sequential plan can include several potentially decomposable concurrent or cyclical plans.

We now examine in more detail each of the knowledge roles of the Asbru ontology.

Preferences Preferences bias or constrain the selection of a plan to achieve a given goal. Examples include:

1. *Strategy*: a general strategy for dealing with the problem (e.g., aggressive, normal);
2. *Utility*: a set of utility measures (e.g., minimize the cost or the patient inconvenience);
3. *Select-method*: a matching heuristic for the applicability of the whole plan (e.g., exact-fit);
4. *Resources*: a specification of prohibited or obligatory resources (e.g., in certain cases of treatment of a pulmonary infection, surgery is prohibited and antibiotics must be used);
5. *Start-conditions*: an indication whether transition from a ready state of the generic plan to an activated state of the plan instance is automatic or requires approval of the user.

Intentions   Intentions are high-level goals at various levels of the plan, an annotation specified by the designer that supports special tasks such as critiquing (mostly as part of a retrospective quality assessment task) and modification. Intentions are temporal patterns of provider actions and patient states, at different levels of abstraction, which should be maintained, achieved, or avoided. We define four categories of intentions:

1. *Intermediate state*: the patient state(s) that should be maintained, achieved, or avoided during the applicability of the plan (e.g., weight gain levels are slightly low to slightly high);
2. *Intermediate action*: the provider action(s) that should take place during the execution of the plan (e.g., monitor blood glucose once a day);
3. *Overall state pattern*: the overall pattern of patient states that should hold after finishing the plan (e.g., patient had less than one high glucose value per week);
4. *Overall action pattern*: the overall pattern of provider actions that should hold after finishing the plan (e.g., patient had visited dietitian regularly for at least three months).

Conditions   Conditions are temporal patterns, sampled at a specified frequency, that need to hold at particular plan steps to induce a particular state transition of the plan instance. They are used for actual execution (application) of the plan. We do not directly determine conditions that should hold during execution; we specify conditions that activate the change of a particular plan state (see Fig. 7.11). A plan instance is completed when the complete conditions become true, otherwise the plan instance's execution suspends or aborts (often, due to failure). Conditions are optional. We distinguish six types of conditions:

1. *filter-preconditions*, which need to hold initially if a generic plan is applicable; these conditions are not goals to be achieved (e.g., patient is a pregnant female), and must be true to achieve a possible state (see Fig. 7.10);
2. *setup-preconditions*, which need to be achieved (usually, within a given time delay relative to the initial time of consideration of the plan) to enable a plan to start (e.g., patent had a glucose-tolerance test) and allow a transition from a possible plan to a ready plan (see Fig. 7.10);
3. *suspend-conditions*, which determine when an active plan instance has to be suspended (e.g., blood glucose has been high for four days); these are informally the inverse of protection conditions in the planning literature, which have to hold during certain time periods (see Fig. 7.11);
4. *abort-conditions*, which determine when an active or suspended plan has to be aborted (e.g., there is an insulin-indicator condition: the patient cannot be controlled by diet) (see Fig. 7.11);
5. *complete-conditions*, which determine when an active plan is completed, typically, but not necessarily, successfully (e.g., delivery has been performed) (see Fig. 7.11);

6. *reactivate-conditions*, which determine when a suspended plan has to be re-activated (e.g., blood glucose level is back to normal or is only slightly high) (see Fig. 7.11).

Effects Effects describe the functional relationship between either (1) each of the relevant plan arguments and measurable parameters it affects in certain contexts (e.g., the dose of insulin is inversely related in some fashion to the level of blood glucose) or (2) the overall plan and the clinical parameters it is expected to effect (e.g., the insulin-administration plan decreases the blood-glucose level). Effects can have a likelihood annotation-a probability of occurrence. Effects can be part of the guideline library (when they annotate plans) and can also be stored in a domain-specific knowledge base (especially for common plans, such as administration of drugs).

Plan-Body The plan body is a set of plans to be executed in parallel, in sequence, in any order, or in some frequency. The Asbru ontology distinguishes among three types of plans: *sequential, concurrent,* and *cyclical (repeating).* Only one type of plan is allowed in a single plan body. A *sequential* plan specifies a set of plans that are executed in sequence; for continuation, all plans included have to be completed successfully. *Concurrent* plans can be executed either together, in parallel, or in any order. Asbru distinguishes two dimensions for classification of sequential or (potentially) concurrent plans: the number of plans that should be completed to enable continuation and the order of plan execution. Table 7.3 summarizes the dimensions of the two plan types. Using the two dimensions, it is possible to define the plan subtypes DO-ALL-TOGETHER, DO-SOME-TOGETHER, DO-ALL-ANY-ORDER, DO-SOME-ANY-ORDER, DO-ALL-SEQUENTIALLY. The continuation condition specifies the names of the plans that must be completed to proceed with the next steps in the plan.

A *cyclical,* or *repeating plan* (a DO-EVERY type) includes a plan that can be repeated, and optional temporal and continuation arguments that can specify its behavior. *Start* and *end* specify a starting and ending time point. *Time base* determines the time interval over which the plan is repeated and the start time, end time, and duration of the particular plan instance in each cycle (e.g., starting with the first Monday's morning, until next Tuesday's morning, perform plan A every morning for 10 minutes). The *times-completed* argument specifies how many times the plan has to be completed to succeed and the times-attempted argument specifies how many attempts are allowed. Obviously, number of attempts must be greater or equal to the number of successful plans. A temporal pattern can be used as a stop condition of the cyclic plan. Finally, the plan itself is associated with its own particular arguments (e.g., dose). The start time, the time base, and the plan name are mandatory to the specification of a cyclic plan; the other arguments are optional.

Table 7.4 summarizes the Asbru knowledge roles and their semantics.

Table 7.3 Categorization of Asbru plan types by continuation conditions and ordering constraints

		Continuation condition	
		All plans should be completed in order to continue	Some plans should be completed in order to continue
Ordering Constraints	Start together	DO-ALL-TOGETHER (no continuation-condition; all plans must complete)	DO-SOME-TOGETHER (continuation-conditions specified as subset of plans)
	Execute in any order	DO-ALL-ANY-ORDER (no continuation-condition; all plans must complete)	DO-SOME-ANY-ORDER (continuation-conditions specified as subset of plans)
	Execute in total order	DO-ALL-SEQUENTIALLY (no continuation-condition; all plans must complete)	

Table 7.4 The tasks involved in guideline-based care and the knowledge required for performing them.

Knowledge role and its subtypes	Semantics	Example
Plan body	The actual actions. Only one subtype is allowed.	See Table 7.6.
Sequential	a set of plans that are executed in sequence; all plans have to be completed to continue the plan	Administer oxygen, administer ventolin, then call an ambulance.
Concurrent: Do-all-together, Do-some-together, Do-all-any-order, Do-some-any-order	Two or more plans that are applied in parallel. The plans either start together or can be executed in any order; some or all of the plans need to complete for continuation of the overall guideline.	Apply in parallel the guidelines (subplans) for glucose monitoring, nutritional management, and monitoring for insulin indication.
Cyclical	Includes a plan that is repeated; specialized temporal and continuation arguments specify its behavior	Administer each day before breakfast 3 units of regular insulin.
Conditions	Temporal patterns, sampled at a specified frequency during guideline application, which need to hold at particular plan steps to induce a particular state transition of the plan.	See Table 7.6.
Filter-preconditions	Must hold initially to decide if a guideline is applicable; allow transition of a generic plan from considered to possible	Patient is female and is pregnant
Setup-preconditions	Need to be achieved to enable a guideline to start; they allow a transition from a possible generic plan to a ready plan	Patent has had a glucose-tolerance test
Complete-conditions	Determine when an active plan is completed	Patient has had delivery
Suspend-conditions	Determine when an active plan instance has to be suspended	Fasting blood glucose has been high for 4 days
Reactivate-conditions	Determine when a suspended plan can be reactivated	blood glucose level is back to normal or is only slightly high
Abort-conditions	Determine when an active/suspended plan has to be aborted	A predefined pattern for insulin-indication has been detected

7.5.2.3 Asbru Example: A Gestational Diabetes Mellitus Guideline

We present, using the original text-based version of Asbru, a Stanford University guideline for controlled observation and treatment of *gestational diabetes mellitus* (GDM) type II (non insulin dependent). The guideline prescribes several concurrent monitoring and management plans following a glucose tolerance test (GTT) between 140 and 200 mg/dl (Table 7.6). The plan body consists of three plans that are executed in parallel (glucose monitoring, nutritional management, and monitoring

Table 7.5

Intentions	High-level goals at various levels of the guideline's plan; support tasks such as quality assessment and modification. Represented as temporal patterns of provider actions or patient states that should be maintained, achieved, or avoided.	See Table 7.6.
Intermediate state	Pattern of patient state(s) that should be maintained, achieved, or avoided during the applicability of the plan	Weight gain levels are slightly low to slightly high during application
Intermediate action	Provider temporal pattern of action(s) that should take place during the application of the guideline's plan	Monitor blood glucose every day
Overall state pattern	The temporal pattern of patient states that should have formed (from start to end of guideline) after finishing the guideline	During the guideline, the patient had up to three episodes of hypoglycemia.
Overall action pattern	The overall pattern of provider actions that should hold after finishing the guideline plan	Patient had visited dietitian regularly once per week, for at least three weeks within each month
Preferences	bias or constrain the selection of a plan to achieve a given goal	See Table 7.6.
Strategy	A general strategy for dealing with the problem	Use aggressive chemotherapy as a general policy
Utility	A set of utility measures for selection of competing options	Minimize the cost
Select-method	Which matching function to use for the filter condition when considering several guidelines	Use only exact fit (versus use the guidelines that has the best fit)
Resources	A specification of prohibited or obligatory resources	Surgery is prohibited in this case, and antibiotics must be used
Start-conditions	Indicate if transition from a ready state of the generic plan to an activated state of the plan instance is automatic or manual	Wait for approval from attending physician to proceed
Effects	Describe functional relationships between actions and patient states; are context-sensitive and have a likelihood annotation	See Table 7.6.
Plan-argument effects	Describe the functional relationship between a plan argument and measurable parameters it affects in certain contexts).	The dose argument of the insulin-administration plan is inversely related to the level of blood glucose
Whole-plan effects	Describe the functional relationship between the overall guideline plan and certain clinical parameters	The insulin-administration plan decreases the blood-glucose level

for insulin indication), exist in the guideline-specification library, and are decomposable into other library plans.

7.5.3 The Importance of Explicit Representation of Guideline Intentions

Intentions, especially in the context of plans, have been examined in philosophy [39] and in artificial intelligence [311]. We view intentions formally as temporally extended goals, comprising constraints on process or outcomes patterns, at various abstraction levels.

The design-time and runtime model underlying the Asgaard project (See Figure 7.8) implies a specific guideline-application critiquing framework. In this framework, five comparison axes exist: the guideline's prescribed actions versus the provider's actual actions; the guideline's intended plan versus the provider's

Table 7.6 A portion of the machine-comprehensible Asbru representation of the guideline for management of non-insulin-dependent gestational diabetes mellitus (GDM) type II. (The expert designer uses a graphical authoring tool).

Semantic role	Asbru Code (ASCII code version)	Explanation
Time assign-ments	(DOMAIN-DEPENDENT TIME-ASSIGNMENT (SHIFTS DELIVERY <- 38 WEEKS) (POINT CONCEPTION <- (ask (ARG "what is the conception-date?"))) (ABSTRACTION-ASSIGNMENT (CYCLIC MIDNIGHTS <- [0, 0 HOURS, 24 HOURS] BREAKFAST-START-TIME <- [0, 7 HOURS, 24 HOURS]))	Time-annotations serve to establish domain-specific temporal terms. Delivery is a point occurring 38 weeks after Conception, whose date is acquired from the user. Midnights and breakfast start times are periodic shifts every 24 hours from the 12:00A.M. time.
Plan body	(DO-ALL-TOGETHER (glucose-monitoring) (nutrition-management) (observe-insulin-indicators)))	The plan body is a concurrent one; All three plans have to start to-gether; all need to be completed to continue the next phase in the plan.
Conditions	(FILTER-PRECONDITIONS (one-hour-GTT-results (140, 200) pregnancy [_, _], [_, _], [_, _], CONCEPTION]) (SETUP-PRECONDITIONS (PLAN-STATE one-hour-GTT COMPLETED [[24 WEEKS, 24 WEEKS], [26 WEEKS, 26 WEEKS], [_, _], CONCEPTION]) (COMPLETE-CONDITIONS (delivery TRUE GDM-Type-II * (SAMPLING-FREQUENCY 24 HOURS))) (SUSPEND-CONDITIONS (STATE(blood-glucose) HIGH GDM-Type-II [[24 WEEKS, 24 WEEKS], [DELIVERY, DELIVERY], [4 DAYS, _], CONCEPTION](SAMPLING-FREQUENCY 24 HOURS))) (REACTIVATE-CONDITIONS (STATE(blood-glucose) (OR NORMAL SLIGHTLY-HIGH) GDM-Type-II [[24 WEEKS, 24 WEEKS], [DELIV-ERY, DELIVERY], [_, _], CONCEPTION] (SAMPLING-FREQUENCY 24 HOURS))) (ABORT-CONDITIONS (insulin-indicator-conditions TRUE GDM-Type-II * (SAMPLING-FREQUENCY 24 HOURS)))	The entry conditions for consider-ing the patient are that the results of a glucose-tolerance test (GTT) are within 140 and 200 gr/dl. To activate the guideline, the patient must have completed the GTT plan within 24 to 26 weeks of preg-nancy.The guideline application is completed on delivery. Suspend the guideline if high blood-glucose lev-els, as defined in the GDM-Type-II context, exist for at least 4 days. Reactivate it if blood glucose levels are normal or slightly high. Abort the guideline if there a pattern of indication for insulin therapy is de-tected. For all conditions, sample the database every 24 hours to de-termine the guideline's status.

(abstracted) plan; the guideline's intended patient state versus the provider's state intention; the guideline's intended state versus the patient's (abstracted) actual state; and the provider's intended state versus the patient's (abstracted) actual state. Com-binations of the comparison results imply a set of different behaviors of the guideline application by the provider. Thus, a care provider might not follow the precise ac-tions, but still follow the intended plan and achieve the desired states. A provider might even not follow the overall plan, but still adhere to a higher-level intention. Alternatively, the provider might be executing the guideline correctly, but the pa-tient's state might differ from the intended, perhaps indicating a complication that needs attention or a failure of the guideline. In theory, there might be up to 32 differ-ent behaviors, or assessment vectors, assuming binary comparisons along five axes. However, the use of consistency constraints prunes this number to approximately 10

Table 7.7

Intentions	(INTENTION:INTERMEDIATE-STATE (MAINTAIN blood-glucose-post-meal (<= 130) GDM-Type-II [[24 WEEKS, 26 WEEKS], [DELIVERY, DELIVERY], [14 WEEKS, 16 WEEKS], CONCEPTION]) (MAINTAIN blood-glucose-fasting (<= 100) GDM-Type-II [[24 WEEKS, 26 WEEKS], [DELIVERY, DELIVERY], [14 WEEKS, 16 WEEKS], CONCEPTION))) (INTENTION:OVERALL-STATE (AVOID STATE(blood-glucose) HIGH GDM-Type-II [[24 WEEKS, 26 WEEKS], [DELIVERY, DELIVERY], [7 DAYS, _], CONCEPTION]))	During application of the guideline (starting around 24 to 26 weeks of conception, lasting until delivery, for a duration of 14 to 16 weeks), the target is to maintain postprandial blood glucose values at less than 130 gr/dl and fasting blood-glucose values at less than 100 gr/dl. An overall intention, to be assessed upon completion of the guideline, is that during the guideline's application, there was never a period of high blood-glucose level (in any context) lasting 7 days.
Preferences	(PREFERENCES (SELECT-METHOD EXACT-FIT) (START-CONDITION AUTOMATIC))	The match in the filter conditions needs to be exact to even consider the guideline for selection. Activation of the guideline by the automated interpreter can start without manual user approval
Effects	(EFFECTS ([GDM-Type-II Weight-at-birth Normal [DELIVERY, DELIVERY], [DELIVERY, DELIVERY], [_, _], CONCEPTION]), 80%)	Expected effects of the guideline include a normal (for this context) weight at birth; estimated likelihood of achievement of the effect is 80%.

major behaviors. (It is also possible to use continuous, rather than binary, measures of pattern matching). Table 7.8 presents several typical behaviors.

Table 7.8 Several of the typical application behaviors defined by the quality-assessment task, based on comparison of guideline specifications and intentions, physician actions and intentions, and patient states. These comparisons result in an *assessment vector*.

Intended action vs. physician action	Intended plan vs. physician plan	Intended state vs. physician intention	Intended state vs. patient state	physician intention vs. patient state	Description of the behavior
+	+	+	+	+	physician executes protocol as specified; protocol succeeds
+	+	+	-	-	physician follows guideline, has the same intentions, but guideline does not work
-	+	+	+	+	overall plan intention followed, albeit through different actions, and it works
-	-	+	+	+	physician follows neither actions nor overall plan; state intentions agree and both succeed
-	-	+	+	+	physician follows neither action nor plan; state intentions differ, and neither materializes

Access to the original process and outcome intentions of the guideline designers supports forming an automated critique of *where, when, how much* the care provider seems to be deviating from the suggested process of applying the guideline, and in *what way* and to *what extent* the care provider's outcome intentions might still similar to those of the author's (e.g., she might be using a different process to achieve the same outcome intention). Thus, effective quality assessment includes searching for a reasonable *explanation* that tries to understand the care provider's rational by comparing it to the design rational of the guideline's author. That design rational needs to be explicitly captured in a set of process and outcome intentions, which can optionally exist at every level (e.g., component) of the guideline. Thus, for example, both false positive and false negative alarms might be prevented during quality assessment time.

Note, however, that such intelligent quality assessment of guideline application requires (1) awareness of the guideline author's intentions, (2) knowledge of the effects of different interventions (e.g., to recognize substitution of a anti-hypertensive drug by a drug that has a similar effect), and (3) a set of general and guideline-specific revision strategies (e.g., stopping administration of a drug that has a negative effect on some clinical parameter, such as Hemoglobin level, is equivalent to administering a drug that has a positive effect on that parameter).

Note also that *intentions* are much more specific than general *themes* such as reducing mortality and morbidity; these cannot be monitored effectively during the lifetime of the guideline's application. Intentions also are not as specific as effects on low-level physiological mechanisms that might lead to reduction in morbidity and mortality we rarely have precise mathematical models for such a complex chain of events. Rather, intentions exist at an intermediate level that captures the guideline designer's constraints on the (temporal) patterns that should emerge from correctly following the guideline's actions and from achieving its expected (short-term) outcomes.

7.5.4 Plan Recognition and Critiquing In the Application of Clinical Guidelines

The following example demonstrates the tasks of *plan-recognition* and *critiquing* in the domain of monitoring and therapy of patients who have insulin-dependent diabetes.

During therapy of a diabetes patient, hyperglycemia (a higher than normal level of blood glucose) is detected for the second time in the same week around bedtime. The diabetes-guideline's prescribed action might be to increase the dose of the insulin the patient typically injects before dinner. However, the provider recommends reduction of the patient's carbohydrate intake during dinner. This action seems to contradict the prescribed action. Nevertheless, the automated assistant notes that increasing the dose of insulin decreases the value of the blood-glucose level directly, while the provider's recommendation decrease the value of the same clinical parameter by reducing the magnitude of an action (i.e., ingestion of carbohydrates) that increases its value. The assistant also notes that the state intention of the guideline was "avoid more than two episodes of hyperglycemia per week." Therefore, the provider is still following the intention of the guideline. By recognizing this high-level intention and its achievement by a different plan, the automated assistant can accept the provider's alternate set of actions, and even provide further support for these actions.

We consider a plan-recognition ability, such as demonstrated in the example, an indispensable prerequisite to the performance of plan critiquing. Such an ability might increase the usefulness of guideline-based decision-support systems to clinical practitioners, who often follow what they consider as the author's intentions rather than the prescribed actions. Note that we must assume knowledge about the effects of interventions on clinical parameters, and knowledge of legitimate domain-independent and domain-specific guideline-revision strategies. Both intervention effects and revision strategies can be represented formally [358].

7.5.5 Acquisition and Maintenance of Asbru Guidelines

As part of the Asgaard project, the developers generated an automated graphical knowledge-acquisition (KA) tool for the object-oriented version of the Asbru language, using the PROTG-II suite of tools, with encouraging results [355]. Domain experts could use the tool to specify and maintain Asbru guidelines. A significant benefit of the KA tool approach is that it detects incorrect syntax while authoring a guideline. Thus, implicit syntactic support for domain experts, who are of course not versatile in the Asbru syntax, is provided at no cost.

However, the complexity of the ontology enforced the automatic generator of the KA tool to produce a user interface with many cascading, small dialogs. Thus, for providing an optimal dialog, more control of the layout of the automatically generated user interface was needed than was possible in early versions of PROTG/Win. Once this additional feature became available, significantly better versions of the KA tool were generated. Another enhancement was the addition of a specialized string editor that accepted as input the BNF syntax of Asbru temporal annotations, and creates automatically a graphical KA tool that acquires legal strings from the user [355]. The string editor was part of the PROTG/Win framework, an early Windows-based version of Protégé.

A significant improvement in the process of acquisition and maintenance of Asbru guidelines was the development, by the Vienna team of the Asgaard project, of the AsbruView visual module for specification and browsing of Asbru guidelines [225]. The AsbruView system enriches the timeline graphic representation by adopting a 3D visualization (especially tailored for representing medical therapy plans): besides the two usual dimensions on which the different (possibly overlapping) parts of plans are temporally laid out, a third dimension is used to add graphic elements which convey further information (e.g., when a plan is completed, or might be suspended, or aborted, ...). The graphic elements are chosen in such a way that the resulting visualization resembles a racing track, complete with green flags (denoting the filter and setup conditions that enable plans to run), yellow flags (denoting the suspend conditions), and red flags (denoting the abort conditions), along which the physician has to run as the treatment of the patient evolves, as depicted in Figure 7.12.

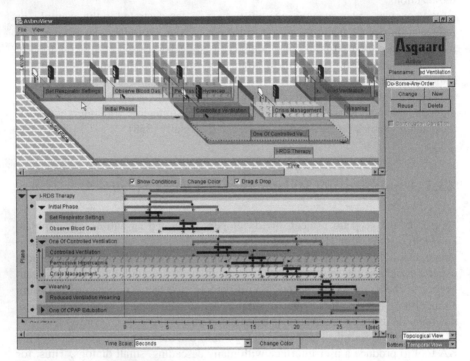

Fig. 7.12 The AsbruView system.

AsbruView uses the metaphor of a racing track to display Asbru plans using two views: a topological view and a temporal view. The topological view mainly depicts relationships among plans (parallel, in sequence, part-of) without a precise time scale. The temporal view focuses on the temporal dimension of plans and various conditions. The racing track metaphor represents a plan as a track along which are laid objects such as sub-plans and conditions.

Another significant development in the transformation of free-text guidelines to Asbru guidelines was the DeGeL project [361, 367], which we discuss in detail in Section 7.6, in which free-text guidelines are incrementally transformed to machine-comprehensible representations in various guideline ontologies. Although multiple guideline ontologies are supported in the DeGeL architecture, Asbru is the default ontology used in the DeGeL architecture, in the sense that in the case of the Asbru ontology, tools exist in the DeGeL architecture not only for generic tasks such as uploading, semi-structuring, indexing, search, and browsing, but also for creation of semi-formal and formal representations.

7.6 Hybrid Guideline Representations and the DeGeL Project

In this section, we will discuss the issue of the transformation of free-text guidelines into increasingly formal representations. We will also discuss in depth the issue of storage and retrieval of clinical guidelines within a digital library. To better clarify this important aspect of the automation of guideline-based care, we shall focus on a particular broad architecture for guideline-based care.

7.6.1 The Guideline Conversion Problem

The existence of automated architectures for guideline representation, such as described in the preceding sections, makes the question "*How will the large mass of free-text guidelines be converted to a formal machine-readable language?*" be a most pertinent one.

The core of the problem is that expert physicians cannot (and need not) program in guideline-specification languages, while programmers and knowledge engineers do not understand the clinical semantics of the guidelines. In addition, some of the guideline's knowledge is of *implicit* nature, clear only to the expert physician authoring the guideline; this knowledge must become *explicit* during the conversion process. Thus, converting guidelines into machine-comprehensible formats must capitalize on the relative strengths of both expert physicians and knowledge engineers.

The conversion process must also support and facilitate collaboration amongst these two very different types of users, and the natural iteration that is likely to be inherent in such a process.

Another aspect we need to consider is that machine-executable representations are crucial for providing computerized assistance, such as automated execution of the guideline, while for some other tasks, such as search and retrieval of relevant guidelines, text-based representations are of a significant importance (and indeed, might be more useful).

Finally, we need to keep in mind that the procedural knowledge encoded in the guideline, such as laboratory tests or clinical procedures, should be truly sharable and generalizable across multiple sites. Thus, the conversion process must support embedding in the guideline's representation terms from *standardized medical vocabularies*, which are well-understood everywhere.

All these considerations have led to the development of the *Digital electronic Guideline Library (DeGeL)* architecture [361, 368]. We will describe the *hybrid* guideline-representation model underlying this framework, then look at the architecture and its tools.

7.6.2 The Hybrid Guideline-Representation Model

To gradually convert clinical guidelines to machine-comprehensible representations, Shahar and his colleagues have developed a *hybrid*, multifaceted representation, an accompanying distributed architecture and a set of Web-based software tools. The specification tools incrementally and in iterative fashion transform a set of clinical guidelines through several intermediate, semi-structured phases, eventually arriving at a fully formal, machine comprehensible representation of the guideline.

The guiding principle in the DeGeL project is that expert physicians (if possible, throughout the world) should be transforming free-text guidelines into intermediate, semantically meaningful representations, while knowledge engineers should be converting these intermediate representations into a formal, executable representation.

7.6.3 The Gradual Conversion Process

Underlying the various modules and tools we will be describing further on, is the guiding principle mentioned above: *Expert physicians* use the tools to classify the guidelines along multiple semantic axes, and to semantically *markup* (i.e., label portions of the text by the semantic labels of the target ontology) existing text-based guidelines, thus creating a *semi-structured format* (which is still text-based). The expert physicians might even further structure the guideline, possibly with a knowledge-engineer's assistance, into a *semi-formal* structure, which includes ontology-specific control-flow knowledge. *Knowledge engineers* convert the marked-up text, or the semi-formal structure, into a *formal, fully structured*, machine-comprehensible representation of the target ontology, using an ontology-dedicated tool. Figure 7.13 summarizes the complete process. All of the hybrid guideline-representation formats co-exist and are organized in the DeGeL library within a unified structure, the *hybrid representation*. Part of the hybrid representation, shared by all hybrid guideline ontologies, is the *hybrid meta-ontology*.

Note that different parts of the same guideline might exist at different levels of specification (e.g., eligibility conditions might include also executable expressions, thus supporting automated eligibility determination, although the guideline's procedural aspect is still only semi-structured or in a semi-formal format). In addition, all specification levels are optional. Finally, if needed, new representation levels can be added.

The hybrid-specification process intertwines the expertise of both the expert physician and knowledge engineer so as to gracefully convert clinical guidelines into a machine-executable format. The conversion process is performed gradually using the following current representation formats:

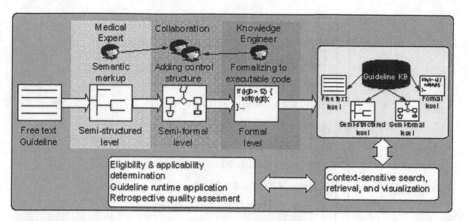

Fig. 7.13 The incremental conversion process in the DeGeL architecture. Input free-text guidelines are loaded into a markup editor; expert physicians index and markup (structure) portions of the guidelines by semantic labels from a chosen target ontology, creating semi-structured and, in collaboration with a knowledge engineer, semi-formal guideline representations. Knowledge engineers use an ontology-specific tool to add executable expressions in the formal syntax of that ontology.

1. *Semi-structured Text* - snippets of text assigned to top-level target-ontology *knowledge-roles*, such as Asbru *filter conditions* (obligatory eligibility criteria) and *outcome intentions*;
2. *Semi-formal representation* - further specification of the structured text, adding more explicit procedural control structures, performed jointly by the knowledge engineer and expert physician, such as specification in explicit fashion of Asbru *sequential* and *concurrent actions.*
3. *Formal representation* - final specification performed by the knowledge engineer, resulting with the guideline converted a machine-comprehensible format, such as an Asbru plan executable by an Asbru runtime execution module. Thus, the output of the DeGeL authoring tool(s) is a *hybrid representation* of a guideline which contains, for each guideline, or even for different sections (knowledge roles) within the same guideline, one or more of the above three formats.

These three current levels of hybrid structuring (or four, including the original free text) are in principle possible within all guideline-representation languages. For example, they were easily implemented within the context of the Asbru language, which is the default guideline ontology in the DeGeL architecture (Figure 7.14).

7.6.4 The DeGeL Architecture

The DeGeL architecture was designed to handle all of the hybrid guideline-representation levels. The DeGeL framework's guideline knowledge-base and various

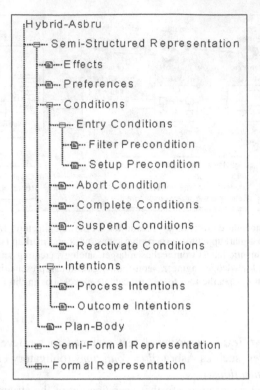

```
Hybrid-Asbru
  ⊟ Semi-Structured Representation
    ▣ Effects
    ▣ Preferences
    ⊟ Conditions
      ⊟ Entry Conditions
        ▣ Filter Precondition
        ▣ Setup Precondition
      ▣ Abort Condition
      ▣ Complete Conditions
      ▣ Suspend Conditions
      ▣ Reactivate Conditions
    ⊟ Intentions
      ▣ Process Intentions
      ▣ Outcome Intentions
    ▣ Plan-Body
  ⊞ Semi-Formal Representation
  ⊞ Formal Representation
```

Fig. 7.14 The semi-structured level of the Asbru hybrid ontology in the DeGeL architecture.

task-specific tools (Figure 7.15) were designed to support all of the design time and runtime tasks involved in guideline-based care [361, 368].

7.6.5 The Hybrid Meta-Ontology

To support the specification of a guideline in one or more different guideline specification languages, the DeGeL architecture includes a *hybrid* guideline *meta-ontology* (*meta* is used here in the sense of "*above*") (Figure 7.16).

The meta-ontology is composed of two components:

1. A *documentation meta-ontology*, which specifies knowledge roles common to all target guideline ontologies, and defines the ontologies of the sources of the guidelines and of the marked-up guidelines (see below).
2. A *specification meta-ontology* for describing a new target ontology, in order to enable adding it into the DeGeL (meta) knowledge base. Thus, we provide an XML schema that describes, for designers of existing or new guideline ontologies, how to generate XML documents that conform to the DeGeL expected

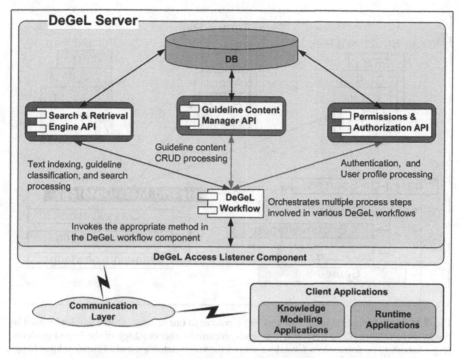

Fig. 7.15 The conceptual architecture of a typical DeGeL server. There are three main components: (1) a *permissions and authorizations* manager component, responsible for generating user-profiles and controlling user access to DeGeL's guideline repository, (2) a *guideline content manager*, responsible for performing *Create, Retrieve, Update* and *Delete* (*CRUD*) operations on all knowledge entities (e.g., guidelines) stored in DeGeL's repository, and (3) a *search & retrieval engine*, responsible for performing text indexing and store semantic classification of guidelines as well as handling search queries processing. The DeGeL *Workflow* component synchronizes all three components during operations that require use of one or more components. The DeGeL architecture has a single conceptual interface that can be accessed through multiple communication methods (e.g., web services, remote procedure calls), the interfaces to which are part of the *Listener* component.

structure. These documents are instances of the specification meta ontology and describe particular target ontologies such as GLIF or Asbru.

The documentary component of the hybrid-meta ontology includes several knowledge-roles, such as *documentation*, common to all guideline ontologies. It distinguishes *source guidelines*, which are free-text guidelines uploaded to DeGeL, from *hybrid guidelines*, which are the output of the gradual hybrid conversion process.

Uploading a guideline into the DeGeL library (e.g., a document published by a professional society) creates a *source guideline*. A guideline source can be named, searched, and retrieved, and is annotated using the dedicated *source-guideline ontology*, which documents the source-guideline's details (e.g., authors, date). However, a source guideline cannot be indexed or applied to a patient.

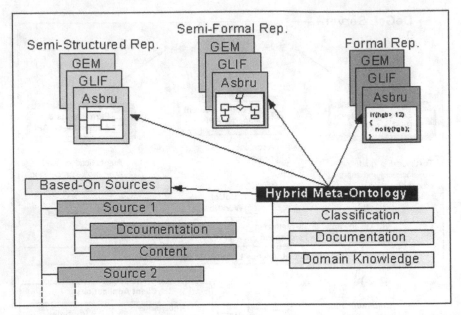

Fig. 7.16 DeGeL's hybrid meta-ontology. (1) a pointer to one or more source ontologies used by the hybrid guideline, (2) a pointer to the semi-structured target ontology of the hybrid guideline (e.g., Asbru), (3) a pointer to a formal version of the target ontology, and (4) several knowledge-roles, independent of the target ontology, that characterize the document (e.g., domain knowledge, semantic indices).

A *hybrid guideline* is a more complex structure, which can be indexed, retrieved, modified, and applied. A hybrid guideline includes one or more source guidelines, additional knowledge roles that are independent of the target ontology, such as documentation; and domain knowledge necessary for guideline application, and the semi-structured, semi-formal, and fully-structured (machine-comprehensible) representations of the guideline using the selected target ontology.

The semi-structured representation of a hybrid guideline will typically exist for all guideline ontologies. This representation can be processed by all DeGeL tools, without adding any ontology-specific extensions; it corresponds roughly to the top level and intermediate concepts of the target ontology. The semiformal and formal representation levels typically need ontology-specific tools for creation and processing. For example, temporal queries to the patient record are specified as semi-structured queries (using the mark-up tool), which are then semi-formalized by the expert physician, or fully formalized by the knowledge engineer, using a tool specific to the target ontology.

Several DeGeL tools are used mostly to specify and retrieve guidelines, irrespective of a particular patient. Other tools are used mostly at runtime and require automated or manual access to patient data. All of the tools were designed to support

the various formats implied by a hybrid representation. Figure 7.17 presents an overall view of the DeGeL architecture.

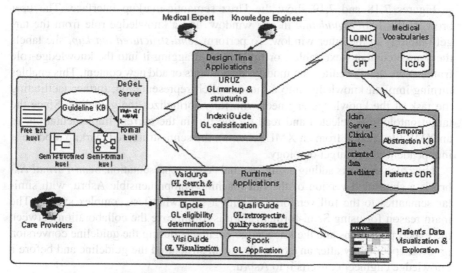

Fig. 7.17 The overall DeGeL guideline-support architecture, showing both the guideline-
representation component (the DeGeL server) and the clinical-data access component (the Idan
server). Both the knowledge modeling time and runtime tools communicate with the guideline
knowledge base, which includes all four hybrid representation formats. The design-time tools, used
by the medical experts and the knowledge engineers, access also several controlled-terminology
medical vocabularies. The runtime tools, used by care providers at the point of care, access also a
mediator to the time-oriented patient data, which enables answering of queries at a varying level
of complexity, regarding the patient data. The runtime tools also have access to a specialized tool
for interactive visualization and exploration of raw patient data and various guideline-specific abstractions derivable from them.

7.6.6 The DEGEL Architecture's Components

We will now examine the various components of the DeGeL architecture, and how
they serve the underlying hybrid, multiple-ontology framework.

7.6.6.1 Uruz: Semantic Markup

The *Uruz* Web-based guideline markup tool (Figure 7.18) enables medical experts
to: create new guideline documents. A source guideline is uploaded into the DeGeL,
and can then be used by Uruz to create a new *guideline document*, marked-up by
the semantic labels of one of the target ontologies available in DeGeL. Uruz can

also be used to create a guideline document *de-novo* (i.e., without using any source) by directly writing into the knowledge roles of a selected target ontology. We are developing an Asbru-dedicated tool to add the formal-specification level.

Figures 7.18 and 7.19 show the Uruz semantic-markup interface. The user browses the *source guideline* in one window, and a knowledge role from the target ontology in the other window. To perform *semi-structured markup*, she labels the source content (text, tables, or figures) by dragging it into the knowledge-role frame. Note that the editor can modify the contents or add new content. This enables turning implicit knowledge into a more explicit representation, further facilitating the task of the knowledge engineer who fully formalizes the guideline. Since the target ontology is selected and read on the fly (in the current implementation, as an XML file created from an XML schema), the semi-structured markup module is independent of the target ontology.

Uruz also supports adding a semi-formal Asbru representation. *Semi-Formal* Asbru is a simplified version of the full machine-comprehensible Asbru, with similar semantics to the full version, but with a somewhat less complex syntax. The main reason for using Semi-formal Asbru is to improve the collaboration between the expert physicians and the knowledge engineers during the guideline conversion process, specifically after an expert physician structured the guideline and before a knowledge engineer converts it to Asbru.

In addition, the semi-formal format still supports text-based retrieval of procedural knowledge, unlike the fully formal format. Finally, a semi-formal structure is obligatory when an electronic medical record is unavailable, since interaction with the clinical user is imperative. This property is exploited to an advantage by our hybrid runtime application module. Semi-formal Asbru has all of Asbru's knowledge-roles, such as conditions (e.g., eligibility, completion, and abort conditions), branching constructs (e.g., *if-then-else* or *switch-case*), various synchronization constraints of sub-guidelines (i.e., do in parallel, do in sequential) and time-annotations for describing temporal constraints.

Instead of using Asbru's complex notion of (plan) arguments, each guideline in semi-formal Asbru has a list of patient-related data, *obtained-values*, defined during design-time. Temporal-patterns, the building blocks of a guideline in Asbru, are expressed with combinations of text and time-annotations instead of Asbru's complicated formal expressions. The semi-formal version syntax is defined using an XML schema.

A list of common clinical actions, such as drug prescription, laboratory observation and physical examination, had been added to semi-formal Asbru. These actions can be used as reusable primitive plans during guideline design-time, thus simplifying the process of guideline structuring.

To create an Asbru semi-formal representation, an Asbru-specific module, the *plan-body wizard* (PBW), had been embedded in Uruz (such modules can be defined also for other ontologies). The PBW is used for defining the guideline's semi-formal control structure (Figure 7.19). The PBW enables a user to decompose the actions embodied in the guideline into atomic actions and other sub-guidelines, and to define the control structure relating them (e.g., sequential, parallel, repeated

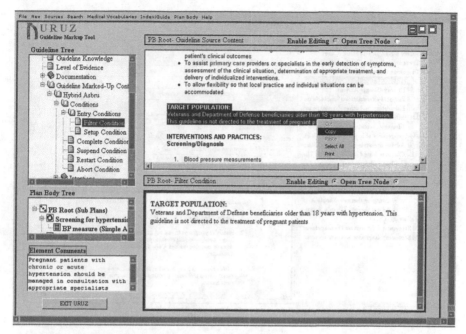

Fig. 7.18 The Uruz Web-based guideline markup tool. The tool's basic semi-structuring interface is uniform across all guideline ontologies. The target ontology selected by the medical expert, in this case, Asbru, is displayed in the upper left tree; the guideline source is opened in the upper right frame. The expert physician highlights a portion of the source text (including tables or figures) and drags it for further modification into the bottom frame labeled by a semantic role chosen from the target ontology (here, *filter condition*). Note that contents can be aggregated from different locations in the source. The bottom left textbox, Element Comments, stores remarks on the current selected knowledge-role, thus supporting collaboration among guideline editors.

application). The PBW, used by medical experts, significantly facilitates the final formal specification by the knowledge engineer.

When a knowledge engineer needs to add a formal, executable expression to a knowledge role, she uses one of the ontology-specific Uruz modules (we are developing one specific to Asbru), which delves deeper into the syntax of the target ontology. For example, in hybrid Asbru, conditions can include temporal patterns in an expressive time-oriented query language used by all application modules.

To be truly *sharable*, guidelines need to be represented in a standardized fashion. Thus, Uruz enables the user to embed in the guideline document, especially when using the PBW, terms originating from standard vocabularies, such as ICD-9-CM for diagnosis codes, CPT for procedure codes, and LOINC for observations and laboratory tests. In each case, the user selects a term when needed, through a uniform, hierarchical search interface to our Web-based *vocabulary server*.

The overall hybrid representation methodology, the functionality of the Uruz module, and the feasibility of using it by clinicians to represent guidelines with

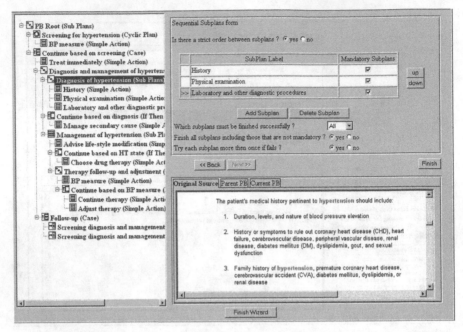

Fig. 7.19 The Asbru plan-body wizard (PBW) module. The module supports creation of an Asbru semi-formal control structure. On the left, the guideline's structure tree is displayed and updated dynamically as the user decomposes the guideline. On the upper right, the user is prompted with wizard-like questions to further specify the selected control structure. In the bottom right, the text of the source, current, or parent guidelines are displayed, as needed by the editor.

adequate completeness and correctness have been rigorously evaluated, using a group of physicians who structured guidelines from several different clinical domains [369].

7.6.6.2 IndexiGuide: Semantic Classification of Guidelines

To facilitate guideline retrieval, the medical expert indexes the guideline document by one or more intermediate or leaf nodes within one or more *external (indexing) semantic axes* trees, using the *IndexiGuide* tool [361].

Currently, the semantic axes include:

1. The Symptoms and Physical Signs axis (e.g., hypertension), which is based on the MeSH standard.
2. The Laboratory and Special Diagnostic Procedures axis (e.g., blood-cell counts), which is based on the CPT standard.The Disorders axis (e.g., endocrine disorders, neoplasms), which is based on the ICD-9 CM standard.
3. The *Disorders* axis (e.g., endocrine disorders, neoplasms), which is based on the ICD-9 CM standard, a version of ICD.

4. The Treatments axis is a combination of a hierarchy of pharmacological treatments (e.g., antibiotic therapy), which is based on the VA-NDF standard, and a hierarchy of other treatments (e.g., Surgery, special therapeutic procedures, anesthesia), which is based on the CPT standard.
5. The Body Systems and Regions axis (e.g., pituitary gland), which is based on the MESH standard.
6. The Guideline Types axis (e.g., screening, prevention, management).
7. The Medical specialties axis (e.g., internal medicine).

It is interesting to note that the determination of the multiple semantic classification indices can be initiated by an automated suggestion, through an analysis of the indexed guideline's text and comparison of its free-text content to a previously classified (using the same semantic hierarchy) set of guidelines [274].

7.6.6.3 Vaidurya: Concept-Based and Context-Sensitive Search and Retrieval of Guidelines

The *Vaidurya* hybrid guideline search and retrieval tool [346] exploits the existence of the free-text source, the semantic indices, and the marked semi-structured-text. Figure 7.20 shows the Vaidurya query interface. The user, performing a search, selects one or more concepts from one or more external (indexing) semantic axes, or scopes, to limit the overall search. (e.g., DISORDERS = *hypertension*; TREATMENTS = *medication*). The tool also enables the user to query marked-up guidelines for the existence of terms within the internal context of one or more target-ontology's knowledge roles (e.g., in the case of Asbru, the FILTER CONDITION context includes the term pregnancy). For search using external scopes, the default constraint is a conjunction (i.e. AND) of all selected axes (e.g., both a Cancer diagnosis within the disorders axis and a Chemotherapy therapy within the TREATMENTS axis) but a disjunction (i.e. OR) of concepts within each axis. For internal contexts, the default semantics are to search for a disjunction of the key words within each context, as well as among contexts (i.e, if using the Asbru target ontology, either finding the term diabetes within the FILTER CONDITION context or the term hypertension within the EFFECTS context). The search results are browsed, both as a set and at each individual-guideline level, using a specialized guideline-visualization tool.

The Vaidurya tool was extensively evaluated on several medical document and guideline repositories, in particular, the National Guidelines Clearinghouse with highly encouraging results, demonstrating the importance of both concept-based and context-sensitive search [275].

7.6.6.4 VisiGuide: Semantic Browsing of Clinical Guidelines

The *VisiGuide* browsing and visualization tool [361] enables users to browse a set of guidelines returned by the Vaidurya search engine and visualize their structure. It is linked to the various DeGeL applications, allowing the user to return one or

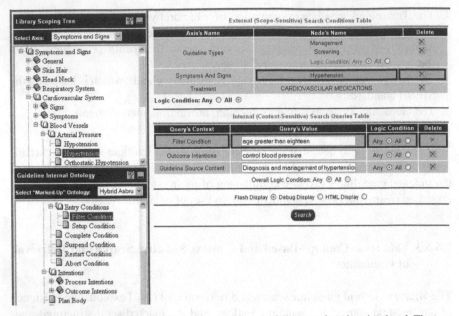

Fig. 7.20 The Vaidurya Web-based, context-sensitive, guideline search and retrieval tool. The user defines the relevant search scope by indicating one or more nodes within the semantic axes (upper left and right frames). The search can be further refined by specifying terms to be found within the source text, and even (after selecting a target ontology such as Asbru), within the context of one or more particular knowledge roles of that ontology (middle right frame).

more selected guideline for use within the Uruz markup tool or the IndexiGuide semantic classifier. Like the Vaidurya search and retrieval tool, VisiGuide makes no assumptions regarding the guideline's ontology, and dynamically parses a guideline ontology expressed as an XML schema, although it can have extensions for specific ontologies (e.g., for display of the Asbru semi-formal plan-body, such as acquired by the PBW).

Visiguide organizes guidelines along the semantic axes in which they were found, distinguishing between axes that were requested in the query (e.g., DISORDERS = *breast carcinoma* and TREATMENTS = *chemotherapy*) and axes that were not requested but which where originally used to classify a retrieved guideline (e.g., TREATMENTS = *radiotherapy*). Axes that were requested in the query but in which no guideline was found are highlighted (differently) as well.

In the *multiple-guideline* display mode (Figure 7.21), a table listing the content of desired semi-structured knowledge roles for all retrieved guidelines or for all guidelines that are indexed by a certain semantic axis can be created on the fly by simply indicating the interesting knowledge roles in the target ontology by which the guideline was marked (semi-structured), thus enabling quick comparison of several guidelines. Several default views exist, such as for eligibility determination or quality assessment. In the *single-guideline* display mode, a listing of the content of each of the knowledge roles or any combination can be requested, thus supporting actual application or quality assessment.

7.6.6.5 DeGeLock: Authorization and Permission in the DeGeL Library

Due to practical and legal considerations, any digital guideline library must include a comprehensive *authorization* model. The hierarchical model used in DeGeL, implemented by *DeGeLock* module, the uses the notions of *virtual expert groups* and of the different *functionalities* inherent in the hybrid meta-ontology model, which imply different levels of authorization. Guideline editors are *members* of one or more *(editing) groups* and have different authorizations in each group [360]. Groups are organized by medical specialty (e.g., oncology). Each *group manager* can accept applications to be a group member, and sets and maintains the authorization configuration of each member in that group. Members of a group can only edit and classify guideline documents based on source guidelines *owned* (uploaded) by a group member, but cannot edit guideline documents owned by another group.

The DeGeL authorization model assumes that each module (e.g., Uruz) enables users to perform several *tasks*. For example, within Uruz, users can View, Edit, Search within guideline documents. Each user is given (within each group) a specific authorization configuration for each module.

To facilitate management, the DeGeL architecture includes several predefined common authorization profiles (more can be constructed in similar fashion).

1. *Searcher* (visits the library, performs searches, views guidelines which have been edited by other users).
2. *Classifier* (classifies guidelines alongside semantic axes).
3. *Expert Editor* (specifies guidelines' content up to the semi-structured level, using DeGeL's hybrid meta-ontology).
4. *Knowledge Engineer* (cannot markup the guideline, but can fully structure the marked-up text up to machine-comprehensible level in the full target ontology).
5. *Group manager* (manages permissions of their group members).
6. *System Administrator* (manages users and groups).

Each user profile targets a specific population of potential users. The majority of physicians will use the library as Searchers; a small number of experts in each specialty will serve as Classifiers or Editors.

The default configuration profiles for each authorization type are predefined. A group manager can easily assign a new member to a predefined authorization type, possibly modifying the configuration if needed, using a Web-based graphical *authorization-management* tool, which the DeGeL developers had developed for that purpose [360]. The management tool is also used by system administrators to manage all DeGeL users, including group managers. Group managers and administrators can view details of group members, authorize addition of new members, and change authorization configurations for existing members. For example, selecting the *Classifier* authorization type defines a particular default configuration, which

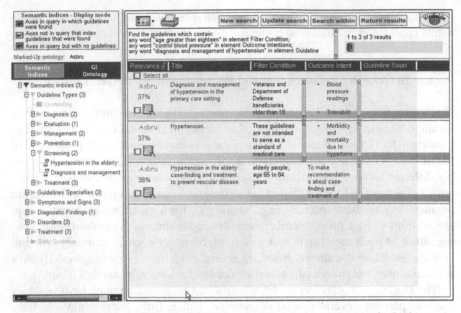

Fig. 7.21 An example of the VisiGuide Interface in the *multiple-ontology* mode. In this mode, multiple guidelines, typically retrieved by Vaidurya search engine, are displayed within the various semantic axes indexing them (left frame); the contents of knowledge roles relevant to the user are displayed and compared as a table (right frame). The "Return Results" button returns selected guideline back to the requesting application (e.g., to the Uruz markup tool). The left-hand icon in the top menu enables the user to select among several preset views, such as an *eligibility view* (which in the case of the Asbru ontology displays, for all guidelines, along with the guideline's title, the filter and *setup* conditions)

authorizes classification in *IndexiGuide*, but not editing in *Uruz*. (When the tool is being used by a system administrator, it displays additional options for extended maintenance).

7.6.6.6 The DeGeL Runtime Tools

In addition to the various DeGeL specification and retrieval tools, several Asbru-specific tools have been developed for runtime guideline application and for retrospective quality assessment of guideline-based care. The *Spock* runtime-application module [432, 433] is a hybrid Asbru runtime application module, which initially focused mainly on the semi-structured and semi-formal representations. *QualiGuide* is a retrospective quality-assessment tool, which uses the concept of intention-based quality assessment and the Asbru *intentions* knowledge role. Besides using the DeGeL knowledge base (i.e., the guideline library itself), both tools use the IDAN mediator [33] and the KNAVE-II visualization tool [354] (described in the chapters discussing temporal abstraction and visualization, respectively) to query and explore the patient's data.

7.7 Additional Research Directions in Clinical Guideline Representation and Application

There exist additional approaches to the problem of guideline representation and application. One approach is treating clinical guidelines as medical business processes, thus enabling their representation using models and languages designed to encode business processes. An example of such a conceptual model is BPMN (Business Process Modeling Notation) [2], while a typical language is BPEL (Business Process Execution Language) [1]. Thus one can either use process management concepts to support representation and application of specific medical processes and in particular clinical guidelines [343] or extend general workflow conceptual models with temporal constructs, allowing a more appropriate representation of temporal aspects of clinical processes and guidelines [92].

In addition, the application of a guidelines creates an execution trace that can be viewed as a set of time oriented measurements and actions. This set can be the basis for inference and reasoning using methodologies such as described in the previous chapter of this book; for example, one can compare the temporal pattern that was expected to be formed by an instance of a correct application of the clinical guideline to the observed pattern formed by the patient actual longitudinal record. The result can be the basis for assigning a quality assessment score to the guideline application process, or support a capability for automated recognition of which guideline, if any, the physician is using [36, 91].

Summary

Clinical guidelines, especially when coupled with the principles of *evidence-based medicine* (*EBM*), are here to stay for a long time, and their efficacy, when followed, is established.

It is not sufficient to use just simple one-time alerts and reminders, although these have their place; frameworks for complex, multi-step, longitudinal care plans, which can handle incremental execution over long time periods, are needed as well. The importance of such frameworks is especially pronounced in the context of caring for chronic patients, the care of which currently is responsible for most of the costs of modern health-care systems. Thus, significant effort must be invested in integration of temporal-reasoning frameworks in platforms for guideline-based care. An example of using a temporal mediator for that purpose was presented in the case of the DeGeL architecture.

However, there is still a clear need for effective integration of automated guideline-support tools at the point of care and at the point of quality assessment, which will relieve the current information overload on both care providers and health-care administrators.

To be effective, these tools will need to be linked to the patient's local record using standardized communication protocols, medical-record schemas, and controlled medical terminologies; must use standard medical vocabularies, should have clear semantics, must facilitate knowledge maintenance and sharing, need to be sufficiently expressive to explicitly capture the design rationale (process and outcome intentions) of the guideline's author, and must be integrated in the overall clinical workflow. At the same time, any approach, to be acceptable to clinicians, must enable flexibility at application time to the attending physicians and their local favorite methods.

Bibliographic Notes

The reader who is interested in learning more about automation of clinical guidelines is well advised to refer to Ten Teije et al.'s primer on computerized guidelines [400]. An excellent case-based comparison of the major early approaches to clinical guideline representation can be found in Peleg et al.'s survey [305].

Problems

7.1. A developer of a set of clinical reminders represented as medical logical modules using the *Arden* syntax needs to reuse in several different modules a certain definition of renal insufficiency in terms of several raw laboratory parameters, such as urine albumin, that exist in the local hospital information system's database schema. Explain what problems might be encountered by the developer who is trying to reuse that knowledge in her local system, and by colleagues who try to share her rules in their own hospital information systems. By what means could they attempt to counter these problems?

7.2. Explain (as an informal architecture and an informal algorithm) how a guideline *eligibility-determination* component in *EON*, which would decide what guidelines might be relevant for each patient, could be constructed from several or all of the EON components (e.g., the knowledge acquisition tool, the temporal mediator, and the ESPR therapy-planning method). (Note: refer to papers such as by Musen et al. [278] or Tu and Musen [407, 408] for further details regarding the EON architecture.)

7.3. Examine the following clinical guideline for management of gestational diabetes (a tendency for high blood-glucose values in certain pregnant women), whose goal is to manage patients without resorting to insulin (whose effect is to reduce blood-glucose level).

This guideline should be considered for every (female) patient who is pregnant. Entry conditions include having a glucose-tolerance test after the 24th and before the 26th week of conception, whose result is between 140 and 400 mg/100cc.

The target of the guideline is to maintain fasting (prebreakfast) blood glucose levels at less than 100 mg/100cc, and postmeal blood glucose values at less than 130 mg/100cc. In general, one of the goals is to avoid periods of more than a week of continuous high fasting blood glucose. Another target is that by the end of the guideline the patient will have had at least 3 visits at the dietician's office.

The guidelines calls for starting on the same visit the monitoring of daily fasting blood glucose, monthly nutrition management, and observation of indications for a need to transfer to an insulin regimen.

The guideline should be continued until delivery, but should be stopped (and insulin therapy started) when an indication for an insulin regimen (as defined elsewhere) is detected. The guideline can be halted temporarily and the patient reevaluated, if the patient has had four consecutive days of high fasting blood glucose.

Describe informally, but using the terms of the *Asbru* ontology, at least three actions, three intentions, and three conditions of the guideline (the more you find, the better). Point out the precise type of each intention, condition, or action(s) you mention. (*Note*: Refer to the Shahar et al. [355] paper or Miksch et al. [263] paper for further details about the Asbru language and its semantics).

Part IV
The Display of Time-Oriented Clinical Information

Chapter 8
Displaying Time-Oriented Clinical Data and Knowledge

Overview

This chapter introduces the reader to the main topics related to the visualization of temporal clinical information. More specifically, the chapter focuses both on displaying raw temporal clinical data and on displaying more abstract temporal clinical data and temporal clinical knowledge. The reader is first introduced to the main aspects of information visualization in medicine and then to the more specific topics related to the visualization of temporal clinical information. Then, several research methodologies and their respective system prototypes are discussed in detail, to demonstrate the different aspects related to the implementation of advanced visualization systems for medical and clinical information, ranging from the visualization of raw temporal data, and the display of more abstract time-oriented data, to the visual composition of temporal queries, and the visual representation of temporal knowledge. The reader learns how to deal with the design and the evaluation of software systems for the visualization of (possibly) complex temporal clinical information. Finally, the reader is guided through the details of several specific systems, designed and implemented for the purpose of visualization and exploration of different kinds of temporal clinical data and knowledge in different real world clinical settings.

Structure of the Chapter

We now move to the task of displaying and visually exploring time-oriented clinical data and knowledge. Information Visualization has become a broad research area within the scientific community of Human-Computer Interaction (HCI). Within the area of information visualization, medical information has attracted some particular attention, since medical data, and in particular clinical data, present several

C. Combi et al., *Temporal Information Systems in Medicine*,
DOI 10.1007/978-1-4419-6543-1_8, © Springer Science+Business Media, LLC 2010

challenging aspects that must be considered when designing systems that allow the visualization and/or visual exploration of (possibly) huge amounts of medical data.

In this chapter, we first give a brief introduction to several basic concepts of information visualization with an emphasis on aspects that are closer to the display of temporal medical data, then we mention several approaches explored by researchers to enable the display of different aspects of temporal medical information, and finally propose a simple taxonomy that we will use in the chapter to describe several real world applications dealing with temporal clinical information visualization. Then, the chapter provides a finer level of detail regarding five research approaches, implemented as prototype systems that deal with different kinds of temporal data and knowledge and with different medical domains. The description of these systems enables the reader to focus in some detail on the difference among different approaches and different application requirements, ranging from (i) techniques for the display of temporal raw data, specifically applied to the visual representation of clinical parameters acquired during hemodialysis sessions, through (ii) the multimodal display of vital signs, symptoms, and visit data, applied to the management of cardiology follow-up patients, (iii) the knowledge-based display and exploration of single-patient data and abstractions, with an emphasis on the monitoring of patients after bone-marrow transplantation, and (iv) the visual composition and definition of complex temporal abstractions to query temporal clinical data with an emphasis on the evaluation of the proposed techniques in a real medical setting, and up to (v) the knowledge-based visualization and exploration of clinical data of multiple patients, with examples from an oncology domain.

The approaches described in this chapter build on the clinical domain representations and applications described and discussed in the previous chapters: in particular, the data models and computational methods adopted by the systems discussed in this chapter were discussed in chapters 4 and 5.

Keywords

Information Visualization, Medical Visualization, Human-Computer Interaction, Timelines, Time relations, Visual temporal queries, Visual temporal patterns, Visual data mining.

8.1 Introduction: When is Picture Worth a Thousand Words?

The amount of clinical data available as part of various medical tasks is increasing every day as a result of technology advancements in computer performance, storage capacity, and networking [67]. It is well known that human cognitive and perceptual capabilities limit the quantity of information a user can observe and manage at a given instant. This becomes more evident as the number of accumulating data items gradually increases over time, and even more so when their temporal dimension must be considered for the purpose of clinical, research, or health-policy decision making. In this situation, members of the clinical staff, such as physicians, technicians, nurses, or students, might not be able to properly assess and analyze the implications of large amounts of time-oriented data and might be overwhelmed by them, if computer applications will not provide suitable solutions for presenting clinical data and for interacting with them.

Answers to these issues are the main theme of the research area called Human-Computer Interaction (HCI).). IV plays an important role in the development of medical information systems. Furthermore, since the medical domain is rather unique in its emphasis on the importance of time and on the need for applying domain-specific knowledge, several specific approaches for the visualization of temporal data and knowledge have been suggested by IV researchers [67, 84, 105, 310].

The importance of information visualization and interactive exploration to the effectiveness of medical decision making can be best appreciated in the larger context of the research on the enhancement of reasoning through an appropriate *diagrammatic representation*.

Larkin and Simon [236], in a highly influential, mostly theoretical, yet quite convincing paper, made an important distinction between *Sentential representations*, which are sequential, such as the propositions in a text, and *Diagrammatic representations*, which are indexed by a two-dimensional location in a plane. Diagrammatic representations also typically display information that is only implicit in sentential representations and that therefore has to be computed, sometimes at great cost, to make it explicit for use.

Larkin and Simon [236] have demonstrated in their paper that the usefulness of visual representation is mainly due to three different aspects of problem solving:

- the reduction of the effort necessary for *search*, through the use of direct perceptual inference, such as localization of all relevant information in one location (e.g., finding the angle between two adjacent sides of a triangle, given a particular vertex);
- facilitation of *recognition* of the conditions necessary for application of certain production (inference) rules, through the use of appropriate diagrammatic representations (i.e., recognizing a certain pattern, such as a very sharp angle, which corresponds to the left hand side, or condition, of the production rule);
- more efficient *computation* of the conclusions of inference rules, through the use of efficient graphical representations (e.g., computing the maxima or minima for each axis, of a set of points in a plane).

As we shall see in the current chapter, different types of IV approaches often try to facilitate one or more of these problem-solving aspects.

8.1.1 Information Visualization Basics

Visualization can be defined as the act or process of interpreting in visual terms or of putting into visible form some concrete or abstract concept [67]. Among the various definitions of *Information Visualization*, we consider here the following two, which underline some basic concepts: IV can be defined as "the process of transforming data, information, and knowledge into visual form making use of humans' natural visual capabilities" [154] or, more concisely, as "the computer-assisted use of visual processing to gain understanding" [49]. It is worth observing that, according to these definitions and in contrast to most medical imaging data, IV focuses also, and mostly, on *abstract* information which cannot be automatically mapped to the physical world because it has no natural and obvious physical representation.

IV aims at reducing the complexity of the examination and understanding of information for humans, by designing proper techniques for the visual display of data. These techniques are aimed at achieving a number of goals, such as:

1. allowing users to explore available data at various levels of abstraction,
2. giving users a greater sense of engagement with data,
3. giving users a deeper understanding of data,
4. encouraging the discovery of details and relations which would be difficult to notice otherwise, and
5. supporting the recognition of relevant patterns by exploiting the visual recognition capabilities of users.

As highlighted by the last two goals, IV aims at making the user an active element in pattern recognition, allowing him/her to detect what would pass unnoticed through automatic recognition systems. From this point of view, IV components and data mining components can have complementary and synergic roles inside the same application. Different research groups in IV have proposed diverse approaches to achieve one or more of the above mentioned goals. Several attempts at proposing taxonomies for the different kinds of IV have been made in the literature. The most cited and influential proposal appears in [374], in which Shneiderman classifies IV approaches along two dimensions: *data type* and *task*.

The task dimension of the classification identifies seven high-level tasks [374]: *overview*: gain an overview of the entire set of data; *zoom*: zoom in on a subset of items of interest; *filter*: filter out uninteresting items; *details-on-demand*: select one or more items and get details; *relate*: view relationships among items; *history*: keep a history of actions to support undo, replay, and progressive refinement; *extract*: allow extractions of subsets of items.

Seven data types for the items to be displayed are identified: *1-dimensional*: linear data organized in a sequential manner, such as alphabetical list of names, program source code, or textual documents; *2-dimensional*: planar or map data covering some part of an area, such as maps, newspaper layouts, or photographs; *3-dimensional*: data with volume and potentially complex relations with each other, such as molecules, the human body, or buildings; *temporal*: data with a start time, finish time, and possible overlaps on a timescale, such as that found in medical records, project management, or video editing; *multi-dimensional*: data with n attributes which become points in an n-dimensional space, such as records in relational and statistical databases), *tree*: collections of items linked hierarchically by a tree structure, such as computer directories, business organizations, genealogy trees; *network*: collections of items linked by a graph structure, such as telecommunication networks, World Wide Web or hypermedia structures.

The data items in each above mentioned category can have multiple attributes (e.g., a 3-dimensional object could have additional attributes such as color, level of transparency, and brightness; a node item in a tree could have additional attributes such as a name, a creation date, a modification date; and so on). Therefore, the separation among the different categories is not always strict, e.g., temporal data can be also seen as an instance of multi-dimensional data. However, this separation is useful for orienting the choice of IV techniques, e.g., when the temporal aspect is dominant in the considered data, display techniques that give a central role to time (such as timelines to visualize personal histories [310], temporal animations to show the evolution of physical phenomena using a familiar VCR metaphor, and so on) can give better results than more general techniques which do not assume specific relations among the multiple attributes.

As one can begin to notice from the previous considerations, *interactivity* is a typical feature of IV systems: an high level of interactivity is indeed important to increase the engagement of the user in the observed data and enhance his/her exploration abilities. More elaborate examples will be described in the next sections, which focus on IV for temporal medical data. For a broad survey of generic IV techniques, we refer the reader to [50]. In spite of the growing number of IV techniques in use in medical applications, a word of caution is needed for the developer who wants to add an IV component to his/her system: since only a few tested guidelines exist [320], and no disciplined design methodologies and engineering principles have been yet identified, special care is needed in order to obtain an effective design.

8.1.2 From Temporal Clinical Data to Temporal Knowledge

Among the data types considered in Shneiderman's taxonomy, the *temporal* data type has a particular relevance to medical IV, as the temporal dimension is always present both in medical data, in the related knowledge, and in the considered decision-oriented tasks: visualizing and exploring temporal clinical information has been widely acknowledged as a relevant need in the medical domain [50]. More precisely, visualization and exploration of information in general, and of large amounts of time-oriented data in particular, is essential for effective medical decision making.

We can now consider several specific IV tasks in the medical domain, and the problems encountered when attempting to tend to all of their requirements.

Visualizing Temporal Histories

Researchers in the areas of *visualization* of *clinical time-oriented data* [105, 319], have developed useful visualization techniques for static presentation of *raw* time-oriented data and for browsing information.

One of the problems that has attracted particular attention since the early '90s [105] has been the visualization of *patient histories*. One of the first systems proposed for this kind of data was the Time Line Browser [105] which visualized instant events (such as the measurement of a clinical parameter with its value) and intervals with duration (such as the status of the patient) on a timeline.

A more elaborate visualization is proposed in Lifelines [310], a system that faces the problem of visualizing patient's histories. In Lifelines, facts are displayed as lines on a graphic time axis, according to their temporal location and extension; color and thickness are then used to represent categories and significance of facts. Lifelines provides a compact overview of the relevant events and intervals, organized into different screen areas, each one devoted to a different aspect of the medical record (such as consultations, conditions, hospitalizations, medications, ...): the user can select items of interest and get details on demand (e.g., a lab report) or perform a temporal zoom-in/zoom-out of the examined range of time, causing a dynamic rearrangement of the events and intervals displayed. Lifelines [310] is one of the best known visualization environment that deals with the history browsing task.

Another interesting problem often appearing in the context of medical temporal data is to augment existing standard representations used by clinicians, by the visualization of additional information. For example, Lowe et al. [249] consider the display methods typical of anesthesia monitors and propose several extensions that convey additional information concerning certainty and vagueness of the displayed data, e.g., color coding is used to give an indication of different levels of certainty by means of various shades of the same color.

Visualizing Knowledge-Based temporal Histories

In previous chapters, especially Chapter 5 we have emphasized the importance of temporal abstraction to the medical domain. In recent years, researchers have investigated various techniques for *information visualization*, including conceptual maps, radar maps, tree maps (conetree/camtree maps, hyperbolic tree), Kohonen maps, fish eye views and dynamic queries interfaces and more [49, 65, 175, 373, 375], as well as using well-known statistical and graphical methodologies, such as three-dimensional representations, scattergrams, pie charts, bar charts, and their derivative techniques. These display methods, however, typically do not focus on visualization of domain-specific *temporal abstractions* and on the issue of *interactive manipulation* and *exploration* of the data and multiple levels of its abstractions, using domain-specific knowledge. Typically, these capabilities have been omitted from several approaches to the visualization of complex information, because they require a formal, domain-independent representation of the domain-specific temporal-abstraction knowledge, considerable effort in modeling the visualized domain, and availability of computational mechanisms for creation of the abstractions.

In this chapter we shall describe, however, several newer approaches that do exploit the existence of a domain ontology and of advanced temporal-abstraction computational capabilities.

Visually Querying Temporal Histories

Although the proposals for visualizing medical histories deal with the representation of intervals and interval relations to some extent, they are not meant to facilitate the visual expression of more complex temporal relations such as those present in temporal patterns that have to be specified for querying the clinical history database. For example, a physician could want to consider only the histories of those patients which were prescribed aspirin and, after the start of the therapy, had an episode of dyspnea followed by headache. In this case, the system has to support the visual definition of a temporal pattern that can be matched by several histories. A suitable graphic definition of such patterns is complementary to the visualization of personal histories, in the context of a more powerful visualization environment for the exploration of large sets of histories.

The definition of temporal patterns is also a basic step in defining *composite temporal abstractions* [347]. While basic temporal abstractions consist of high level information derived from several time-stamped raw data, composite temporal abstractions are obtained from other previously defined basic and/or composite abstractions. Previous attempts at providing user interfaces for the definition of clinical abstraction [81, 352] have assumed a skilled user, able to manage all the technical details of the definition process, and have used the visualization capabilities of graphic interfaces only in a very limited way. Furthermore, for most of the above contributions, the experimental evaluation of the proposed visualizations is either missing/limited or performed without any rigorous statistical backing.

Visual Analysis of Temporal Histories and of Derived Concepts

As discussed in previous chapters, management of patients, especially chronic patients, requires collection, interpretation and exploration of large amounts of time oriented data. One of the ways to support care providers performing tasks such as diagnosis, therapy, and research is by supplying them with the technology for on-the-fly visualization, interpretation and exploration of the data and of the higher level, knowledge-based, concepts (abstractions) that can be derived from these data. Indeed, one of the benefits of creating temporal abstractions is that domain-specific, meaningful, interval-based characterizations of time-oriented data are a prerequisite for effective *visualization* and *exploration* of these data [105, 351, 352].

Temporal Knowledge Visualization

The visualization of temporal knowledge is a step forward: the focus in this case is to provide a visualization of intensional information, as for example, that underlying several different therapy plans, that can be adapted and further specified for single patient. An example is the AsbruView system [225] for specification of longitudinal guidelines using the ASBRU ontology, which was described in Chapter 7. Another example, although a less visual one, is the temporal-abstraction knowledge-acquisition tool, used to acquire and maintain knowledge needed by the KBTA method, which was described in Chapter 5. An example that will described at length in this chapter is a system using a paint-roller metaphor for specification of complex temporal patterns [68].

8.1.3 Desiderata and a Taxonomy of Dimensions for Systems Performing Visualization of Temporal Data and Knowledge

Considering the data-, information-, and knowledge-visualization tasks discussed so far in this chapter, and in particular, the special needs of various types of clinicians, researchers, and administrators in the medical domain, the desiderata for provision of a software system for interactive exploration of time oriented clinical data may be summarized through the following requirements:

1. *Accessibility.* A modular, scalable architecture to enable the application of diverse (and preferable distributed) types of knowledge to the same data and different data bases to the same knowledge; the architecture should support access to both the data and the abstractions derivable from it.
2. *Visualization of raw data and abstractions.* Effective visualization and exploration, which should include both raw data and the abstract concepts derived from those data.
3. *Temporal granularity.* The visualization should support interactive exploration of time oriented data at different levels of temporal granularities.
4. *Absolute and Relative time lines.* The system should support both a calendar-based time line as well as a relative time line, which refers only to clinically significant events (e.g., start of therapy).
5. *Intelligent exploration of raw data and abstractions.* Effective exploration of both raw data and their abstractions, using meaningful domain-specific semantic relations.
6. *Explanation.* The system should enable an explanation of abstractions from data and knowledge.
7. *Statistics.* The system should provide statistics on both the raw data and the abstracted concepts.
8. *Search and Retrieval.* The system should support easy and fast search and retrieval of clinically significant concepts.

9. *Dynamic Sensitivity Analysis.* The system should include capabilities for interactive exploration of the effects of simulated hypothetical modifications of raw data on the derived concepts.
10. *Clinical-Task support.* The system should be customizable for a specific clinical task (e.g., monitoring of diabetic patients).
11. *Collaboration.* The system should support collaboration between different clinicians and researchers on the same patient data and abstracted knowledge.
12. *Documentation.* The system should support documentation of the exploration process.

Furthermore, according to the previous discussion, it should be noted that two major types of visualization can be distinguished:

- visualization of the time-oriented *data* or of the concepts derivable from these data;
- visualization of *knowledge* regarding time-oriented information.

In some more detail, one can distinguish at least four dimensions of visualization and interactive exploration of time-oriented data or knowledge, in particular in medical domains:

- subject cardinality (SC): individual versus multiple patients;
- abstraction level (AL): raw data versus abstract concepts versus knowledge;
- concept cardinality (CC): one concept versus multiple concepts;
- temporal granularity (TG): single versus multiple time units.

In the rest of this chapter, we will discuss in detail several system prototypes and their related design aspects, using the suggested dimensions taxonomy.

We will move from proposals dealing with raw data to other ones related to the visualization of temporal knowledge, according to the classification provided in Table 8.1 and considering the highlighted desiderata.

Table 8.1 A characterization of the information visualization prototypes described in this chapter according to the proposed taxonomy of dimensions. SC = subject cardinality; AL = Abstraction level; CC = concept cardinality; TG = temporal granularity.

Prototype	SC	AL	CC	TG
IPBC	single	raw data	single	single
KHOSPAD	single	abstract data	multiple	multiple
KNAVE II	single	raw data and abstractions	multiple	single but variable
Paint rollers	single	abstractions	multiple	single
VISITORS	multiple	raw data, abstractions and knowledge	multiple	multiple

¢

8.2 Interactive Exploration of Clinical Raw-Data Time Series: the IPBC System

The system proposed by Chittaro et al. [69], called IPBC (*Interactive Parallel Bar Charts*) connects to the hemodialysis clinical database, produces a visualization that replaces tens of separate screens used in traditional hemodialysis systems, and extends them with a set of interactive tools. The data visualization is based on bar charts. Historically, bar charts are a widely adopted approach to display time-series [410]. Unfortunately, while a bar chart allows for an easy comparison among the data values for a single time-series, when the considered task requires to compare a *collection* of time-series (such as a monitored signal from the same patient in different sessions of the same clinical treatment), traditional bar charts (as other historical approaches) become unfeasible. IPBC visually represents each time-series in a bar chart format where the X axis is associated with time and the Y axis with the value (height of a bar) of the series at that time. Then, the obtained bar charts are laid out side by side, using an additional axis to identify the single time-series, in a 3D space. It must be noted that also the additional axis has typically a further temporal dimension, e.g., it is important to order the series by date of the hemodialysis session to analyze the evolution of the state of a patient. An example is shown in Figure 8.1, that illustrates a visualization of 47 time-series of about 240 values each, resulting in a total of more 11000 values (the axis on the right is the time axis, while the axis on the left identifies the different time-series). Hereinafter, we will refer to this visual representation as a *parallel bar chart*.

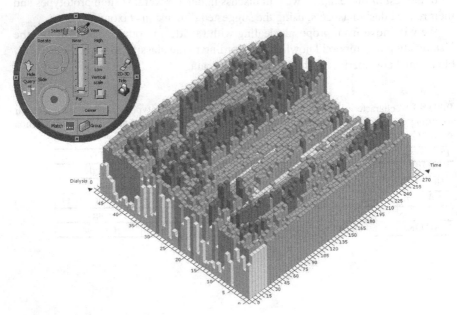

Fig. 8.1 A Parallel Bar Chart.

While a 3D space can significantly increase both the number of time-series that can be simultaneously analyzed in a convenient way and the number of values associated with each time-series, it poses well-known problems such as occlusions, 3D navigation, difficulties in comparing heights, proper use of space, and the need for effective interaction techniques to aid the user in the analysis of large datasets (e.g., highlighting interesting patterns, checking trends, ...).

In designing how the different interactive functions of IPBC should be invoked by the user, two different problems have been considered: (i) one well-known limitation of many 3D visualizations is the possible waste of screen space towards the corners of the screen; (ii) an approach based on traditional menus would require long mouse movements from the visualization to the menu bar and vice versa, for the interactive analysis of data.

A specific round-shaped pop-up menu (see Figure 8.2), called *RoundToolbar* (RT), is the proposed solution to the mentioned issues. RT can appear anywhere the user clicks on the screen with the right mouse button and can be easily positioned in the unused screen corners, thus allowing a better usage of the screen space (e.g., see Figure 8.1) and a reduction of the distance between the visualization and the menu. Moreover, to further improve the selection time of functions with respect to a traditional menu, the organization of modes in the toolbar is inspired by Pie Menus: in particular, the main modes are on the perimeter of the RT, and when a mode is selected, the center of the RT contains the corresponding tools (which are immediately reachable by the user, who can also quickly switch back from the tools to a different mode).

It is well-known that free navigation in a 3D space is difficult for the average user, because (s)he has to control 6 different degrees of freedom and can follow any possible trajectory. To make 3D navigation easier, when the *View* mode is selected in the RT (as in Figure 8.2), the proposed controls (called *Rotate*, *High-Low* and *Near-Far*) for viewpoint movement cause movement along limited pre-defined trajectories which can be useful to examine the visualization. The *Slide* control simply allows the user to place the entire visualization in a specific screen position: to do so, (s)he has to drag the circle within the widget in the desired direction. The *Vertical scale* control in the *View* mode is used to scale the bars on the Y axis. Vertical scaling has been included in the *View* mode, because it has been observed that when users scaled the bars, they typically changed the viewpoint as the following operation. Finally, the *Center* button is used to restore the default viewpoint position which points to the middle of the bar chart visualization.

IPBC uses color to classify time-series values into different ranges. In particular, at the beginning of a session, the user can define her general *range of interest* for the values, specifying its lowest and highest value. These will be taken as the lower and upper bounds for an IPBC dynamic query control (shown in Figure 8.3) that allows the user to interactively partition the specified range into subranges of interest. Different colors are associated to the subranges and when the user moves the slider elements, colors of the affected bars in the IPBC change in real-time. For example, Figure 8.1 shows a partition that includes the three subranges corresponding to the colors shown by the slider in Figure 8.3. Possible values outside the specified

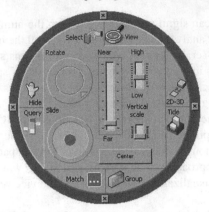

Fig. 8.2 The View mode.

general range of interest are highlighted with a proper single color and (optionally) the corresponding bars are reduced to a zero height (this feature is typically used by clinicians to highlight values which are physiologically impossible and are due to measurement errors, e.g., disconnected or malfunctioning sensors). The entire color coding scheme can be personalized by the user. The dynamic query control allows the user to move the two slider elements *independently* (to change the relative size of adjacent subranges). For example, in Figure 8.3, one has been set to 140 mmHg and the other to 175 mmHg. This can be done both by dragging the edge or (more easily) the tooltips which indicate the precise value. Up and down arrow signs in the tooltips also allow for a fine tuning of the value. It is also possible to move the two slider elements together by clicking and dragging the area between the two bounds. This can be particularly useful (especially when the other areas are associated to the same color), because it will result in a "spotlight" effect on the visualization: as we move the area, our attention is immediately focused on its corresponding set of bars, highlighted in the visualization. Although it would not be difficult from a technical point of view to offer a dynamic query control with more than three variable subranges, it is interesting to note that the proposed organization into 4 categories of values (the 3 variable subranges inside the general range, plus the out of general range category) was the one considered most useful from the clinical user's point of view (introducing more subranges would thus only make the system more complicated and difficult to use).

When the time-series are generated from an high frequency sampled signal or, more generally, when the user wants to temporally aggregate data to obtain a more abstract view, (s)he can use the *Group By* mode of RT. In this mode, bar charts display the result of an aggregation function applied to raw data. In the current version of IPBC, aggregate functions allow to compute either the mean or the median of parameter values grouped according to their time of occurrence. Thus, the user can visualize (i) the *mean* calculated over intervals of a given duration; (ii) the *median* calculated over intervals of a given duration. Dynamic queries can obviously be applied also to this abstract data. Figure 8.4 shows a visualization obtained after

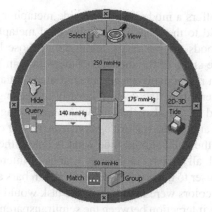

Fig. 8.3 The Query mode.

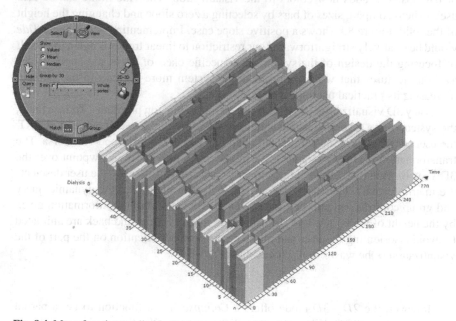

Fig. 8.4 Mean function applied to data grouped over 30 minutes intervals.

applying the mean function over intervals of 30 minutes to the dataset of Figure 8.1. Although the set of available functions can be easily extended, it is interesting to note that median and mean are the only two functions that were specifically requested by clinicians during the evaluation of preliminary versions of the system.

A frequent need in visual data mining is to quickly perceive how many and which values are below or above a given threshold. This can be easily done with the previously described dynamic queries when the threshold is constant. However, the required threshold is often time-varying, e.g., one can be interested in knowing how many and which values are not consistent with an increasing or decreasing trend.

For this need, IPBC offers a mode based on a tide metaphor. The *Tide* mode adds a semitransparent solid to the visualization: the solid metaphorically represents a mass of water that floods the bar chart, highlighting those bars which are above the level of water. The slope of the top side of the solid can be set by moving two tooltips in the RT (that respectively specify the initial and final values for the height of the solid), thus determining the desired linearly increasing or decreasing trend. The height of the solid can be also changed without affecting the slope by clicking and dragging the blue area in the RT. An *opaque/transparent* control allows the user to choose how much the solid should hide what is below the threshold. When the *Tide* mode is activated, all the bars in the user's range of interest are turned to a single color to allow the user to more easily perceive which bars are above or below the threshold; if multiple colors were maintained, the task would be more difficult, also because the chromatic interaction between the semitransparent surface and the parts of bars inside it adds new colors to the visualization. The *Tide* mode can be also used to help compare sizes of bars by selecting a zero slope and changing the height of the solid. Figure 8.8 shows a positive slope case. Implementing a non-linear *Tide* would be relatively straightforward (our restriction to linear trends is again the result of focusing the design of the system on a specific class of users and applications, avoiding features that would only make the system more difficult to use without increasing its practical usefulness).

As any 3D visualization, IPBC can suffer from occlusion problems. To face them, the system offers several possible solutions. By using the $2D - 3D$ mode on the RT, the user can transform the parallel bar chart into a matrix format and vice versa. The transformation is simply obtained by automatically moving the viewpoint over the 3D visualization (and taking it back to the previous position when the user deselects the matrix format). This can solve any occlusion problem (and the dynamic query and group controls can still be used on the matrix cells), but the information given by the height of the bars is lost. Transitions to matrix format and back are animated to avoid disorienting the user and allow her to keep her attention on the part of the visualization (s)he was visually focusing on.

Moreover, the $2D - 3D$ mode offers a *Collapse series* function to get a partial flattening of the IPBC. When the user clicks on a specific time-series (which can possibly occlude an interesting one), this bar chart and all the possibly occluding ones are collapsed (through a smooth animation) into a flat representation analogous to the matrix one, as illustrated in Figure 8.5. When the user selects a flat bar chart, all the flattened bar charts between the considered one and the first non-flat bar chart are raised up. Finally, the $2D - 3D$ mode offers another *Collapse* function to obtain a flattening of every bar that can possibly occlude a chosen bar, as shown in Figure 8.6. With this function, users can also collapse all bar charts before a given time, allowing a comparison of time-series at the same instant of time.

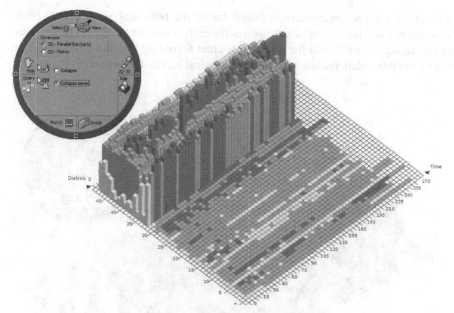

Fig. 8.5 Removing occlusions by flattening entire time-series.

8.2.1 Mining Hemodialytic Data with IPBC

In the following, we will show how IPBC can be used during real clinical tasks, to help clinicians in evaluating the quality of care given to single patients, on the

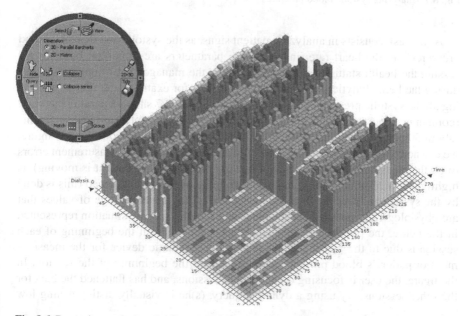

Fig. 8.6 Removing occlusions by flattening portions of time-series.

basis of the clinical parameters acquired during the hemodialysis sessions. Each hemodialysis session returns a time-series for each parameter; different time-series are displayed side by side in the parallel bar chart according to date (in this case, the axis on the left orders the sessions in chronological increasing order).

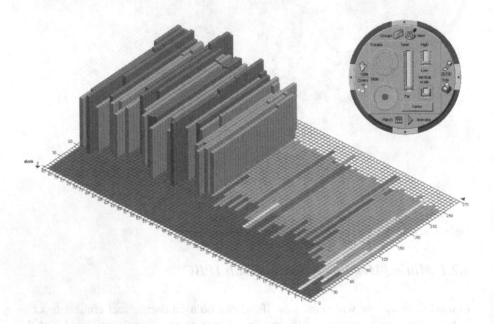

Fig. 8.7 Analyzing systolic blood pressures.

A first task consists in analyzing patient signs, as the systolic and diastolic blood pressures and the heart rate; indeed, these parameters are important both for assessing the health status of the patient and for the management of device settings during the hemodialytic sessions. Let us consider, for example, the task of analyzing all the systolic pressures of a given patient: Figure 8.7 shows a parallel bar chart (containing more than 5000 bars), representing the systolic pressure measurements (about 50 per session) during more than 100 hemodialytic sessions. In this figure, we can notice that the presence of out-of-scale values, related to measurement errors (e.g., the measurement device cannot operate properly if the patient is moving), is highlighted in a suitable color and the corresponding bars are flattened (this is done by the system because the clinician has specified the general range of values that are physiologically possible for the parameter). In the specific situation represented in the figure, the presence of several out-of-scale values at the beginning of each session is due to the fact that nurses apply the automatic device for the measurement of patient's blood pressure some time after the beginning of the session. In the figure, the user is focusing on a group of sessions, and has flattened the bars for the other sessions. By using a dynamic query, (s)he is visually distinguishing low (yellow bars), normal (orange bars), and high (red bars) blood pressures. By looking at the visualization, one can immediately notice that some sessions are red for the most part: in those sessions the patient suffered from hypertension, i.e. a clinically undesired situation. Moreover, the clinician can quickly determine how frequent the problem is, how long does it last in a session, and if the possible therapeutic actions taken over time had the desired effect on the subsequent sessions (e.g., the visualization in the figure clearly shows that the situation has worsened in the most recent sessions).

Another task is related to observing the percentage of reduction of the blood volume (mainly due to the removal of water in excess from the patient's blood) during hemodialysis. The needed amount and the speed of this reduction depend on the conditions of individual patients (e.g., sometimes the process is slowed down to avoid situations in which the blood pressure of the patient is too low). For this task, the *Tide* mode in IPBC is useful for clinicians. As an example, Figure 8.8 shows a visualization with more than 9000 bars, representing 36 hemodialytic sessions, containing about 250 values each. The clinician has set the desired increasing linear trend for the percentage of reduction of the blood volume during a session. The resulting visualization allows her to distinguish those (parts of) sessions characterized by a percentage of reduction above or below the desired trend. For example, the selected session (highlighted in a lighter shade of grey) has a first part emerging from the tide, while the last part is below. At the same time, it is possible to observe that one of the last sessions has the percentage of reduction above the tide during almost the entire session. The *Tide* mode can be combined with the *Group* mode to study trends of derived abstract data.

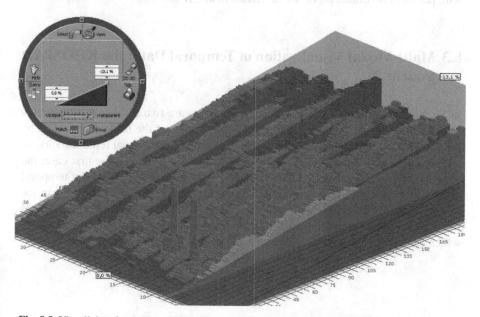

Fig. 8.8 Visualizing the time-varying reduction of the blood volume in the *Tide* mode.

The last task is related to the coordinated analysis of systolic and diastolic blood pressures (measured on the patient) and of the blood flow (QB) entering the hemodialyzer. QB is initially set by the hemodialyzer, but it can be manually changed by nurses: for example, it may be reduced when the patient becomes hypotensive, according to the physician's evaluation. Since the quality of care is affected negatively by suboptimal QBs, it could be important to visually relate QB and blood pressures, to check whether suboptimal QBs are due to low pressures. Indeed, it could be that suboptimal values of QB are due to human errors during the manual setting of the hemodialyzer. Figure 8.9 depicts the coordinated visualization of the three clinical parameters for a given patient: the diastolic blood pressure is shown in the window in the lower left part, the systolic blood pressure is shown in the window in the upper left part, and QB is shown in the right window. The user is allowed to freely organize the visualization, switching the different charts from the left windows to the right window. In the figure, the clinician is focusing on a session where the QB is below the prescribed value during the first two hours of hemodialysis (yellow color for QB) and (s)he has selected a specific value (the system highlights that value and the corresponding values in the other windows with black arrows). It is easy to observe that the suboptimal QB is justified by low blood pressures (yellow bars in the corresponding time-series in the upper left window): QB was set to the correct value by nurses (see black arrow in the right window) only after blood pressures reached normal values (orange color in the blood pressures charts). In this case, the clinician can conclude that the suboptimal QB is not due to human errors in the treatment, but was correctly set by nurses because of the patient's hypotension.

Other advanced features of IPBC, as pattern matching capabilities and integration with parallel coordinate plots, are discussed in detail in [69].

8.3 Multi-Modal Visualization of Temporal Data: the KHOSPAD System

In visualizing complex temporal information there are two different requirements to be dealt with: the first is related to the visualization of the history, described by the considered temporal objects; the second is related to the visual representation of temporal relations existing among different temporal objects. In the first case, the focus is on providing users with a synthetic visual description of a set of temporal objects (e.g., related to a given patient), according to the absolute time axis; in the second case, the attention is on allowing users to explore the different temporal relations existing among temporal objects (e.g., among symptoms and therapies), which can not be observed with the previous history-oriented representation.

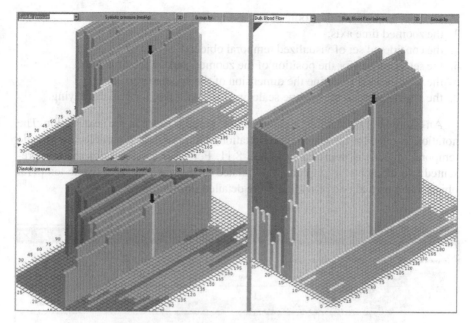

Fig. 8.9 Coordinated analysis of three related hemodialysis parameters.

8.3.1 Displaying Temporal Objects on the Time Axis in KHOSPAD

This history-oriented visual representation of temporal data is provided to the user with graphical tools easily displaying the relative position among temporal objects and allowing a quick identification of the temporal extension, and of its indeterminacy, of the object valid intervals. The following features have been identified as important ones when designing the graphical interface devoted to the representation of histories:

1. the selection of subsets of temporal objects, to allow the user to focus on a subpart of the history, described by the whole set of temporal objects;
2. the visualization on the considered screen window of the minimal interval containing the whole set of temporal objects, to allow the user to have an overall view of the considered history;
3. the displaying of a reference time axis, to locate temporal data;
4. the selection of the preferred time unit to be used in displaying information;
5. the selection of the part of the time axis to be displayed;
6. the zooming in and out on the time axis.

An example of the layout of a visualization system, namely the KHOSPAD system [87], is represented in Figure 8.10, where from the top to the bottom and from the left to the right are shown:

1. the starting instant and the scale for the displayed time axis;

2. the time axis, where the zoomed part is highlighted;
3. the zoomed time axis;
4. the considered set of visualized temporal objects;
5. the selection bar for the position of the zoomed part on the time axis;
6. the selection bar to define the dimension of the zoomed part;
7. the selection menu of the time scale and the check box for grid displaying.

An example of displaying of clinical information is given in the next section. The notation adopted for the visual representation of temporal data, modeled by suitable temporal objects, is depicted in Figure 8.11. For each temporal object are represented two elements: a graphical one related to the valid interval of the temporal object and a textual one providing more detailed information.

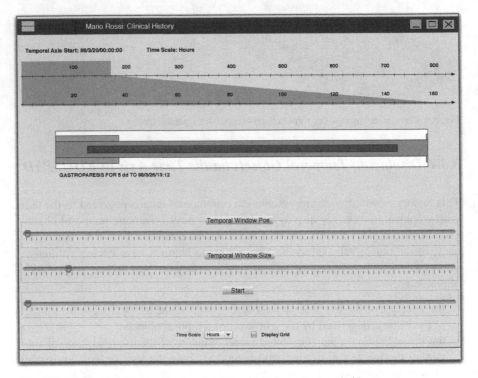

Fig. 8.10 The layout of the system for visualizing histories on temporal objects.

The graphical representation is strictly related to the GCH-OODM model, described in Chapter 4, and allows the visualization of the indeterminacy related to the object valid interval. The color of the box is associated to the specific temporal type of the database schema, the considered temporal object is instance of. The box represents the extension of the object valid interval, i.e. the segment on the time axis having as lower and upper bounds the smallest time point possibly being the start of the interval and the greatest time point possibly being the end of the interval, respectively. For example, in Figure 8.10 the considered box, related to the valid interval FOR 5 dd TO 1998/3/26/13:12 of the temporal object labeled as GASTROPARE-SIS, has the lower bound 1998/3/20:13:12:01 (coming from the subtraction of the upper distance of the duration 5 dd, i.e. 5 dd 23 hh 59 mi 59 ss, to the lower bound of the final instant 1998/3/26/13:12, i.e. 1996/3/26:13:12:00) and the upper bound 1996/3/26:13:12:59.

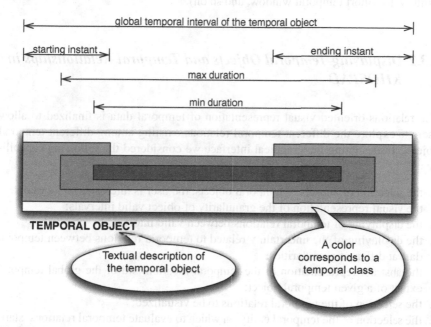

Fig. 8.11 The graphical notation adopted for the valid interval of temporal objects.

Subparts of the box represent, the starting instant, the duration, and the ending instant of the interval, respectively: in the example of layout depicted in Figure 8.10, the ending instant visualized is the minute 1998/3/26/13:12; the visualized duration is 5 dd; the visualized starting instant is evaluated by subtracting a duration of five days to March 26, 1998 at 13:12. Subparts representing starting and ending instants are as extended as the indeterminacy related to these instants is high: the subparts representing these instant are, in fact, all the time points on the time axis, possibly being the start (end) of the considered interval: the sub-box related to the ending instant in Figure 8.10, for example, is bounded by the time points 1998/3/26/13:12:0 and 1998/3/26/13:12:59 on the basic time axis (having the granularity of *seconds*). The duration is represented by two sub-boxes having a dimension respectively related to the lower and the upper time span that the valid interval can have: in our example, we have 5 dd 0 hh 0 mi 0 ss as lower time span and 5 dd 23 hh 59 mi 59 ss as upper time span. The two sub-boxes are centered in respect with the box representing the valid interval of the considered temporal object: this way, the visual notation underlines that, while starting and ending instants of the valid interval are anchored temporal concepts (i.e., related to an absolute position on the time axis), the duration of the valid interval is an unanchored temporal concept (i.e., related to the possible distances between the starting and ending time points of the valid interval). Suitable graphical representations are defined also for instantaneous events or, more generally, for temporal objects for which it is not possible to visualize in a correct way the valid interval in some displaying conditions (e.g., too big temporal window, too short temporal window, and so on).

8.3.2 Displaying Temporal Objects and Temporal Relationships in KHOSPAD

The relations-oriented visual representation of temporal data is finalized to allow users to explore the different temporal relations existing among different temporal objects. In designing the graphical interface we considered the following capabilities:

1. the selection of subsets of temporal objects, the user is interested to focus on;
2. the visual representation of the granularity of object valid intervals;
3. the displaying of temporal relations between valid intervals;
4. the displaying of the uncertainty related to temporal relations between temporal data at different granularities;
5. the abstract representation of the temporal location and of the global temporal extent of a given temporal object;
6. the selection of the temporal relations to be visualized;
7. the selection of the temporal entity on which to evaluate temporal relations: starting instant, ending instant, duration, or valid interval itself.

The notation adopted for the visual representation of temporal objects is depicted in Figure 8.12. The valid interval of a temporal object is represented by a node, composed by three circular sectors: each sector corresponds to one of the dimensions of the valid interval (starting instant, ending instant, and duration, as in Figure 8.12). The label at top of each node represents, after a system-assigned number, the atemporal content of the considered temporal object (e.g., a brief description of a therapy, or of a pathology) and an abstract description of the valid interval. The valid interval is represented by the granule containing it at the finest allowed granularity. The colors of the sectors are related to the granularity used in specifying the parts of the valid interval: the color ranges from cool (blue) to warm (red) tones for granularities ranging from coarser to finer ones. Radial and angular dimensions of each sector are related to the granularity used in specifying the considered temporal entity: the radial dimension increases with the granularity (big radial dimensions for coarse granularities) and the correspondence between radial dimension and granularity is the same for each node. This way, the radial dimension is useful to compare granularities of different temporal objects. The angular dimensions depend on the different granularities used in expressing the valid interval related to the given node. The angular dimensions of the temporal entities given at coarser granularity are greater than the angular dimensions of the temporal entities given at finer granularity for the same interval. Angular dimensions are useful in comparing granularities used for the same valid interval. For temporal objects having an instantaneous validity, a special notation has been defined, as depicted in Figure 8.12.

The user can also display the temporal object with some representation of the temporal location. Around the sectors, a watch-like face is visualized, showing a suitable time scale according to the given valid interval: on this scale the position of the valid interval is highlighted. For example, the node labeled as "(3) Gastroparesis on Mar 1998" in Figure 8.12 refers to the temporal object, previously stored in the database, having FOR 5 dd TO 1998/3/26:13:12 as valid interval: "Gastroparesis" is the part of the label related to the description of the atemporal data of the object; "on Mar 1998" represent the minimal granule (March 1998) containing the whole valid interval; the smaller sector is in green (i.e., granularity of minutes) and represent the finest granularity, i.e. that for the ending instant, used for the given valid interval; the sector at the bottom of the node is in cyan and represent the duration (given at the granularity of days, corresponding to cyan); the leftmost sector of the node stands for the starting instant and is depicted in magenta (i.e., granularity of months), having an indeterminacy in its temporal location due to granularities of both the ending instant and the duration. On the external circular scale, March days are represented and the days containing the valid interval are highlighted with the color corresponding to the granularity of days.

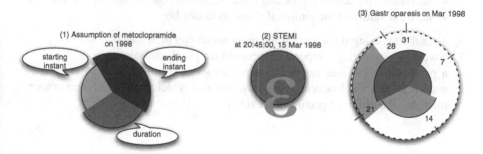

Fig. 8.12 Graphical notation for temporal objects (from left to right): main notation, instantaneous temporal object notation, extension of the main notation to represent temporal location.

Temporal relations are visualized by edges between the nodes representing temporal objects; different edge colors are associated to different truth values. Edges are labeled; labels represent the different relations existing between nodes. The edges have a direction from the first operand to the second one of the relation. Bidirectional edges stand for two directional edges with the suitable labels. For example, in Figure 8.13, two nodes are displayed, related to a pathology and to a therapy, respectively. The first edge, labeled by "<" (before) and "O" (overlaps), stands for undefined relations (blue color); the second one, labeled by "M(M)" (meets at granularity of months) in one direction and "F(M)" (finishes at granularity of months) in the other direction, stands for true relations (black color); the third one, labeled by "F(Mi)" (finishes at granularity of minutes) "M(M)", "O", and "<" in one direction and by "F(Mi)" in the other direction, stands for false relations (red color).

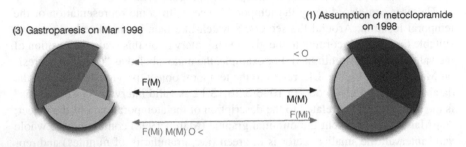

Fig. 8.13 Graphical notation for relations between temporal objects.

The overall structure of the designed visualization system is represented in Figure 8.14, where from the top to the bottom are shown:

1. a panel containing some controls for the visualization:

   - the selection of the temporal entities (starting instant, ending instant, duration, or valid interval) to consider,
   - the buttons for defining how and where temporal objects must be represented,
   - the truth values of the temporal relations to display;

2. a panel containing the temporal relations which can be selected;
3. a window visualizing temporal objects and temporal relations;
4. a panel where the user can select granularities at which each relation must be evaluated (this window remembers also the user which are the different colours related to the different granularity levels.)

8.3.3 Visualizing Temporal Clinical Data in KHOSPAD

The history-oriented visualization provides a graphical and expressive abstract representation of the patient's clinical history, thus facilitating the process of data overview performed by the physicians. An example is shown in Figure 8.15: temporal objects are displayed on a time scale having days as units. The first three objects (from top to bottom) are related to pathologies: their valid intervals are given by the ending instant and the duration (FOR ... TO notation), by the starting instant and the duration (FROM ... FOR notation) and by the starting and ending instant (FROM ... TO notation). The other visualized temporal objects are related to a therapy and

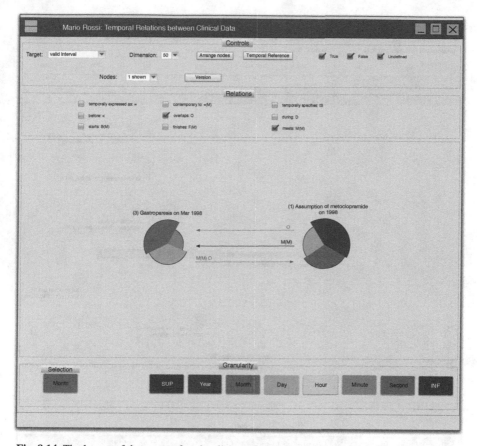

Fig. 8.14 The layout of the system for visualizing relations between temporal objects.

to follow-up visits (the last two temporal objects): their valid intervals are given by the FROM .. FOR or FOR .. TO notations. The displayed temporal dimension of these objects is not always the real one: a box having a fixed dimension displays valid intervals, having a too short span on the zoomed time axis (in this case, the first months of year 1998) for a suitable visualization. The box with an "ε" inside displays a temporal object having a valid interval with duration zero; this way, instantaneous events can be visually identified.

The relations-oriented visualization, focusing on the representation of the different relations between temporal clinical objects, facilitates the process of data analysis performed by physicians. Figure 8.16 shows an example of the user interface, where the focus is on the precedence relation (<). Only the true relations have been visualized. Nodes are related to pathologies (2, 3, 5), therapies (1), and follow-up visits (4, 6). Granularities of temporal clinical objects range from years and more (blue) to seconds (red); the temporal object related to a myocardial infarction is instantaneous and is displayed with the suitable notation. The small window in right

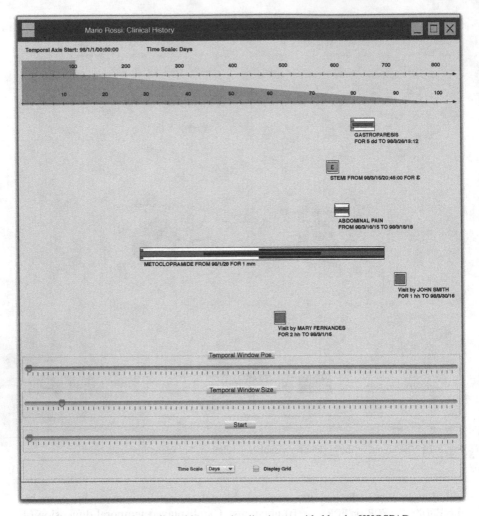

Fig. 8.15 An example of the clinical history visualization provided by the KHOSPAD prototype.

part of the figure shows detailed data related to the node 1 about the assumption of metoclopramide: the visualization of this kind of window, as the position of nodes on the panel, can be decided by the user simply by clicking and/or by drop-and-drag operations.

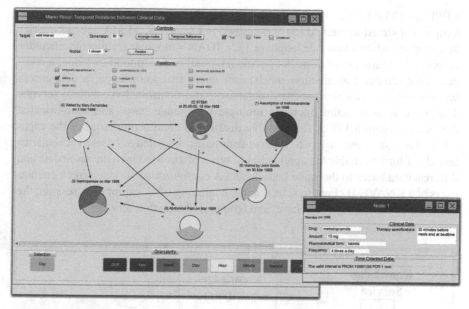

Fig. 8.16 An example of the relations-oriented visualization provided by the KHOSPAD prototype.

8.4 Knowledge-Based Visualization and Exploration of Single-Patient Time-Oriented Data: the KNAVE II System

In previous research, Shahar et al. have implemented a stand-alone prototype module of the visualization service, called *knowledge-based navigation of abstractions for visualization and explanation* (KNAVE) [351, 352, 366]. Preliminary assessments of KNAVE in the oncology domain, were highly encouraging, and demonstrated the feasibility of the whole architecture and in particular of the knowledge-based exploration concept [351, 352]. KNAVE-II is a significantly enhanced version of KNAVE, and is an intelligent interface designed to fulfill all the above mentioned desiderata. KNAVE-II is a knowledge-based interactive visualization and exploration intelligent user interface. The interface is used to explore a single patient record. KNAVE-II enables interactive, dynamic exploration by the user of raw data and their abstractions. Figure 8.18 shows the main interface of a KNAVE-II visualization and exploration client: it demonstrates the design for the integration of a knowledge browser (which reflects the contents of the domain's temporal-abstraction knowledge base, that is, its temporal-abstraction *ontology*) and data panels (which show the contents of the database or of the results of a temporal-abstraction process applied on such contents). In this section, we explain in detail how the KNAVE-II architecture supports all desiderata previously introduced. In order to support the *accessibility*, KNAVE-II uses a knowledge-based distributed temporal-abstraction mediator IDAN [32]. IDAN uses a modern version of the

RÉSUMÉ [359] problem solver, ALMA, which implements the knowledge-based temporal-abstraction method [347] described in Chapter 5. ALMA accepts queries in a temporal-abstraction rule language [32]. IDAN's modular architecture includes automated acquisition of domain-specific temporal-abstraction knowledge, a computational temporal-abstraction mechanism using that knowledge, a data-access service that accesses time-oriented databases, and controlled-vocabulary servers. The modular architecture includes multiple knowledge bases and time-oriented databases (Figure 8.17). A full scalable distributed architecture requires the capability of *remote connectivity* to diverse data bases, knowledge bases, vocabularies and algorithms to enable the application of types of knowledge to the same data and different data bases to the same knowledge. A *configuration* service screen enables users of a KNAVE-II client to select, at the beginning of an exploration session, the desired data base, knowledge base, and temporal-abstraction service.

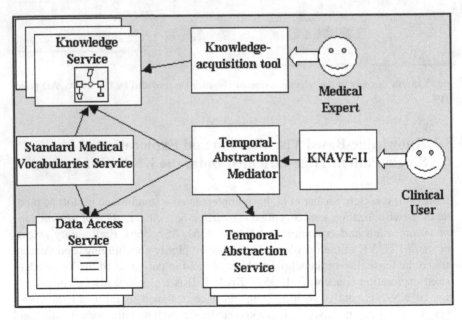

Fig. 8.17 The knowledge based distributed architecture for visualization and exploration of individual patient records, as used by the KNAVE-II system.

End users interact with KNAVE-II to submit time-oriented queries. The temporal-abstraction mediator, using data from the appropriate local data-source through the data-access service, and temporal-abstraction and visualization knowledge from the appropriate domain-specific knowledge base through the knowledge service, answers these queries. KNAVE-II enables users to visually and dynamically explore resultant abstractions, using a specialized graphical display. Arrows indicates a "uses" relation.

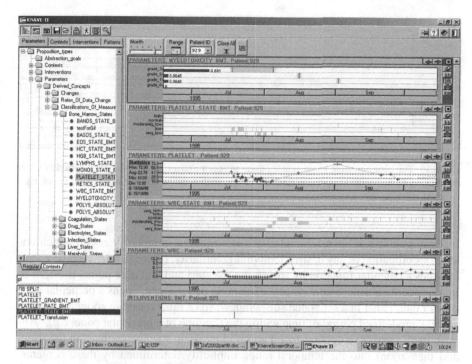

Fig. 8.18 A view of an individual patient's data in the KNAVE-II prototype (in an oncology domain).

Figure 8.18 provides a view of an individual patient's data in the KNAVE-II prototype within an oncology domain. On the left hand side, a browser to the clinical domain's ontology, coming from the ontology knowledge base, is shown. The user selects a raw data (3rd 5th and 6 th panels from the top) or abstract concept (1st 2nd and 4th) from the ontology tree or by asking a query from the left hand bottom panel, which is then retrieved or computed on the fly and displayed as a panel on the right hand side. Operators represented as icons, in each panel, enable the user to perform actions. The time-synchronization function [pin-shaped] icon, enables to synchronize the display of the panels according to the specified time in the selected panel; the "kb" icon allows one to query the knowledge used to derive the concept through; it is also possible to add statistics, clicking on the stats icon, about raw and derived concepts (see the statistics displayed on 3rd and 1st panels from the top, respectively); semantic exploration of concepts is accessed by clicking on the cross icon; finally, the right and left direction arrows enable to skip to the nearest period (at the selected direction) data was found.

The content-based zoom enables users to mark specific contents in the panel whether within a complete temporal-granularity unit or not (see shadowed area), and then zoom into the temporal range implicitly determined by these contents. KNAVE-II also supports the exploration of time oriented data at different levels of temporal granularities. KNAVE-II implements five operators (zoom-in functions)

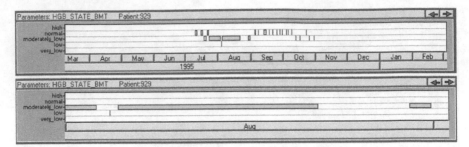

Fig. 8.19 The time-sensitive zoom enables users to zoom into a specific predefined period of time, by clicking on a specific temporal granularity button within the timeline of a particular panel, e.g., "August 1995".

for manipulating temporal granularity: (a) a random granularity zoom, which enables specification of any desired temporal granularity; (b) a calendaric-range zoom, operates a calendar function by specifying the start and end time points to zoom-in into a specific absolute time range; (c) a single panel zoom, opens a particular concept or data panel in a separated sub-window; (d) a time-sensitive zoom, to select a specific predefined period of time within the timeline of a particular panel e.g., "August 1995"; and (e) a content-based zoom to mark-up specific contents in the panel and then zoom into the temporal range implicitly determined by these contents, whether within a complete temporal-granularity unit or not. Figure 8.19 demonstrates the time-sensitive zoom and the content-based zoom. Changing dynamically the point of view from an *absolute* (calendar-based) time line to a *relative* time line is another KNAVE-II specific capability. The relative time line is set by identifying clinically significant events (e.g., start of therapy, birth of the child) which serve as a date of reference to all the other displays. Once the relative time-line was selected the time display will change to +/- time units starting from that event, based on the time granularity selected (hours, days, months, years). The user can select the event to be used as the zero-time reference, through access to a predefined list of events, defined within the knowledge-base, that can be used as reference points, as in Figure 8.20: once the relative time-line was selected the time display will change to +/- units starting from that event. The selection of the time reference event can be done by direct manipulation or by selecting a predefined reference event from the knowledge-base, in which case KNAVE-II will show the nearest event enabling direct browsing between events (in the case that there were more than one such points in the patient's record).

Exploration of raw data and abstract concepts includes navigation along semantic links in the domain's temporal-abstraction ontology, such as *abstracted-from* relations, using the *semantic explorer* (Figure 8.21). The semantic explorer is evoked by clicking on the exploration (cross icon) button of the panel (see Figure 8.18). The user uses the semantic relationships of a concept, which depend on its type (e.g., abstracted-from, abstracted-into, sibling-argument, generated-context, and is-a, in the case of a raw or abstract data type) to navigate to other concepts semanti-

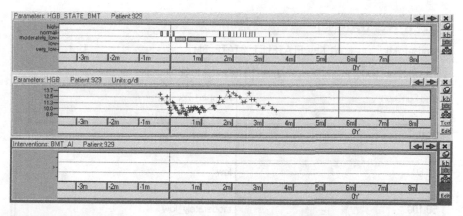

Fig. 8.20 Relative time lines in KNAVE-II. Note that time units are no longer calendaric but are rather relative to the zero-time milestone (or any arbitrary time point) chosen by the user.

cally related to the original concept. During exploration, the user is able to obtain context-sensitive explanations to questions such as "From which data is this concept abstracted?" and "What classification function defines this abstraction?" (Figure 8.21): the knowledge-based explanation is evoked by clicking on the knowledge button (kb) in a panel (see Figure 8.18). The user queries the temporal-abstraction knowledge that was used to derive a specific displayed concept. For any domain, the semantics of the query, visualization, and exploration processes are the same, since these processes use the terms of the domain-independent knowledge-based temporal-abstraction ontology. However, the exploration operators use, for any domain, the domain-specific temporal-abstraction properties of that particular domain. The result is a uniform-behavior, but context-sensitive (with respect to the knowledge) visualization and exploration interface in all time-oriented domains. To support clinical research, it is imperative to provide several types of descriptive statistics as part of the interactive visualization and exploration. Statistics in KNAVE-II can be computed and displayed for either raw data or abstracted parameters. The computation of statistics is, sensitive to the particular time window displayed in each panel, and thus changes dynamically when the contents of the panel are changed. Default statistics for raw data types include descriptive statistics such as mean, maximum, minimum, standard deviations, etc.(see 3rd panel from the top in Figure 8.18). In the case of abstract data types, the default statistics displayed are a detailed distribution of the duration of each value of the abstraction (e.g., GRADE-II bone-marrow toxicity) (see 1st panel from the top in Figure 8.18). KNAVE-II supports easy and fast search and retrieval of ontology-based clinically-significant concepts. The search supports easy ordering of concepts according to the type and the related context, opening the found concept in the knowledge browser, and (optional) opening the matching panel for further visualization and exploration (see Figure 8.18).

The exploration supports, among other features, dynamic simulation of hypothetical raw data or domain-specific knowledge. The user is able to simulate the effect of modifying the data or the knowledge, thus adding a dynamic sensitivity

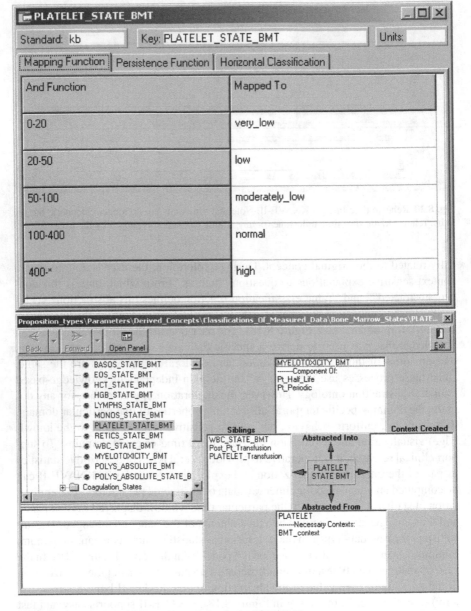

Fig. 8.21 Exploration of data and knowledge in the KNAVE-II prototype.

analysis capability (What-if dynamic simulation) by changing or adding values in a specific panel. The modified values are kept in the cache and do not affect the real patients' data in the database. Dynamic sensitivity analysis enables the following actions: adding the hypothetical raw data, modifying the existing data by

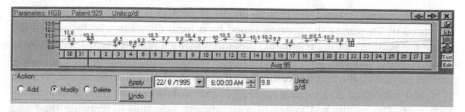

Fig. 8.22 Dynamic sensitivity analysis. The user can modify the time and value of any data point (including adding or deleting points) and apply the changes, to explore the resulting modifications to the temporal abstractions derivable from the data.

changing their values and time-stamps; or deleting selected data. The apply button triggers the computation of abstractions derived from the modified raw data, while the undo button cancels the modifications and their effects. The display reflects the computational implications of these modifications. Exploiting the direct access to the domain's temporal-abstraction ontology and to the temporal abstraction server will provide these explanations. Clinical-Task support is achieved by enabling the physicians and medical researchers to select several raw-data and abstract-concept panels that are all related to a specific clinical task (e.g., monitoring of diabetic patients). KNAVE-II enables customization of the displays for a specific clinical task by opening the selected panels and saving them as a *profile* (e.g., a diabetes profile) in the knowledge base through the profile-save icon in the top-level menu. Clicking on the profile-load icon, another icon in the top-level menu, retrieves all the profiles, enables selection of a specific profile, and applies it to the current patient record. Clinicians and researchers usually like to consult or share the result of the exploration of data and abstractions with colleagues. KNAVE-II enables collaboration by saving the selected data and abstractions of a particular exploration to a special exploration file format, through another one of the top-level menu icons. An exploration file can be saved to a shared directory or sent by email to a collaborating colleague. The collaborator can open the file, and explore the same data, starting from the same point in which the image of the exploration was saved. She also can add other raw-data and abstract-concept panels that seem to be relevant to the case in discussion, and continue the visualization and exploration session (and even send it back). Standard clinical, research and administrative procedures require documentation of the patient's clinical data to the patient's file, thus showing the exploration that supported a clinical decision. Clicking on the camera icon at the top-level menu enables taking a snapshot of the exploration process. Another icon enables printing the exploration for further reference.

An extensive evaluation of the KNAVE-II system has been performed by researchers from Stanford University in the Veterans Administration Palo Alto Health Care System, in the domain of monitoring patients after bone-marrow transplantation [255]. The use of the KNAVE-II system has been compared, in a cross-over study, to the use of an electronic spreadsheet (ESS) and of paper charts (in a later study, the comparison was just with the ESS), in the context of the task of answering a set of time and value queries, at varying levels of difficulty, regarding the patient's course of disease, taken from typical oncology protocols. The measures included the speed of answering the queries, timed by a local clinician who monitored the users; the accuracy of the responses, measured using a predetermined set of scores to all possible answers with respect to time or value; the usability of each interface as judged by the users, measured on the Standard Usability Score (SUS) scale [41], and a directed comparative ranking of the usability of the three interfaces by each user. The results had indicated a significant reduction in response time and enhancement in accuracy for users of the KNAVE-II system, which became more evident as queries were more difficult. Even a highly conservative methodology had been used by the investigators in the second evaluation to assign the [finite] time of the considerably less difficult previous query for users who ran out of time on the hardest query, the mean difference in time for answering queries in the second evaluation was 155 seconds faster when using KNAVE-II ($p = 0.0003$) than when using the ESS. At the same time, 91.6% (110/120) of the total number of questions posed within queries of all levels of difficulty that were asked in the second evaluation, produced correct answers when using KNAVE-II, as opposed to only 57.5% (69/120) when using the ESS ($p < 0.0001$). The average SUS questionnaire score for KNAVE-II was 64 (range 45-72.5, median 67.5), and for the ESS 45 (range 42.5-50, median 42.5). Based on a paired t-test of the difference in mean SUS scores, KNAVE-II's usability was significantly superior to the ESS ($p = 0.011$; 95% CI = 7 to 31) . All subjects ranked KNAVE-II first with respect to overall preference, easiness to find answers to clinical queries, and time to find answers to clinical queries.

8.5 Specification of Temporal Patterns and Queries Using Visual Metaphors: the Paint-Roller System

One of the approaches proposed for the interactive, graphical specification of temporal patterns in the medical domain was suggested by Combi and Chittaro [68]. We shall refer to this approach as using the *Paint-rollers metaphor*. We will focus in this section on the specification of composite abstractions, which, in Shahar's KBTA terminology, can also be referred to as *patterns*, using the Paint roller metaphor. In the following, we assume that for all composite abstractions to be specified, the necessary component abstractions (hereinafter, concisely referred to as *components*) have been already suitably defined. Two different requirements arise when dealing with composite abstractions (hereinafter, concisely referred to as *abstractions*):

- the first one involves *temporal* aspects and concerns the definition of temporal relations between the components (e.g., "headache overlapping analgesics" could be the definition of the (composite) abstraction *headache therapy pattern*) and the definition of the interval associated to the abstraction (e.g., "3 days before the start of analgesics up to the end of analgesics");
- the second one involves *logical* aspects and deals with the composition through the usual connectives - AND, OR, NOT - of the specified abstractions in order to

query the temporal abstraction database (e.g., "*headache therapy pattern* AND *increase of SBP and DBP*", or "*headache therapy pattern* AND NOT *increase of SBP and DBP**").

The proposed user interface supports the definition process by allowing the user to: (i) easily set the relative temporal positions among components, (ii) define the temporal interval associated to the abstraction, on the basis of the intervals of components, and (iii) logically relate the components through the standard connectives which can be combined to define more complex expressions. The interface has been designed with the following features in mind: (i) use of simple graphical metaphors related to the physical world, (ii) visual separation of temporal and logical aspects in two different graphic windows, (iii) point-and-click selection of the components and the various graphic operators (the abstraction can be interactively defined without resorting to the keyboard), (iv) use of different colors to highlight different kinds of abstractions, (v) clear connection among graphic objects in the temporal and logical parts of a given abstraction. In the following, we first describe the metaphors for the definition of the temporal and logical parts of abstractions, and then provide an overview of the interface.

8.5.1 Temporal Aspects: Displaying Intervals and Temporal Relations Using the Paint-Roller Metaphor

Intervals for abstractions are visualized as paint strips. The temporal location of these strips can be specified in different ways (see Figure 8.23):

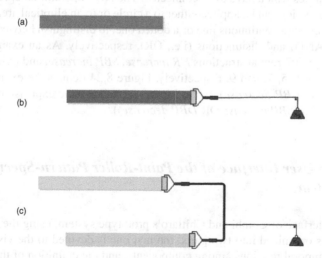

Fig. 8.23 A graphical notation for the interval representation.

1. Paint strips can be represented plainly without any attached object (Figure 8.23, example *a*). In this case, the goal is to represent intervals' ends that have a precisely set position with respect to other intervals. The commonsense reasoning motivating this choice is that "the end of a paint strip cannot move by itself".

2. Alternatively, an end of a strip can be attached to a paint roller, connected to a weight by means of a wire as shown in Figure 8.23 (example *b*). This notation expresses that the end of the interval can take different positions on the time axis: the roller can extend the end of the paint strip up to the wall, which stops the roller.

3. Finally, a weight can be connected to more than one roller simultaneously to represent intervals' ends which can move but keep their relative position. For example, the left ends of the two intervals in Figure 8.23 (example *c*) can move, but the lower interval will always terminate after the upper one.

8.5.2 Logical Aspects: Displaying Logical Expressions as Part of the Paint-Roller Metaphor

To visualize and interactively compose logical expressions, each involved component is first graphically associated to a circle. Every circle is filled with the same color associated to the abstraction. Moreover, a numeric ID is used as an additional mean to relate a single abstraction to its two visual representations (one for its related temporal interval and one for the propositional part), as shown in the following section. The AND and the OR operators are represented by elliptical areas, containing all the circles which have to be connected. The NOT operator is represented as a diagonal line which can be applied either to a circle or to an elliptical area. The edge of areas are either a continuous line or a dotted one, to distinguish between conjunctions (i.e., AND) and disjunctions (i.e., OR), respectively. As an example, let us consider three different abstractions *HR increase*, *SBP increase*, and *DBP decrease*, identified by IDs 5, 7, and 9, respectively. Figure 8.24 depicts the expression (*HR increase* AND *DBP decrease*), while Figure 8.25 shows the expression (NOT(*HR increase*) AND (*SBP increase* OR *DBP decrease*)).

8.5.3 The User Interface of the Paint-Roller Pattern-Specification System

The user interface of Combi and Chittaro's prototype system, using the Paint-roller metaphor, is organized into two parts: the first one is devoted to the visual specification of temporal relations among components, and the definition of the name and

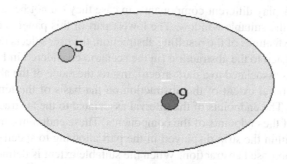

Fig. 8.24 Graphical representation of the expression (5 AND 9).

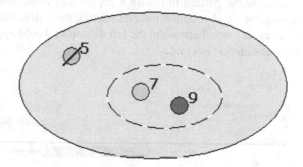

Fig. 8.25 Graphic representation of the expression (NOT 5 AND (7 OR 9)).

validity interval of the abstraction; the second one is devoted to the visual definition of the logical expression on the components.

Figure 8.26 is a screenshot of the user interface during the definition of the abstraction *flu* involving *fever*, *headache*, and *sore throat* as components. Just below the menu items, the interface show the abstractions opened/defined by the user in the current session. Each abstraction is identified by a numerical ID. In the considered scenario, for example, the user is dealing with the specification and the further query related to abstractions *flu*, identified by the ID 0 and displayed in detail with its components, *pneumonia*, with ID 12, and *viral gastroenteritis*, with ID 20. The user can switch from the temporal specification of an abstraction to that of another abstraction by simply clicking on the suitable abstraction label. In the panel displayed in the upper part of the screen window, the system displays the temporal intervals for the components, according to their relative positions, following the previously described metaphors. For a given abstraction, a label (e.g., *headache (2)* in Figure 8.26) describing its propositional (atemporal) content and the assigned ID is displayed at the beginning of the same line. The color of paint strips is related to the considered component. Insertion, modification, deletion, and connection of different graphic objects can be performed by switching among the different options either through the menu items and its sub-items or through a suitable toolbar. A scrollbar allows one to display different components, in case they cannot be displayed at the same time into the suitable window. The lowest part of this panel is devoted to the definition of the features of the resulting abstraction: the user selects the color which has to be associated to the abstraction (in the scenario considered in Figure 8.26 the abstraction *flu* is associated to a dark green), inserts the name of the abstraction, and defines the temporal extent of the abstraction on the basis of the temporal extents of components. The endpoints of the interval associated to the abstraction can coincide with any of the endpoints of the components. These endpoints are displayed as vertical bars within the strip displayed in the part allowing to specify the temporal extent of the composed abstraction. When the suitable extent is defined through the pointing device, the corresponding part of the strip assumes the chosen color. A further possibility is to add (subtract) some fixed time span to the chosen endpoints: for example, with respect to the pattern in Figure 8.26, the user could define, through a suitable dialog window, an abstraction interval starting three days before the start of headache. In this case, a small arrow (in the left direction) would appear near the left endpoint of the abstraction interval.

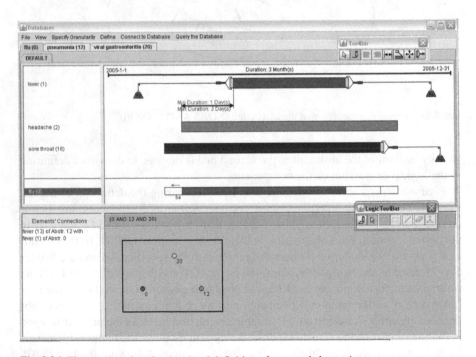

Fig. 8.26 The user interface for the visual definition of temporal abstractions.

The panel displayed in the lower part of the window allows the user to define the logical expressions for querying the abstraction database. Abstractions are displayed as previously described. Insertion, modification, and deletion of different graphic objects can be performed by switching among different options either in the menu items and its sub-items or through a toolbar (named *Logic toolbar* in the interface). As an example, in Figure 8.26 the AND between abstractions 0, 12, and 20 is being selected, as highlighted by the continuous-line square representing the AND. The same abstraction can appear more than once in the logical expression: this is graphically achieved by allowing to duplicate the circles associated to a specific ID. The expression the user is defining appears at the top of the panel in textual form using IDs (as depicted in Figure 8.26) or abstraction labels.

Further features of the interface are related to the selection of a time frame to restrict the evaluation of the query on the patients' abstraction database and to the connection of specific components of different abstractions when evaluating the query: indeed, up to this point abstractions have been defined in isolation and thus there are no connections between the components of different abstractions.

For example, considering the query in Figure 8.26, we are looking for patients who had *viral gastroenteritis* (20) AND *pneumonia* (12) AND *flu* (0); in this case, we could be not interested in the period when these abstractions occurred (i.e., a patient suffering both from flu in 1980 and from pneumonia in 1990 and from viral gastroenteritis in 2000 would be selected as well as a patient suffering from intestinal flu, pneumonia and viral gastroenteritis in the same month). When we need to identify patients who are involved in all the considered abstractions in the same period, we could use a suitable fixed time frame, restricting the query evaluation to only a portion of patient histories. The defined time frame is displayed for each abstraction in the upper part of the panel devoted to the temporal specification of the abstraction components; in the given case, only abstractions occurring in 2005 are considered.

Moreover, the physician could be interested in identifying the patients satisfying several abstractions sharing some of their components. For example, let us consider the query depicted in Figure 8.26 involving three abstractions: *flu* (0), *viral gastroenteritis* (20), and *pneumonia* (12). As we already explained, we are looking for patients who had pneumonia and viral gastroenteritis, and flu (i.e., (0 AND 12 AND 20). However, we are interested in selecting only those patients who had pneumonia and episodes of flu related to the same symptoms and/or signs. In this case, we have to connect the components of the considered abstractions; more precisely, we need that the query to be evaluated with the fever component set to be the same for the two abstractions flu and pneumonia. This is specified in the *Connection* area, in the left part of the *Logical* area, as shown in Figure 8.26, where the component *fever* (13) of abstraction *pneumonia* has to be the same of component *fever* (1) of abstraction *flu*.

8.5.4 User-Based System Design and Evaluation of the Paint-roller Metaphor System

The development of the Paint-roller metaphor system's interface has been carried out in three different steps, each one followed by proper tests with actual users.

First, a low-fidelity paper prototype was built to test preliminary ideas and better understand user needs; second, different metaphors were proposed and evaluated; third, different versions of the overall user interface were implemented and tested. Results obtained with each step has been used to guide the next one. The design and evaluation of the first prototype has been carried out through the well-known *storyboarding* approach. Paper figures representing screenshots or possible interface elements were used to present possible ideas to users and obtain their feedback. The paper prototype was tested on 10 subjects: after a brief presentation about temporal abstractions, people were gradually introduced to the different graphic objects, and then asked about their interpretation of some predefined scenes. Finally, people were asked to compose specific abstractions using the paper elements. Several issues and new ideas came out from this preliminary evaluation (e.g., the need for a proper metaphor to visually specify non fixed endpoints) which were dealt with in the next two steps. Three different graphic vocabularies were evaluated for representing temporal aspects. Besides the already introduced paint strips vocabulary, two other possible design choices have been considered. One used elastic strips for visualizing temporal intervals, where strips' ends can also be either fixed by screws or attached (possibly together with another strip) to an horizontally moving mass system (as those seen in physics textbooks), to express the same relations already described for paint strips. The other vocabulary used springs for representing intervals, and any end of a spring can also be either fixed with a screw or connected (possibly together with another spring) to a weight by means of a wire.

The evaluation was based on a questionnaire containing 4 exercises for each of the 3 proposals (i.e., a total of 12 exercises): the first part (a total of 3 exercises) of the questionnaire aimed at assessing which objects are correctly perceived by users as having some freedom of movement; the purpose of the second part (a total of 9 exercises) was to assess how much the possible temporal locations and respective temporal relations of intervals are correctly perceived. The questionnaire was administered to 30 subjects (13 females and 17 males). Age ranged from 24 to 37, averaging at 27. Nine subjects were physicians, while 17 were university students (1 in Mathematics, 2 in Computer Science, 3 in Engineering, 8 in Business Administration, 3 in Agricultural Sciences), 3 persons recently completed their Master (2 in Philosophy, 1 in Biology), and 1 subject held a secretarial position. The majority of subjects (19 people) had taken a course in Physics in their university curriculum (and were thus expected to be familiar with the moving mass notation adopted in the Elastic Strips proposal). Subjects were not given any information about the meaning of the specific graphic elements in the three proposals. Each subject was first asked to fill the first part of the questionnaire, then (s)he was provided with the second. Since each part contained exercises for all three proposals, the order in which they were presented was changed for each subject to minimize learning effects.

Statistical analysis pointed out that the correctness results obtained with Elastic Strips were significantly lower than those obtained both with Springs and with Paint Strips, while the difference between Springs and Paint Strips turned out not to be statistically significant. Elastic Strips were discarded as a possible design choice, while the other two vocabularies were more thoroughly evaluated in the next step.

In the last step, different Visual C++ implementations of the user interface were implemented. Interaction with the interfaces was analogous to common drawing applications: the user can create and position the available graphic objects in the window and then set their desired size/position by using only the mouse. The difference between the interfaces was the available set of graphic objects for specifying temporal and logical relations. In particular: (i) for expressing temporal relations, the Paint Strips and the Springs vocabularies were contrasted, (ii) for logical expressions, the previously described approach (hereinafter, Ellipses) for the visual definition of the expressions was compared with the recent Tabular Query Form (TQF) [294] proposal. TQF allows one to visually define expressions as disjunctions of conjunctions: each conjunction is displayed as a separate box, containing the propositions which have to be considered. The experimental task consisted of two parts. In the first part, subjects were asked to solve some *interpretation exercises* on paper. Each exercise showed a temporal pattern (or a logical expression), and asked to choose its correct interpretation among 3 possibilities. In the second part of the task (*definition exercises*), the subject used instead one of the graphic interfaces to visually define some temporal patterns (or logical expressions) described (only in natural language) on a sheet given to him/her. Subjects were recruited at the Medical Clinic Department of our University. None of them had been involved in the first experiment described in the previous section.

A total of 31 people (15 males and 16 females) were recruited: 6 students in Medicine, 4 medical doctors (MDs) who had just earned their degree, 18 MDs specializing in various subfields (2 in Clinical Pharmacology, 10 in Public Health, 5 in Clinical Psychiatry, 1 in Surgery), 1 psychologist, 1 pharmacologist, and 1 physician employed in the Public Health Department. Age ranged from 23 to 44, averaging at 30. With respect to computer usage, only one subject never used a computer; the others were equally split among those who use it for 5 or more hours per week and those that use it for a couple of hours per week. Each subject performed the task in two different sessions. Each session evaluated separately one design choice for temporal aspects and one for logical aspects: the task was thus performed two times in each session. Before using an interface, a training phase was carried out: subjects were told about the meaning of the graphic elements; then, they were guided to directly interact with the system: for each graphic object, they learned how to insert, modify, and delete it; finally, the meaning of each graphic object in the context of temporal patterns was explained. When subjects felt ready, they were introduced to the task. After task completion, they filled a 28-questions questionnaire inspired by QUIS (Questionnaire for User Interaction Satisfaction) [66].

There are no imposed limits on the duration of any part of the session. Since each session lasted about 50 minutes, the two sessions for the same subject were scheduled in different days to avoid excessive tiredness. To minimize learning effects, different users carried out sessions on different pairs of temporal and logical interfaces. We collected the following quantitative data: time spent to complete the interpretation exercises, number of correct answers to interpretation exercises, time spent to complete the definition exercises, and number of correct answers to interpretation exercises. Qualitative impressions were recorded with the satisfaction questionnaire. Data concerning correct answers to exercises were analyzed by using the Wilcoxon test for two related samples. With respect to temporal aspects, the following results were obtained: average number of correct answers was 3.39 for Springs and 3.52 for Paint Strips in the 4 given interpretation exercises; while it was 2.55 for Springs and 2.81 for Paint Strips in the 3 definition exercises. With respect to logical expressions, the average number of correct answers was 4.32 for Ellipses and 4.06 for TQF in the 5 given interpretation exercises; while it was 3.77 for Ellipses and 3.55 for Paint Strips in the 4 definition exercises. Although numerical results are slightly better for Paint Strips and Ellipses, the p=0.05 threshold for statistical significance was not met in the comparisons of means.

Data concerning time spent to complete the two parts of the task (analyzed with the paired-samples $t$test) yielded analogous results. However, Paint Strips obtained statistically significant results better than those of Springs when considering the final qualitative questions: for example, 22 of the 31 subjects preferred Paint Strips to Springs for the visual definition of temporal relations (χ_r^2=6.25, p<0.025, Pearson's Chi-Square test for one-way tables). No statistically significant differences were instead found in the questionnaires concerning Ellipses and TQF. The last evaluation step, carried out on more than 30 physicians and based on rigorous statistical techniques, helped the developers taking the final design choices, which were adopted in the interface. The choice of Paint Strips over Springs was relatively straightforward: while quantitative data showed only very slight differences in user performance with the two interfaces, qualitative data indicated a statistically significant user preference for Paint Strips. The choice of Ellipses over TQF was less easy. In general, Ellipses performed similarly to TQF, which has been shown to improve (compared to textual definition) the capability of untrained users of defining logical expression [294]. At the end, the choice of Ellipses for the interface was based on the observation that Ellipses allow users to visually compose arbitrary expressions in a more simple way, while TQF forces the user to represent any expression as a disjunction of conjunctions.

8.6 Knowledge-Based Visualization and Exploration of Time-Oriented Data of Multiple Patients: the VISITORS System

Due to the large volume of time-stamped information, clinicians and medical researchers alike require useful, intuitive, intelligent tools to process large amounts of time-oriented data of multiple patients from multiple sources. For example, to analyze the results of clinical trials, to perform large-scale assessment of the quality of care, or to make health-care policy decision, an aggregated view of a group or even a whole population of patients is required. To meet this need, Klimov and Shahar have designed and developed the *VISualizatIon of Time-Oriented RecordS* (VISITORS) system, which combines intelligent temporal analysis and information visualization techniques ([218]. The VISITORS system includes tools for intelligent querying, visualization, exploration, and analysis of raw time-oriented data and derived (abstracted) concepts for multiple patient records. To derive meaningful interpretations from raw time-oriented data (known as temporal abstractions), the researchers used the knowledge-based temporal-abstraction (KBTA) method [347] described in detail in Chapter 5. The main module of the VISITORS system is an interactive, ontology-based exploration module, which enables users to visualize multiple raw data and abstract (derived) concepts at multiple levels of temporal granularity, to explore them, and to display associations among raw and abstract concepts, as depicted in Figure 8.27. A knowledge-based delegate function is used to convert the multiple data points of each patient into one delegate value for each temporal granule. The function is specific to each concept type and temporal granularity level, although several default functions exist.

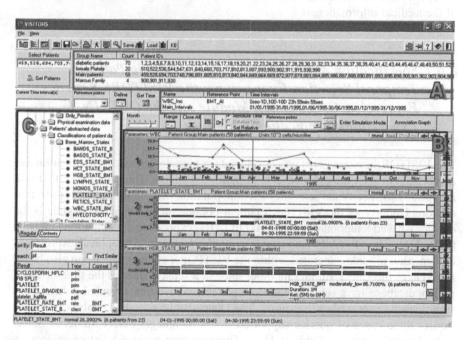

Fig. 8.27 The VISITORS main interface. In this case, the knowledge base used to abstract the data and browse the resulting abstractions is from the oncology domain. The two top panels (denoted by A) display lists of patients and lists of time intervals, returned as answers to previous queries. The graphs (denoted by B) show the data, for a group of 58 patients, of the white blood cell count (WBC) raw concept (graph 1) and of the monthly distribution of the abstract concept Platelet-state values during 1995 (graph 2). In the case of raw data, a line connects the minimal, maximal, and mean values for each time granule of the current selected time granularity (e.g., the WBC of the patient whose value was maximal for a particular month among all patients). Graph 3 shows the monthly distribution of the Hemoglobin-state abstract concept values during the first year (relative time line) following bone-marrow transplantation. The left panel (denoted by C) of the interface includes a browser to the knowledge-base in the oncology domain.

Functionality and usability evaluations of the VISITORS interactive exploration
module were performed, using a database of more than 1, 000 oncology patients
and a group of 10 users–five clinicians and five medical informaticians. Both types
of users were able in a short time (mean of 2.5±0.27 min per question) to answer
a set of clinical questions, given the patient groups selected by the queries, including questions that require the use of specialized operators for finding associations
among derived temporal abstractions, with high accuracy (mean of 98.7±2.3 on a
predefined scale from 0 to 100). There were no significant differences between the
response times and between accuracy levels of the exploration of the data using
different time lines, i.e., absolute (i.e., calendaric) vs. relative (which refers to some
clinical key event). A standard usability score (SUS) [41] questionnaire filled by the
users demonstrated the VISITORS system to be usable (mean score for the overall
group: 68), but the clinicians' usability assessment was significantly lower than that
of the medical informaticians.

8.6.1 Specification of Knowledge-Based, Multiple-Patient Temporal Selection Expressions: the VISITORS System's Expression-Construction Module and the OBTAIN Expression-Specification Language

The VISITORS system [218] was developed for knowledge-based visualization and
exploration of the time-oriented data of multiple patients. Thus, the user's first task
is typically to specify the relevant patient population that the user intends to further
explore. Underlying the VISITORS patient-population specification module is the
ontology-based temporal-aggregation (OBTAIN) expression-specification language
[220, 221]. The OBTAIN language was implemented by a graphical expression-
construction module integrated within the VISITORS system, which can also be referred to as the Query-Builder module, since in a sense it queries the time-oriented
patient database for a sub-population of patients whose time-oriented data and abstractions the user wishes to explore. The Query-Builder module enables construction of three types of queries supported by the language: Select Subjects, Select
Time Intervals and Get Subjects Data. These queries retrieve a list of subjects, a
list of relevant time intervals, and a list of time-oriented subjects' data sets, respectively. The OBTAIN language enables population querying, through the Select Subjects query, by using an expressive set of time and value constraints. For example:
"Find patients who had, during the first month following BMT, at least one episode
of bone-marrow toxicity (an abstraction defined by the relevant clinical protocol)
of grade I or higher that lasted at least two days.", as shown in Figures 8.28 and
8.29. The language enables also querying for temporal intervals that fulfil certain
knowledge-based statistical constraints (e.g., "locate periods during which at least
15% of the patients had a bone-marrow state that could be abstracted as Grade-II
toxicity or higher, during the period following their bone-marrow transplantation."),

as depicted in Figure 8.30. To evaluate the Query Builder, five clinicians and five medical informaticians defined queries, using the query builder, on a database of more than 1000 oncology patients. After a brief training session, both user groups were able in a short time (mean: 3.3 ± 0.53 min) to construct ten complex queries using the query builder, with high accuracy (mean: 95.3 ± 4.5 on a predefined scale of 0 to 100). When grouped by time and value constraint subtypes, five groups of queries emerged. Only one of the five groups (queries using fuzzy time constraints) resulted in a significantly lower accuracy of constructed queries. The five groups of queries could be clustered into four homogeneous groups, ordered by increasing query-construction time. A standard usability scale questionnaire filled by the users demonstrated the query builder to be usable (mean score for the overall group: 68), but the clinicians' usability assessment was lower than that of the medical informaticians.

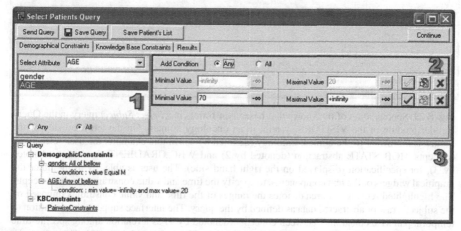

Fig. 8.28 Specification of the demographic constraints in a *Select Subject* query in the Query Builder module of the VISITORS system. The left-hand list includes the demographic parameters that are constrained (denoted by 1). In this example, the user specifies the two constraints on the age parameters (displayed as two panels on the right hand side, and denoted by 2). The relationship between two constraints is "Any", i.e., denotes the "OR" Boolean relation. The relationship between two demographic parameters is "All", i.e., denotes the "AND" Boolean relation. The bottom part of the interface displays the query that is created automatically and incrementally from the user's specification (denoted by 3).

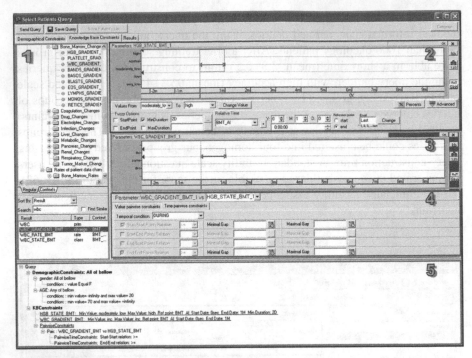

Fig. 8.29 Specification of the knowledge-based constraints in a *Select Subject* query in the Query Builder module of the VISITORS system, in an oncology domain. On the left-hand side of the Figure is a browser to the clinical ontology (denoted by 1). In this example, the user selects two concepts: HGB_STATE abstraction (denoted by 2) and WBC_GRADIENT abstraction (denoted by 3), for specification (displayed on the right hand side). The user is able, by using either the graphical widgets or the text components, to specify the time and value constraints on the concepts. The highlighted rectangular area denotes the ranges of the time and value constraints imposed on the subject's raw or abstracted data as defined by the query. The interface supports the definition of temporal pairwise constraints between concepts (denoted by 4) as a relation among specific time points (i.e., start/end) according to the 13 temporal logic relations of Allen [8]. Thus, once the user has selected the necessary temporal pairwise constraints, the panel of pairwise constraints is fulfilled according to the interrelation (in this example, the interrelation is during). The bottom part of the interface displays the query that is created automatically and incrementally from the user's specification (denoted by 5).

8.6.2 Visual Data Mining of Temporal Data

Klimov and Shahar developed a user-driven interactive knowledge-based visualization technique, called a Temporal Association Chart (TAC), for the investigation of temporal and statistical associations within multiple patient records among raw concepts, and among the temporal abstractions derived from them [220, 222]. The TAC visualization technique was implemented as part of the VISITORS system [218], which supports intelligent visualization and exploration of longitudinal data of multiple patients (Figure 8.31).

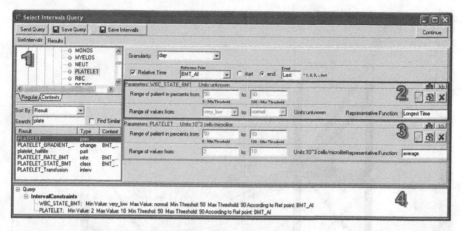

Fig. 8.30 Specification of the constraints in a *Select Time Intervals* query in the Query Builder module of the VISITORS system, in an oncology domain. On the left hand side is a browser to the clinical ontology (denoted by 1). In this example the user selects two concepts: WBC_STATE abstraction (denoted by 2) and Platelet raw data (denoted by 3) for specification (displayed on the right hand side). The bottom part of the interface displays the query that is created automatically and incrementally from the user's specifications (denoted by 4).

The TAC module was evaluated by a group of ten users, five clinicians and five medical informaticians [222]. Users were asked to answer ten complex questions using the VISITORS system, five of which required the use of TACs. Both types of users were able to answer the questions in reasonably short periods of time (a mean of 2.5±0.27 minutes) and with high accuracy (95.3±4.5 on a 0-100 scale), without a significant difference between the two groups. All five questions requiring the use of TACs were answered with similar response times and accuracy levels. Similar accuracy scores were achieved for questions requiring the use of TACs and for questions requiring the use only of general exploration operators. Although response times when using TACs were slightly (but significantly) longer.

Summary

The chapter presented some basics on information visualization with some remarks on those notions more useful for the task of visualizing temporal clinical information. Then, after introducing a small taxonomy allowing to identify, together with some general requirements for the medical domain, the specific features of the different proposals dealing with IV of temporal medical information, we discussed five different visualization systems focusing on visualization of temporal information in medicine. We started with the description of a system dealing with the visualization of multiple raw data time series related to a given patient; then, we introduced a

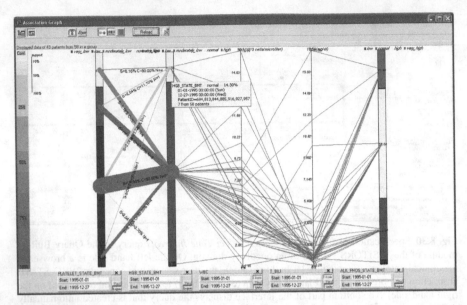

Fig. 8.31 An example of using the *Temporal Association Chart* module within the VISITORS system. In this case, associations among three haematological concepts (two abstract concepts and one raw-data concept) and two hepatic concepts (one abstract and one raw data) are visualized and explored for 49 patients during the year 1995. Association rules are displayed between the Platelet-state and HGB-state abstract concepts. The confidence and support scales are represented on the left. The width of the link among two values signifies the amount of support (prevalence of patients within the explored group) to that association; a deeper shade signifies a higher confidence (probability that, given the left-hand abstract-concept value, the patient will have the right-hand abstract-concept value).

system for a multi-modal visualization of time-oriented data at higher abstraction level, given with indeterminacy and at different granularities; a further system deals with the knowledge-based visual representation of temporal abstractions and their related raw data/abstractions; then, we considered a proposal for visually expressing new composite abstractions and for querying a temporal clinical abstraction database; finally, we moved to a system for the knowledge-based visualization and exploration of abstractions and related raw data for multiple patients.

Bibliographic Notes

Information visualization in medicine, with some special focus on temporal data, receives attention both on journals and conferences proceedings in the computer science area (more particularly in the HCI community) and in the medical informatics/AI in medicine areas. At the best of our knowledge the only special issue devoted to IV in medicine is that edited by Luca Chittaro [67], containing discussions and papers close to the visualization of temporal clinical data. A survey paper on visualization of temporal information has been published by et al. [6] and consider several examples taken from the medical domain. The visual display of temporal clinical data is founded on several concepts that are discussed and summarized in widely acknowledged readings book by Card et al. [67]. Moreover, an excellent treatise on visualization is the series of books by Edward Tufte on methods to display information [410, 411, 412].

Problems

8.1. Larkin and Simon [1987] have demonstrated that the usefulness of visual (diagrammatic) representations is mainly due to the reduction of the efforts involved in (1) the search for information, (2) the recognition of relevant elements, and (3) logical computation.

Although not absolutely necessary for answering the following questions, We highly recommend that you read Larkin and Simon's original excellent paper.

1. Explain these three objectives in your own words.
2. Provide at least 2 examples for each of them in the medical domain.
3. Explain how each of these objectives might be accomplished or assisted, in the case of each of the KNAVE-II and VISITORS systems, when a clinician or a clinical researcher needs to answer a particular question regarding a patient, in the first case, or a group of patients, in the second case.

8.2. Consider the KNAVE system's architecture in Figure 8.17. A user who uses a knowledge-based visualization and exploration distributed architecture (such as KNAVE) starts a session by formulating a query about a patient. Which of the following resources does she most need to formulate that initial query: (1) the patient database, (2) the knowledge base, (3) the data display module, (4) the temporal mediator? Please explain.

8.3. Consider systems such as KNAVE-II and VISITORS, which enable users to explore time-oriented data with the help of domain ontology. What is the difference between a syntactic temporal zoom, which uses only the data and timeline, and a semantic temporal zoom, which exploits domain knowledge? Give examples of each. Consider both the various types of zooming described in the chapter, as well as potential new ones, such as automated summarization of the data.

8.4. What does *dynamic sensitivity analysis* mean in the context of the KNAVE system? What might it mean in the context of the VISITORS system? How might a Truth Maintenance System (TMS) (see Chapter 5) be useful for such a task? Explain. Consider both knowledge and data aspects.

8.5. Interactive exploration of large sets of time-oriented data using distributed data and knowledge and perhaps also computational resources might be computationally expensive. For example, the user might be exploring the white blood cells state abstraction and suddenly asking for abstractions of the Hemoglobin gradient abstraction. The KNAVE and VISITORS systems have in practice used several strategies to alleviate this problem. Try to suggest several potential strategies to facilitate the runtime response performance.

References

1. Bpel. In Liu and Özsu [245], page 266.
2. Bpmn. In Liu and Özsu [245], page 266.
3. Serge Abiteboul, Peter Buneman, and Dan Suciu. *Data on the Web: From Relations to Semistructured Data and XML*. Morgan Kaufmann, 2000.
4. Tamas Abraham and John F. Roddick. Survey of spatio-temporal databases. *GeoInformatica*, 3(1):61–99, 1999.
5. Klaus-Peter Adlassnig, Carlo Combi, Amar K. Das, Elpida T. Keravnou, and Giuseppe Pozzi. Temporal representation and reasoning in medicine: Research directions and challenges. *Artificial Intelligence in Medicine*, 38(2):101–113, 2006.
6. Wolfgang Aigner, Silvia Miksch, Wolfgang Müller, Heidrun Schumann, and Christian Tominski. Visual methods for analyzing time-oriented data. *IEEE Trans. Vis. Comput. Graph.*, 14(1):47–60, 2008.
7. C.F. Aliferis and G.F. Cooper. A new formalism for temporal modeling in medical decision-support systems. In R.M. Gardner, editor, *19. annual symposium on computer applications in medical care*, pages 213–217, Philadelphia, 1995. Hanley & Belfus.
8. J. Allen. Towards a General Theory of Action and Time. *Artificial Intelligence*, 23:123–154, 1984.
9. James F. Allen and Patrick J. Hayes. A common-sense theory of time. In *IJCAI*, pages 528–531, 1985.
10. J.F. Allen. Maintaining Knowledge About Temporal Intervals. *Communications of the ACM*, 26:832–843, 1983.
11. T. Amagasa, M. Yoshikawa, and S. Uemura. A Bitemporal XML Data Model. In *IPSJ SIGNotes DataBase Systems*, volume 125. Japanese National Institute of Informatics, 2001.
12. T. Amagasa, M. Yoshikawa, and S. Uemura. Realizing Temporal XML Repositories using Temporal Relational Databases. In *Proceedings of the Third International Symposium on Cooperative Database Systems and Applications*, pages 63–68. IEEE Computer Society, 2001.
13. S. Andreassen, M. Woldbye, B. Falck, and S.K. Andersen. MUNIN – A causal probabilistic network for interpretation of electromyographic findings. In *International Joint Conference on Artificial Intelligence (IJCAI)*, pages 366–372, 1987.
14. G.E.M. Anscombe. Before and after. *The Philosophical Review*, 73:3–24, 1964.
15. R. Arlein, J. Gava, N. Gehani, and Lieuwen D. *ODE 4.2 User Manual*. AT&T Bell Laboratories Murray Hill, New Jersey 07974, 1996.
16. P. Atzeni, S. Ceri, S. Paraboschi, and R. Torlone. *Database Systems: Concepts, Languages, and Architecture*. McGraw-Hill, 1999.
17. Juan Carlos Augusto. Temporal reasoning for decision support in medicine. *Artificial Intelligence in Medicine*, 33(1):1–24, 2005.

C. Combi et al., *Temporal Information Systems in Medicine*,
DOI 10.1007/978-1-4419-6543-1, © Springer Science+Business Media, LLC 2010

18. John Bair, Michael H. Böhlen, Christian S. Jensen, and Richard T. Snodgrass. Notions of upward compatibility of temporal query languages. *Wirtschaftsinformatik*, 39(1):25–34, 1997.
19. M. Balaban, D. Boaz, and Y. Shahar. Applying temporal abstraction in medical information systems. *Annals of Mathematics, Computing, & Teleinformatics*, 1(1):54–62, 2003.
20. M. Barnes and G. O. Barnett. An architecture for a distributed guideline server. In Gardner [151], pages 233–237.
21. R. Bellazzi. Drug Delivery Optimization through Bayesian Networks: An application to Erythropoietin Therapy in Uremic Anemia. *Computers and Biomedical Research*, 26:274–293, 1993.
22. J. H. Van Bemmel. Medical infornatics, art or science? *Methods of Information in Medicine*, 35:157–117, 1996.
23. Elisa Bertino and Elena Ferrari. Temporal synchronization models for multimedia data. *IEEE Trans. Knowl. Data Eng.*, 10(4):612–631, 1998.
24. Elisa Bertino, Elena Ferrari, and Marco Stolf. Mpgs: An interactive tool for the specification and generation of multimedia presentations. *IEEE Trans. Knowl. Data Eng.*, 12(1):102–125, 2000.
25. C. Berzuini. Representing time in causal probabilistic networks. In *Proc. UAI89*, pages 15–28, 1989.
26. C. Bettini, C. Dyreson, W. Evans, R. Snodgrass, and X. Sean Wang. *A Glossary of Time Granularity Concepts*, pages 406–413. In Etzion et al. [139], 1998.
27. C. Bettini, X. S. Wang, and S. Jajodia. A general framework for time granularity and its application to temporal reasoning. *Annals of Mathematics and Artificial Intelligence*, 22:29–58, 1998.
28. C. Bettini, X. S. Wang, and S. Jajodia. *Time Granularities in Databases, Data Mining, and Temporal Reasoning*. Springer, 2000.
29. I. Bichindaritz and E. Conlon. Temporal knowledge representation and organization for case-based reasoning. In Chittaro et al. [71], pages 152–159.
30. M. Blaha and W. Premerlani. *Object-Oriented Modeling and Design for Database Applications*. Prentice Hall, 1998.
31. R.L. Blum. *Discovery and Representation of Causal Relationships from a Large Time-Oriented Clinical Database: The RX Project*. Lecture Notes in Medical Informatics. Springer-Verlag, 1982.
32. D. Boaz and Y. Shahar. Idan: A Distributed Temporal-Abstraction Mediator for Medical Databases. In M. Dojat, E.T. Keravnou, and P. Barahona, editors, *Proceedings of the 95th Conference on Artificial Intelligence in Medicine Europe (AIME)*, LNAI 2780, pages 21–30. Springer, 2003.
33. David Boaz and Yuval Shahar. A framework for distributed mediation of temporal-abstraction queries to clinical databases. *Artificial Intelligence in Medicine*, 34(1):3–24, 2005.
34. M. H. Boehlen. Temporal database system implementations. *SIGMOD Record*, 24(4):53–60, 1995.
35. Michael H. Böhlen, Richard T. Snodgrass, and Michael D. Soo. Coalescing in temporal databases. In T. M. Vijayaraman, Alejandro P. Buchmann, C. Mohan, and Nandlal L. Sarda, editors, *VLDB*, pages 180–191. Morgan Kaufmann, 1996.
36. I. Boldo, Y. Shahar, A. Laor, and O. Etzion. Automated recognition of clinical guidelines from electronic medical records. In *The 8th International Workshop on Intelligent Data Analysis in Medicine and Pharmacology (IDAMAP-2004)*, pages 103–110, Stanford, CA, 2003.
37. G. Bortolan, C. Cavaggion, R. Degani, M. Bressan, and P. Marinato. The Role of Patient History in a Decision Support System. In *Proceedings of Computers in Cardiology*, pages 357–360. IEEE Computer Society Press, 1989.
38. A.A Boxwala, M. Peleg, S. Tu, O. Ogunyemi, Q.T. Zeng, D. Wang, V.L. Patel, R. Greenes, and E.H. Shortliffe. Glif3: A representation format for sharable computer-interpretable clinical practice guidelines. *Journal of Biomedical Informatics*, 37(3):147–161, 2004.

39. M. E. Bratman. *Intention, Plans and Practical Reason.* Harvard University Press, Cambridge, MA, 1987.
40. E. Braunwald, R.H. Jones, D.B. Mark, and et al. Diagnosing and Managing Unstable Angina. Agency for Health Care Policy and Research. *Circulation*, 90:613–622, 1994.
41. J. Brooke. *SUS: A 'quick and dirty' usability scale*, pages 189–194. Taylor & Francis, 1996.
42. V. Brusoni, L. Console, P. Terenziani, and D. Theseider Dupre. A spectrum of definitions for temporal model-based diagnosis. *Artificial Intelligence*, 102:39–79, 1998.
43. B.G. Buchanan and E.H. Shortliffe, editors. *Rule-based expert systems: the MYCIN experiments of the Stanford Heuristic Programming Project.* Addison-Wesley, 1984.
44. Alex A. T. Bui, Denise R. Aberle, and Hooshang Kangarloo. Timeline: Visualizing integrated patient records. *IEEE Transactions on Information Technology in Biomedicine*, 11(4):462–473, 2007.
45. P. Buneman, S. B. Davidson, G. G. Hillebrand, and D. Suciu. A Query Language and Optimization Techniques for Unstructured Data. In *Proceedings of the 1996 ACM SIGMOD International Conference on Management of Data*, pages 505–516. ACM Press, 1996.
46. Peter Buneman, Sanjeev Khanna, Keishi Tajima, and Wang Chiew Tan. Archiving scientific data. In Michael J. Franklin, Bongki Moon, and Anastassia Ailamaki, editors, *SIGMOD Conference*, pages 1–12. ACM, 2002.
47. T. Bylander, D. Allemang, M. C. Tanner, and J. R. Josephson. The computational complexity of abduction. *Artificial Intelligence*, 49:25–60, 1991.
48. P.V. Caironi, L. Portoni, C. Combi, F. Pinciroli, and S. Ceri. HyperCare: A prototype of an active database for compliance with essential hypertension therapy guidelines. In *Proceedings of the 1997 AMIA Annual Fall Symposium (formerly the Symposium on Computer Applications in Medical Care)*, pages 288–292, Philadelphia, PA, 1997. Hanley & Bclfus.
49. S. Card and J. MacKinlay. The Structure of the Information Visualization Design Space. In *InfoVis '97 IEEE Symposium on Information Visualization*, pages 92–99. IEEE Computer Society Press, 1997.
50. S. Card, J. MacKinlay, and B. Shneiderman, editors. *Readings in Information Visualization: Using Vision to Think.* Morgan - Kaufmann, 1998.
51. A.F. Cardenas, I.T. Ieong, R.K. Taira, R. Barcher, and C.M. Breant. The Knowledge - Based Object-Oriented PIQUERY+ Language. *IEEE Transactions on Knowledge and data Engineering*, 5:644–657, 1993.
52. Alfonso F. Cardenas, Ion Tim Ieong, Ricky K. Taira, Roger Barker, and Claudine M. Breant. The knowledge-based object-oriented picquery+ language. *IEEE Trans. Knowl. Data Eng.*, 5(4):644–657, 1993.
53. G. Carrault, M. O. Cordier, R. Quiniou, and F. Wang. Temporal abstraction and inductive logic programming for arrythmia recognition from electrocardiograms. *Artificial Intelligence in Medicine*, 28:231–263, 2003.
54. R.G.G. Cattel and D.K. Barry, editors. *The Object Database Standard: ODMG 2.0.* Morgan - Kaufmann, 1997.
55. R.G.G. Cattel and D.K. Barry, editors. *The Object Data Standard: ODMG 3.0.* Morgan - Kaufmann, 2000.
56. R.G.G. Cattell. *The Object Database Standard: ODMG-93.* Morgan Kaufmann, San Francisco, 1996.
57. R.G.G. Cattell. *The Object Database Standard: ODMG 3.0.* Morgan Kaufmann, San Francisco, 2000.
58. S. Chakravarty and Y. Shahar. A constraint-based specification of periodic patterns in time-oriented data. *Annals of Mathematics and Artificial Intelligence*, 30(1-4), 2000.
59. S. Chakravarty and Y. Shahar. Specification and detection of periodicity in clinical data. *Methods of Information in Medicine*, 40(5):410–420, 2001.
60. B. Chandrasekaran. Generic tasks in knowledge-based reasoning: High-level building blocks for expert system design. *IEEE Expert*, 1:23–30, 1986.
61. B. Chandrasekaran and S. Mittal. Conceptual representation of medical knowledge for diagnosis by computer: MDX and related systems. *Advances in Computers*, 22:217–293, 1983.

62. S. S. Chawathe, S. Abiteboul, and J. Widom. Representing and Querying Changes in Semistructured Data. In *Proceedings of the Fourteenth International Conference on Data Engineering*, pages 4–13. IEEE Computer Society, 1998.

63. S. S. Chawathe, S. Abiteboul, and J. Widom. Managing historical semistructured data. *Theory and Practice of Object Systems*, 5(3):143–162, 1999.

64. T.S. Cheng and S.K. Gadia. An object-oriented model for temporal databases. In *Proceedings of the International Workshop on an Infrastructure for Temporal Databases*, pages 1–19, Arlington, TX, 1993.

65. E.H. Chi. *A Framework for Visualizing Information*. Kluwer Academic Publishers, 2002.

66. J.P. Chin, V.A. Diehl, and K.L. Norman. Development of an instrument measuring user satisfaction of the human-computer interface. In *Proceedings of the CHI '88 Conference on Human Factors in Computing Systems*, pages 213–218. ACM Press, 1988.

67. L. Chittaro. Information visualization and its application to medicine. *Artificial Intelligence in Medicine*, 22:81–88, 2001.

68. L. Chittaro and C. Combi. Visualizing queries on databases of temporal histories: new metaphors and their evaluation. *Data and Knowledge Engineering*, 44:239–264, 2003.

69. L. Chittaro, C. Combi, and G. Trapasso. Data mining on temporal data: a visual approach and its clinical application to hemodialysis. *Journal of Visual Languages and Computing*, 14:591–620, 2003.

70. L. Chittaro and M. Dojat. Using a General Theory of Time and Change in Patient Monitoring: Experiment and Evaluation. *Computers in Biology and Medicine*, 27:435–452, 1997.

71. L. Chittaro, S. Goodwin, H. Hamilton, and A. Montanari, editors. *Third International Workshop on Temporal Representation and Reasoning (TIME '96)*, Los Alamitos, CA, 1996. IEEE Computer Society Press.

72. L. Chittaro and A. Montanari. Efficient Temporal Reasoning in the Cached Event Calculus. *Computational Intelligence*, 12:359–382, 1996.

73. L. Chittaro, A. Montanari, and E. Peressi. *An Integrated Framework for Temporal Aggregation and Omission in the Event Calculus*, pages 47–54. Computational Mechanics, Boston, 1995.

74. L. Chittaro, M. Del Rosso, and M. Dojat. Modeling medical reasoning with the event calculus: an application to the management of mechanical ventilation. In P. Barahona, M. Stefanelli, and J. Wyatt, editors, *Proceedings of the 5th Conference on Artificial Intelligence in Medicine Europe (AIME)*, LNAI 934, pages 79–901. Springer, 1995.

75. W.J. Clancey. Heuristic classification. *Artificial Intelligence*, 27:289–350, 1985.

76. W.J. Clancey and R. Letsinger. NEOMYCIN: Reconfiguring a rule-based expert system for application to teaching. In *Proceedings of the International Joint Conference on Artificial Intelligence (IJCAI)*, pages 829–836, 1981.

77. J. Clifford. Formal semantics and pragmatics for natural language querying. Cambridge tracts in theoretical computer science. Cambridge University Press, 1990. Cambridge.

78. J. Clifford and A. Rao. A simple, general structure for temporal domains. In C. Rolland, F. Bodart, and M. Leonard, editors, *Temporal Aspects in Information Systems*, Amsterdam: NorthHolland, 1988.

79. J. Clifford and A. Tuzhilin, editors. *Recent Advances in Temporal Databases*. Springer Verlag, 1995.

80. C. Combi, , L. Missora, and F. Pinciroli. Supporting Temporal Queries on Clinical Relational Databases: the S-WATCH-QL Language. In J.J. Cimino, editor, *1996 AMIA Annual Fall Symposium*, pages 527–531. Hanley & Belfus, 1996.

81. C. Combi and L. Chittaro. Abstraction on clinical data sequences: an object-oriented data model and a query language based on the event calculus. *Artificial Intelligence in Medicine*, 17:271–301, 1999.

82. C. Combi, C. Cucchi, and F. Pinciroli. Applying Object-Oriented Technologies in Modeling and Querying Temporally-Oriented Clinical Databases Dealing with Temporal Granularity and Indeterminacy. *IEEE Transactions on Information Technology in Biomedicine*, 1:100–127, 1997.

83. C. Combi, F. Pinciroli, M. Cavallaro, and G. Cucchi. Querying temporal clinical databases with different time granularities: the GCH-OSQL language. In R.M. Gardner, editor, *19. annual symposium on computer applications in medical care*, pages 326–330. Hanley & Belfus, 1995.

84. C. Combi, F. Pinciroli, G. Musazzi, and C. Ponti. Managing and Displaying Different Time Granularities of Clinical Information. In J.G. Ozbolt, editor, *18th Annual Symposium on Computer Applications in Medical Care*, pages 954–958. Hanley & Belfus, 1994.

85. C. Combi, F. Pinciroli, and G. Pozzi. Temporal Clinical Data Modeling and Implementation for PTCA Patients in an OODBMS Environment. In *Proceedings of Computers in Cardiology*, pages 505–508, Los Alamitos, 1994. IEEE Computer Society Press.

86. C. Combi, F. Pinciroli, and G. Pozzi. Managing different time granularities of clinical information by an interval-based temporal data model. *Methods of Information in Medicine*, 34:458–474, 1995.

87. C. Combi, L. Portoni, and F. Pinciroli. Visualizing Temporal Clinical Data on the WWW. In *Joint European Conference on Artificial Intelligence in Medicine and Medical Decision Making, AIMDM'99*, LNCS 1620, pages 301–314. Springer, 1999.

88. C. Combi and G. Pozzi. HMAP - A Temporal Data Model Managing Intervals with Different Granularities and Indeterminacy from Natural Language Sentences. *The VLDB Journal*, 9:294–311, 2001.

89. Carlo Combi. Modeling temporal aspects of visual and textual objects in multimedia databases. In *TIME*, pages 59–68, 2000.

90. Carlo Combi, Sara Degani, and Christian S. Jensen. Capturing temporal constraints in temporal er models. In Qing Li, Stefano Spaccapietra, Eric S. K. Yu, and Antoni Olivé, editors, *ER*, volume 5231 of *Lecture Notes in Computer Science*, pages 397–411. Springer, 2008.

91. Carlo Combi, Matteo Gozzi, José M. Juárez, Roque Marín, and Barbara Oliboni. Temporal similarity measures for querying clinical workflows. *Artificial Intelligence in Medicine*, 46(1):37–54, 2009.

92. Carlo Combi, Matteo Gozzi, José M. Juárez, Barbara Oliboni, and Giuseppe Pozzi. Conceptual modeling of temporal clinical workflows. In *TIME*, pages 70–81. IEEE Computer Society, 2007.

93. Carlo Combi and Angelo Montanari. Data models with multiple temporal dimensions: Completing the picture. In Klaus R. Dittrich, Andreas Geppert, and Moira C. Norrie, editors, *CAiSE*, volume 2068 of *Lecture Notes in Computer Science*, pages 187–202. Springer, 2001.

94. Carlo Combi, Angelo Montanari, and Giuseppe Pozzi. The t4sql temporal query language. In Mário J. Silva, Alberto H. F. Laender, Ricardo A. Baeza-Yates, Deborah L. McGuinness, Bjørn Olstad, Øystein Haug Olsen, and André O. Falcão, editors, *CIKM*, pages 193–202. ACM, 2007.

95. Carlo Combi and Barbara Oliboni. Managing valid time semantics for semistructured multimedia clinical data. In Torsten Grust, Hagen Höpfner, Arantza Illarramendi, Stefan Jablonski, Marco Mesiti, Sascha Müller, Paula-Lavinia Patranjan, Kai-Uwe Sattler, Myra Spiliopoulou, and Jef Wijsen, editors, *EDBT Workshops*, volume 4254 of *Lecture Notes in Computer Science*, pages 375–386. Springer, 2006.

96. Carlo Combi, Barbara Oliboni, and Elisa Quintarelli. A graph-based data model to represent transaction time in semistructured data. In Fernando Galindo, Makoto Takizawa, and Roland Traunmüller, editors, *DEXA*, volume 3180 of *Lecture Notes in Computer Science*, pages 559–568. Springer, 2004.

97. Carlo Combi, Barbara Oliboni, and Elisa Quintarelli. Specifying temporal data models for semistructured data by a constraint-based approach. In Hisham Haddad, Andrea Omicini, Roger L. Wainwright, and Lorie M. Liebrock, editors, *SAC*, pages 1103–1108. ACM, 2004.

98. Carlo Combi, Barbara Oliboni, and Rosalba Rossato. Merging multimedia presentations and semistructured temporal data: a graph-based model and its application to clinical information. *Artificial Intelligence in Medicine*, 34(2):89–112, 2005.

99. L. Console, D. Theseider Dupré, and P. Torasso. A theory of diagnosis for incomplete causal models. In *Proc. IJCAI-89*, pages 1311–1317, 1989.

100. L. Console, D. Portinale, D. Theseider Dupré, and P. Torasso. *Combining heuristic and causal reasoning in diagnostic problem solving*, pages 46–68. Springer-Verlag, 1993.

101. L. Console and P. Torasso. Integrating models of the correct behavior into abductive diagnosis. In *Proc. ECAI-90*, pages 160–166, 1990.

102. L. Console and P. Torasso. A spectrum of logical definitions of model-based diagnosis. *Computational Intelligence*, 7:133–141, 1991.

103. L. Console and P. Torasso. On the co-operation between abductive and temporal reasoning in medical diagnosis. *Artificial Intelligence in Medicine*, 3:291–311, 1991.

104. Luca Console, Anna Furno, and Pietro Torasso. Dealing with time in diagnostic reasoning based on causal models. In Z. Ras and L. Saitta, editors, *Methodologies for Intelligent Systems 3*, pages 230–239, NorthHolland, 1988.

105. S.B. Cousins and M.G. Kahn. The visual display of temporal information. *Artificial Intelligence in Medicine*, 3:341–357, 1991.

106. P. Dagum and A. Galper. Forecasting sleep apnea with dynamic network models. In *Proc. UAI93*, pages 64–71, 1993.

107. P. Dagum and A. Galper. Sleep apnea forecasting with dynamic network models. Knowledge Systems Laboratory Report KSL9320, Stanford University, Stanford, CA, May 1993.

108. P. Dagum and A. Galper. Time-series prediction using belief network models. *International Journal of Human-Computer Studies*, 42:617–632, 1995.

109. P. Dagum, A. Galper, E. Horvitz, and A. Seiver. Uncertain Reasoning and Forecasting. *International Journal of Forecasting*, 11:73–87, 1993.

110. P. Dagum, A. Galper, and E. J. Horvitz. Dynamic network models for forecasting. In *Proceedings of the Eight Conference on Uncertainty in Artificial Intelligence*, pages 41–48. Morgan Kaufmann, 1992.

111. A.K. Das and M.A. Musen. A Temporal Query System for Protocol-Directed Decision Support. *Methods of Information in Medicine*, 33:358–370, 1994.

112. A.K. Das and M.A. Musen. A comparison of the temporal expressiveness of three database query methods. In R.M. Gardner, editor, *19. annual symposium on computer applications in medical care*, pages 331–337. Hanley & Belfus, 1995.

113. A.K. Das and M.A. Musen. A Formal Method to Resolve Temporal Mismatches in Clinical Databases. In *AMIA Annual Symposium*. Hanley & Belfus, 2001.

114. A.K. Das, Y. Shahar, S.W. Tu, and M.A. Musen. A temporal-abstraction mediator for protocol-based decision support. In J.G. Ozbolt, editor, *18. annual symposium on computer applications in medical care*, pages 320–324, Philadelphia, 1994. Hanley & Belfus.

115. C.J. Date, Hugh Darwen, and Nikos Lorentzos. *Temporal Data and the Relational Model*. Morgan Kaufmann, 2002.

116. J. de Kleer and B. C. Williams. Diagnosing multiple faults. *Artificial Intelligence*, 32:97–130, 1987.

117. T. Dean and D. McDermott. Temporal Data Base Management. *Artificial Intelligence*, 32:1–55, 1987.

118. T.L. Dean and K. Kanazawa. Probabilistic causal reasoning. In *Proceedings Conference on Uncertainty in Artificial Intelligence (UAI)*, pages 73–80, 1988.

119. R. Dechter, I. Meiri, and J. Pearl. Temporal Constraint Networks. *Artificial Intelligence*, 49:61–95, 1991.

120. P. Degoulet, C. Devries, J. Chantalou, E. Klinger, D. Sauquet, P. Zweigenbaum, and F. Aime'. LIED: A Temporal Database Management System. In *Medinfo 86*, pages 532–536. North-Holland, 1986.

121. P. R. Dexter, S. Perkins, M. J. Overhage, K. Maharry, R. B. Kohler, and C. J.McDonald. A computerized reminder system to increase the use of preventive care for hospitalized patients. *New England Journal of Medicine*, 345(13):965–970, 2001.

122. John D. N. Dionisio and Alfonso F. Cardenas. Unified data model for representing multimedia, timeline, and simulation data. *IEEE Trans. Knowl. Data Eng.*, 10(5):746–767, 1998.

123. R.H. Dolin. Modeling the temporal complexities of symptoms. *Journal of AMIA*, 2:323–331, 1995.

124. R.H. Dolin. Expressiveness and query complexity in an electronic health record data model. In J.J. Cimino, editor, *1996 AMIA Annual Fall Symposium*, pages 522–526. Hanley & Belfus, 1996.

125. W. Dorda. WAREL: A System for Retrieval of Clinical Data Considering the Course of Diseases. *Methods of Information in Medicine*, 28:133–141, 1989.

126. W. Dorda. Data-Screening and Retrieval of Medical Data by the System WAREL. *Methods of Information in Medicine*, 29:3–11, 1990.

127. W. Dorda, W. Gall, and G. Duftschmid. Clinical data retrieval: 25 years of temporal query management at the university of vienna medical school. *Methods of Information in Medicine*, 41(2):89–97, 2002.

128. S.M. Downs, M.G. Walker, and R.L. Blum. Automated summarization of on-line medical records. In R. Salamon, B. Blum, and M. Jorgensen, editors, *MEDINFO '86: Proceedings of the Fifth Conference on Medical Informatics*, pages 800–804, North-Holland, Amsterdam, 1986.

129. G. Duftschmid, S. Miksch, and W. Gall. Verification of temporal scheduling constraints in clinical practice guidelines. *Artificial Intelligence in Medicine*, 25(2):93–121, 2002.

130. C. Dyreson. Towards a temporal world-wide web: a transaction-time web server. In *In Proceedings of the Australian Database Conference (ADC '01)*, pages 169–175, 2001.

131. C. E. Dyreson, M. H. Böhlen, and C. S. Jensen. Capturing and Querying Multiple Aspects of Semistructured Data. In *VLDB'99, Proceedings of 25th International Conference on Very Large Data Bases*, pages 290–301. Morgan Kaufmann, 1999.

132. Curtis E. Dyreson. Temporal coalescing with now, granularity, and incomplete information. In Alon Y. Halevy, Zachary G. Ives, and AnHai Doan, editors, *SIGMOD Conference*, pages 169–180. ACM, 2003.

133. Andrew Eisenberg and Jim Melton. SQL standardization: The next steps. *SIGMOD Record*, 29(1):63–67, 2000.

134. R. Elmasri and S. Navathe. *Fundamentals of Database Systems, 2nd Edition*. The Benjamin/Cummings Publishing Company, Redwood City, CA, 1994.

135. R. Elmasri and S. Navathe, editors. *Fundamentals of Database Systems, Third Edition*. Addison-Wesley, 2000.

136. E. A. Emerson. Temporal and Modal Logic. In J. Van Leeuwen, editor, *Handbook of Theoretical Computer Science, Vol. B*, pages 995–1072. Elsevier Science Publishers, North-Holland, 1990.

137. O. Etzion, A. Gal, and A. Segev. Temporal Support in Active Databases. In *Workshop on Information Technologies & Systems (WITS)*, pages 245–254, 1992.

138. O. Etzion, A. Gal, and A. Segev. *Extended Update Functionality in Temporal Databases*, pages 56–95. In Etzion et al. [139], 1998.

139. O. Etzion, S. Jajodia, and S. Sripada, editors. *Temporal Databases - Research and Practice*. LNCS 1399. Springer, 1998.

140. L. M. Fagan, J. C. Kunz, E. A. Feigenbaum, and J. J. Osborn. *Extensions to the rule-based formalism for a Monitoring task*, pages 397–423. In Buchanan and Shortliffe [43], 1984.

141. M. J. Field and K. N. Lohr. Clinical practice guidelines: Directions for a new program. Committee to Advise the Public Health Service on Clinical Practice Guidelines, Institute of Medicine, 1992.

142. M. Fisher, D. Gabbay, and L. Vila, editors. *Handbook of Temporal Reasoning in Artificial Intelligence*. Elsevier, The Netherlands, 2005.

143. K.D. Forbus. Qualitative process theory. *Artificial Intelligence*, 24:85–168, 1984.

144. J. Fox and S. Das. *Safe and Sound: Artificial Intelligence in Hazardous Applications*. AAAI and MIT Press, 2000.

145. J. Fox, N. Johns, and A. Rahmanzadeh. Disseminating medical Knowledge: the PROforma approach. *Artificial Intelligence in Medicine*, 14:157–181, 1998.

146. P. Friedland and Y. Iwasaki. The Concept and Implementation of Skeletal Plans. *Journal of Automated Reasoning*, 1(2):161–208, 1985.

147. J.F. Fries. Time oriented patient records and a computer databank. *Journal of the American Medical Association*, 222:1536–1543, 1972.

148. A. Gal, O. Etzion, and A. Segev. Representation of highly-complex knowledge in a database. *Journal of Intelligent Information Systems*, 3(2):185–203, 1994.

149. A. Galton, editor. *Temporal Logics and Their Applications*. London: Academic Press, 1987.

150. J. Gamper and W. Nejdl. Abstract temporal diagnosis in medical domains. *Artificial Intelligence in Medicine*, 10:209–234, 1997.

151. R. M. Gardner, editor. *Proceedings of the Annual Symposium on Computer Applications in Medical Care (SCAMC-95)*, New Orleans, LA, 1995. Hanley & Belfus.

152. A. Gerevini, L. Schubert, and S. Schaeffer. The temporal reasoning tools TIMEGRAPH I-II. *International Journal of Artificial Intelligence Tools*, 4(1-2):281–299, 1995.

153. P. Gershkovich and R.N. Shiffman. An implementation framework for gem encoded guidelines. In *Proceedings of the 2001 AMIA Annual Symposium*, Philadelphia, 2001. Hanley & Belfus.

154. N. Gershon, S.G. Eick, and S. Card. Information Visualization. *ACM Interactions*, 5:9–15, 1998.

155. M. K. Goldstein, B. B. Hoffman, R. W. Coleman, M. A. Musen, S. W. Tu, A. Advani, R. D. Shankar RD, and M. O'Connor. Implementing clinical practice guidelines while taking account of evidence: ATHENA, an easily modifiable decision-support system for management of hypertension in primary care. In *Proceedings of the 2000 AMIA Annual Symposium*, pages 300–304, Philadelphia, 2000. Hanley & Belfus.

156. M. K. Goldstein, B. B. Hoffman, R. W. Coleman, M. A. Musen, S. W. Tu, A. Advani, R. D. Shankar RD, and M. O'Connor. Patient safety in guideline-based decision support for hypertension management: ATHENA DSS. *Journal of the American Medical Informatics Association*, 9(6Suppl):S11–16, 2002.

157. Matteo Golfarelli and Stefano Rizzi. A survey on temporal data warehousing. *nternational Journal of Data Warehousing and Mining*, 5(1):1–17, 2009.

158. I.A. Goralwalla, M.T. Özsu, and D. Szafron. Modeling Medical Trials in Pharmacoeconomics Using a Temporal Object Model. *Computers in Biology and Medicine*, 27:369–387, 1997.

159. Iqbal A. Goralwalla, Yuri Leontiev, M. Tamer Özsu, Duane Szafron, and Carlo Combi. Temporal Granularity for Unanchored Temporal Data. In Georges Gardarin, James C. French, Niki Pissinou, Kia Makki, and Luc Bouganim, editors, *Proceedings of the 1998 ACM CIKM International Conference on Information and Knowledge Management*, pages 414–423, Maryland, USA, 1998.

160. Iqbal A. Goralwalla, Yuri Leontiev, M. Tamer Özsu, Duane Szafron, and Carlo Combi. Temporal Granularity: Completing the Puzzle. *J. Intell. Inf. Syst.*, 16(1):41–63, 2001.

161. C. Gordon and M. Veloso. The PRESTIGE Project: Implementing Guidelines in Healthcare. *Medical Informatics Europe '96*, pages 887–891, 1996. IOS Press.

162. F. Grandi and F. Mandreoli. The Valid Web: an XML/XSL infrastructure for Temporal Management of Web Documents. In *Advances in Information Systems, First International Conference, ADVIS 2000*, volume 1909 of *Lecture Notes in Computer Science*, pages 294–303. Springer-Verlag, Berlin, 2000.

163. H. Gregersen and C.S. Jensen. Temporal Entity-Relationship Models: a Survey. *IEEE Transactions on Knowledge and Data Engineering*, 11:464–497, 1999.

164. Heidi Gregersen and Christian S. Jensen. Temporal entity-relationship models - a survey. *IEEE Trans. Knowl. Data Eng.*, 11(3):464–497, 1999.

165. J. M. Grimshaw and I. T. Russel. Effect of clinical guidelines on medical practice: A systematic review of rigorous evaluations. *Lancet*, 342:1317–1322, 1993.

166. W. E. Grosso, H. Eriksson, R. Fergerson, J. H. Gennari, S. W. Tu, and M. A. Musen. Knowledge modeling at the millennium (the design and evolution of Protg 2000). In *Proceedings of the 12th Banff Knowledge Acquisition for Knowledge-Based Systems Workshop*, pages 7–4–1 to 7–4–36, Banff, Canada, 1999.

167. P. Haddawy, J.W. Helwig, L. Ngo, and R.A. Krieger. Clinical simulation using context-sensitive temporal probability models. In R.M. Gardner, editor, *19. annual symposium on computer applications in medical care*, pages 203–207, Philadelphia, 1995. Hanley & Belfus.

168. I. J. Haimowitz and I. S. Kohane. An epistemology for clinically significant trends. In *Proceedings of the Tenth National Conference on Artificial Intelligence*, pages 176–181. AAAI Press, 1993. Menlo Park, CA.

169. I. J. Haimowitz and I. S. Kohane. Automated trend detection with alternate temporal hypotheses. In *Proceedings of the Thirteenth International Joint Conference on Artificial Intelligence, San Mateo*, pages 146–151. Morgan Kaufman, 1993.

170. I. J. Haimowitz and I. S. Kohane. Managing temporal worlds for medical trend diagnosis. In *Artificial Intelligence in Medicine* [206], pages 299–322.

171. I.J. Haimowitz. *KnowledgeBased Trend Detection and Diagnosis*. PhD thesis, Department of Electrical Engineering and Computer Science, Massachusetts Institute of Technology, 1994.

172. J.Y. Halpern and Y. Shoham. A propositional modal logic of time intervals. In *Proceedings of the Symposium on Logic in Computer Science*, New York, 1986. IEEE. Boston, Mass.

173. Keith W. Hare. JCC's SQL standards page. (accessed: September 1, 2006). http://www.jcc.com/sql.htm, February, 2006.

174. P. Hayes. *The second naive physics manifesto*. MIT Press, Cambridge, MA, 1985.

175. M. Hearst. *User Interfaces and Visualization*, pages 257–324. ACM Press, 2000.

176. D.E. Heckerman and R.A. Miller. Towards a better understanding of the INTERNIST-1 knowledge base. In R. Salamon, B. Blum, and M. Jorgensen, editors, *MEDINFO '86: Proceedings of the Fifth Conference on Medical Informatics*, pages 27–31, 1986.

177. S. I. Herbert, C. J. Gordon, A. Jackson-Smale, and J. L. Renaud Salis. Protocols for clinical care. *Computer Methods and Programs in Biomedicine*, 48:21–26, 1995.

178. W. Horn. AI in medicine on its way from knowledge-intensive to data-intensive systems. *Artificial Intelligence in Medicine*, 23:3–12, 2001.

179. W. Horn, S. Miksch, G. Egghart, C. Popow, and F. Paky. Effective Data Validation of High-Frequency Data: Time-Point-, Time-Interval-, and Trend-Based Methods. *Computer in Biology and Medicine, Special Issue: Time-Oriented Systems in Medicine*, 27(5):389–409, 1997.

180. R. Horne, D. James, K. Petrie, J. Weinman, and R. Vincent. Patients' interpretation of symptoms as a cause of delay in reaching hospital during acute myocardial infarction. *Heart*, 83:388–393, 2000.

181. G. Hripcsak, P. Ludemann, T. A. Pryor, O. B. Wigertz, and P. D. Clayton. Rationale for the Arden Syntax. *Computers and Biomedical Research*, 27:291–324, 1994.

182. A. Hunter. Merging structured text using temporal knowledge. *Knowledge and Data Engineering*, 41(1):29–66, 2002.

183. R. S. Irwin, L. S. Boulet, and M. M. Cloutier. Managing cough as a defense mechanism and as a Symptom: a consensus panel report of the American College of Chest Physicians. *Chest*, 114(2):133S–181S, 1998.

184. C. Jensen and C. Dyreson (Eds.) et al. *Consensus Glossary of Temporal Database Concepts - February 1998 Version*, pages 367–405. In Etzion et al. [139], 1998.

185. C. Jensen and R. Snodgrass. Temporal Specialization and Generalization. *IEEE Transactions on Knowledge and Data Engineering*, 6:954–974, 1994.

186. C. Jensen and R. Snodgrass. Semantics of Time-Varying Information. *Information Systems*, 21:311–352, 1996.

187. C. Jensen and R. Snodgrass. Temporal Data Management. *IEEE Transactions on Knowledge and Data Engineering*, 11:36–44, 1999.

188. Christian S. Jensen and Richard T. Snodgrass (eds.). Temporal Database Entries for the Springer Encyclopedia of Database Systems. Technical Report TR-90, TimeCenter, 2008.

189. Christian S. Jensen and Richard T. Snodgrass. Temporal data management. *IEEE Trans. Knowl. Data Eng.*, 11(1):36–44, 1999.

190. F.V. Jensen. *Bayesian Networks and Decision Graphs*. Springer, 2001.

191. Haitao Jiang and Ahmed K. Elmagarmid. Spatial and temporal content-based access to hypervideo databases. *VLDB J.*, 7(4):226–238, 1998.

192. P. D. Johnson, S. W. Tu, N. Booth, B. Sugden, and I. N. Purves. Using scenarios in chronic disease management guidelines for primary care. In *Proceedings of the 2000 AMIA Annual Symposium*. Hanley & Belfus, 2000.

193. P. D. Johnson, S. W. Tu, M. A. Musen, and I. Purves. A virtual medical record for guideline-based decision support. In *Proceedings of the 2000 AMIA Annual Symposium*. Hanley & Belfus, 2001.

194. D. Jung, R. Brennecke, M. Kottmeyer, R. Hering, W. Clas, R. Erbel, and J. Meyer. Tree-base, a query language for a time-oriented cardiology data base system. In *Proceedings of Computers in Cardiology*, pages 381–384. IEEE Computer Society Press, 1986.

195. K. Kahn and G.A. Gorry. Mechanizing temporal knowledge. *Artificial intelligence*, 9:87–108, 1977.

196. M. Kahn, L.M. Fagan, and S. Tu. Extensions to the Time-Oriented Database Model to Support Temporal Reasoning in Medical Expert Systems. *Methods of Information in Medicine*, 30:4–14, 1991.

197. M. Kahn, L.M. Fagan, and S. Tu. TQuery: A Context-Sensitive Temporal Query Language. *Computers and Biomedical Research*, 24:401–419, 1991.

198. M.G. Kahn. *ModelBased Interpretation of TimeOrdered Medical Data*. PhD thesis, Section on Medical Information Sciences, University of California, San Francisco, CA, 1988.

199. M.G. Kahn. Combining physiologic models and symbolic methods to interpret time-varying patient data. *Methods of Information in Medicine*, 30:167–178, 1991.

200. M.G. Kahn and K.A. Marrs. Creating temporal abstractions in three clinical information systems. In R.M. Gardner, editor, *19. annual symposium on computer applications in medical care*, pages 392–396. Hanley & Belfus, 1995.

201. K. Kanazawa. A logic and time nets for probabilistic inference. In *Proceedings, Ninth National Conference on Artificial Intelligence*, pages 360–365, Cambridge, MA, 1991. MIT Press. Los Angeles, CA.

202. B. T. Karras, S. D. Nath, and R. N. Shiffman. A Preliminary Evaluation of Guideline Content Mark-up Using GEM-An XML Guideline Elements Model. In M. J. Overhage, editor, *Proceedings of the 2000 AMIA Annual Symposium*, Philadelphia, 2000. Hanley & Belfus.

203. E. T. Keravnou. Special Issue on Temporal Reasoning in Medicine. *Artificial Intelligence in Medicine*, 3(6), 1991.

204. E. T. Keravnou. Engineering time in medical knowledge-based systems through time-axes and time-objects. In *Proc. TIME-96*, pages 160–167, New York, 1996. IEEE Computer Society Press.

205. E. T. Keravnou. An ontology of time using time-axes and time-objects as primitives. Technical report, Department of Computer Science, University of Cyprus, 1996.

206. E. T. Keravnou. Special Issue on Temporal Reasoning in Medicine. *Artificial Intelligence in Medicine*, 8(3), 1996.

207. E. T. Keravnou. *Temporal abstraction of medical data: deriving periodicity*, pages 61–79. In Lavrač et al. [238], 1997.

208. E. T. Keravnou. A multidimensional and multigranular model of time for medical knowledge-based systems. *Journal of Intelligent Information Systems*, 13:73–120, 1999.

209. E. T. Keravnou. A time ontology for medical knowledge-based systems. In *Proc. EMCSR'98*, 831–835.

210. E. T. Keravnou, F. Dams, J. Washbrook, C. M. Hall, R. M. Dawood, and D. Shaw. Background knowledge in diagnosis. *Artificial Intelligence in Medicine*, 4:263–279, 1992.

211. E. T. Keravnou, F. Dams, J. Washbrook, C. M. Hall, R. M. Dawood, and D. Shaw. Modelling diagnostic skills in the domain of skeletal dysplasias. *Comput. Methods Progr. Biomed.*, 45:239–260, 1994.

212. E. T. Keravnou and J. Washbrook. A Temporal Reasoning Framework Used in the Diagnosis of Skeletal Dysplasias. *Artificial Intelligence in Medicine*, 2:239–265, 1990.

213. E. T. Keravnou and J. Washbrook. Abductive diagnosis using time-objects: criteria for the evaluation of solutions. *Computational Intelligence*, 17(1), 2001.

214. E.T. Keravnou. Temporal diagnostic reasoning based on time-objects. In *Artificial Intelligence in Medicine* [206], pages 235–266.

215. S.K. Kim and S. Chakravarthy. Modeling Time: Adequacy of Three Distinct Time Concepts for Temporal Databases. In *12th International Conference of the Entity-Relationship Approach*, LNCS 823, pages 475–491. Springer, 1993.

216. S.K. Kim and S. Chakravarthy. Resolution of Time Concepts in Temporal Databases. *Information Sciences*, 80:91–125, 1994.

217. W. Kim. *Modern Database Systems*. Addison Wesley, New York, 1995.

218. D. Klimov and Y. Shahar. A framework for intelligent visualization of multiple time-oriented medical records. In *Proceedings of the 2005 AMIA Annual Fall Symposium*, 2005.

219. D. Klimov and Y. Shahar. Intelligent Visualization of Temporal Associations for Multiple Time-Oriented Patient Records. In *Proceedings of the 11th International Workshop on Intelligent Data Analysis in Medicine and Pharmacology (IDAMAP-2007)*, 2007.

220. Denis Klimov and Yuval Shahar. Intelligent querying and exploration of multiple time-oriented medical records. In Klaus A. Kuhn, James R. Warren, and Tze-Yun Leong, editors, *MedInfo*, volume 129 of *Studies in Health Technology and Informatics*, pages 1314–1318. IOS Press, 2007.

221. Denis Klimov, Yuval Shahar, and Meirav Taieb-Maimon. Intelligent selection and retrieval of multiple time-oriented records. *J. Intell. Inf. Syst.*, 2009.

222. Denis Klimov, Yuval Shahar, and Meirav Taieb-Maimon. Intelligent visualization of temporal associations for multiple time-oriented patient records. *Methods of Information in Medicine*, 48(3):254–262, 2009.

223. I. S. Kohane. Temporal reasoning in medical expert systems. Technical Report 389, Laboratory of Computer Science, Massachusetts Institute of technology, Cambridge, MA, 1987.

224. I.S. Kohane. Temporal reasoning in medical expert systems. In R. Salamon, B. Blum, and M. Jorgensen, editors, *MEDINFO '86: Proceedings of the Fifth Conference on Medical Informatics*, pages 170–174, Amsterdam: North-Holland, 1986.

225. R. Kosara and S. Miksch. Metaphores of movement: A visualization and user interface for time-oriented, skeletal plans. *Artificial Intelligence in Medicine*, 22(2):111–131, 2001.

226. Manolis Koubarakis, Timos K. Sellis, Andrew U. Frank, Stéphane Grumbach, Ralf Hartmut Güting, Christian S. Jensen, Nikos A. Lorentzos, Yannis Manolopoulos, Enrico Nardelli, Barbara Pernici, Hans-Jörg Schek, Michel Scholl, Babis Theodoulidis, and Nectaria Tryfona, editors. *Spatio-Temporal Databases: The CHOROCHRONOS Approach*, volume 2520 of *Lecture Notes in Computer Science*. Springer, 2003.

227. V. Kouramajian and J. Fowler. Modeling past, current, and future time in medical databases. In J.G. Ozbolt, editor, *18th Annual Symposium on Computer Applications in Medical Care*, pages 315–319. Hanley & Belfus, 1994.

228. R. Kowalski and M. Sergot. A Logic-Based Calculus of Events. *New Generation Computing*, 4:67–95, 1986.

229. David Krieger and Tim Andrews. C++ bindings to an object database. In *Modern Database Systems*, pages 89–107. 1995.

230. M.M Kuilboer, Y. Shahar, D.M. Wilson, and M.A. Musen. Knowledge reuse: Temporal-abstraction mechanisms for the assessment of children's growth. In *Proceedings of the Seventeenth Annual Symposium on Computer Applications in Medicine*, pages 449–453, Washington, DC, 1993.

231. P. Ladkin. Primitives and units for time specification. In *Proceedings of the Sixth National Conference on Artificial Intelligence*, pages 354–359, Philadelphia, PA, 1986. Morgan Kaufmann.

232. P. Ladkin. Time representation: A taxonomy of interval relations. In *Proceedings of the Sixth National Conference on Artificial Intelligence*, pages 360–366, Philadelphia, PA, 1986. Morgan Kaufmann.

233. C. Larizza, R. Bellazzi, and A. Riva. Temporal abstractions for diabetic patients management. In *Proc. AIME-97, Lecture Notes in Artificial Intelligence*, volume 1211, pages 319–330. Springer, 1997.

234. C. Larizza, C. Berzuini, and M. Stefanelli. A general framework for building patient monitoring systems. In P. Barahona, M. Stefanelli, and J. Wyatt, editors, *5 Artificial Intelligence in Medicine*, LNAI 934, pages 91–102. Springer, 1995.

235. C. Larizza, A. Moglia, and M. Stefanelli. M-HTP: A System for Monitoring Heart Transplant Patients. *Artificial Intelligence in Medicine*, 4:111–126, 1992.

236. J. Larkin and H. Simon. Why a diagram is (sometimes) worth ten thousand words. *Cognitive science*, 11:65–99, 1987.

237. S.L. Lauritzen and D.J. Spiegelhalter. Local computations with probabilities on graphical structures and their application to expert systems. *Journal of the Royal Statistical Society Ser. B*, 50:157–224, 1987.

238. N. Lavrač, E. T. Keravnou, and B. Zupan, editors. *Intelligent Data Analysis in Medicine and Pharmacology*. Kluwer Academic Publishers, Boston, 1997.

239. John Zhong Li. *Modeling and querying multimedia data*. PhD thesis, Edmonton, Alta., Canada, 1998. Advisor-Özsu, M. Tamer.

240. Yingjiu Li, Peng Ning, Xiaoyang Sean Wang, and Sushil Jajodia. Discovering Calendar-based Temporal Association Rules. In *TIME 2001: Eigth International Symposium on Temporal Representation and Reasoning*, pages 111–118, Civdale del Friuli, Italy, 2001. IEEE Computer Society.

241. Yingjiu Li, Peng Ning, Xiaoyang Sean Wang, and Sushil Jajodia. Discovering calendar-based temporal association rules. *Data Knowl. Eng.*, 44(2):193–218, 2003.

242. Daniel F. Lieuwen and Narain H. Gehani. Versions in ode: Implementation and experiences. *Softw., Pract. Exper.*, 29(5):397–416, 1999.

243. Daniel F. Lieuwen, Narain H. Gehani, and Robert M. Arlein. The ode active database: Trigger semantics and implementation. In Stanley Y. W. Su, editor, *ICDE*, pages 412–420. IEEE Computer Society, 1996.

244. Thomas D. C. Little and Arif Ghafoor. Interval-based conceptual models for time-dependent multimedia data. *IEEE Trans. Knowl. Data Eng.*, 5(4):551–563, 1993.

245. Ling Liu and M. Tamer Özsu, editors. *Encyclopedia of Database Systems*. Springer US, 2009.

246. Ling Liu and Tamer M. Özsu, editors. *Encyclopedia of Database System*. Springer-Verlag, 2009.

247. W. Long. Temporal reasoning for diagnosis in a causal probabilistic knowledge base. *Artificial Intelligence in Medicine*, 8:193–215, 1996.

248. W.J. Long and T.A. Russ. A control structure for time dependent reasoning. In A. Bundy, editor, *Proceedings of the Eighth International Joint Conference on Artificial Intelligence*, pages 230–232. William Kaufmann, 1983.

249. A. Lowe, R.W. Jones, and M.J. Harrison. The graphical presentation of decision support information in an intelligent anaesthesia monitor. *Artificial Intelligence in Medicine*, 22:xx–xx, 2001.

250. P.J.F. Lucas. Analysis of notions of diagnosis. *Artificial Intelligence*, 105:295–343, 1998.

251. A.K. Mackworth. Consistency in networks of relations. *Artificial Intelligence*, 8:99–118, 1977.

252. Paolo Magni, Silvana Quaglini, Monia Marchetti, and Giovanni Barosi. Deciding when to intervene: a markov decision process approach. *International Journal of Medical Informatics*, 60(3):237–253, 2000.

253. R. Maiocchi, B. Pernici, and F. Barbic. Automatic Deduction of Temporal Information. *ACM Transactions on Database Systems*, 17(4):647–698, 1992.

254. S.B Martins, Y. Shahar, M. Galperin, D. Goren-Bar, D. Boaz, G. Tahan, H. Kaizer, L.V. Basso, and M.K. Goldstein. Evaluation of KNAVE-II: A tool for intelligent query and exploration of patient data. In *Proceedings of MEDINFO-04, the Eleventh World Congress on Medical Informatics*, San Francisco, CA., 2004.

255. S.B. Martins, Y. Shahar, D. Goren-Bar, M. Galperin, H. Kaizer, L.V. Basso, D. McNaughton, and M.K. Goldstein. Evaluation of an architecture for intelligent query and exploration of time-oriented clinical data. *Artificial Intelligence in Medicine*, 43:17–34, 2008.

256. S. M. Maviglia, R. D. Zielstorff, M. Paterno, J. M. Teich, D. W. Bates, and G. Kuperman. Automating complex guidelines for chronic disease: Lessons learned. *Journal of the American Medical Informatics Association*, 10(2):154–165, 2003.

257. J. McCarthy. Situations, actions and causal rules. AIMemo 1, Artificial Intelligence Project, Stanford University, Stanford, CA, 1957.

258. J. McCarthy and P. Hayes. *Some philosophical problems from the standpoint of artificial intelligence*, pages 463–502. Edinburgh University Press, Edinburgh, UK, 1969.

259. D. V. McDermott. A temporal logic for reasoning about processes and plans. *Cognitive Science*, 6(2):101–155, 1982.

260. J. Melton and A.R. Simon. *Understanding the New SQL: A Complete Guide*. Morgan - Kaufmann, 1993.

261. A.O. Mendelzon, F. Rizzolo, and A.A. Vaisman . 2004:. Indexing temporal xml documents. In *VLDB'04, Proceedings of 30th International Conference on Very Large Data Bases*, pages 216–227. Morgan Kaufmann, 2004.

262. S. Miksch, W. Horn, C. Popow, and F. Paky. Utilizing temporal data abstraction for data validation and therapy planning for artificially ventilated newborn infants. *Artificial Intelligence in Medicine*, 8(6):543–576, 1996.

263. S. Miksch, Y. Shahar, and P. Johnson. Asbru: A task-specific, intention-based, and time-oriented language for representing skeletal plans. In *Proceedings of the Seventh Workshop on Knowledge Engineering Methods and Languages (KEML-97)*, pages 9–20, Milton Keynes, UK, 1997.

264. P. L. Miller. *Expert Critiquing Systems: Practice-Based Medical Consultation by Computer*. Springer-Verlag, New York, NY, 1986.

265. R. A. Miller. Internist-1/caduceus: Problems facing expert consultant programs. *Methods Info. Med.*, 23:9–14, 1984.

266. R.A. Miller, H.E. Pople, and J.D. Myers. INTERNIST-1: an experimental computer-based diagnostic consultant for general internal medicine. *New England Journal of Medicine*, 307:468–476, 1982.

267. S. Mittal and B. Chandrasekaran. Conceptual representation of patient data bases. *Journal of Medical Systems*, 4:169–185, 1980.

268. M. J. O'Connor M.J., W. E. Grosso, S. W. Tu, and M. A. Musen. RASTA: A Distributed Temporal Abstraction System to Facilitate Knowledge-Driven Monitoring of Clinical Databases. In *Proceedings of MEDINFO-2001, the Tenth World Congress on Medical Informatics*, pages 508–512, London, UK, 2001.

269. R. Mohr and T.C. Henderson. Arc and path consistency revisited. *Artificial Intelligence*, 28:225–233, 1986.

270. A. Montanari and B. Pernici. *Temporal Reasoning*, pages 534–562. In Tansell et al. [399], 1993.

271. U. Montanari and F. Rossi. Constraint relaxation may be perfect. *Artificial Intelligence*, pages 143–170, 1991.

272. Fabian Mörchen. Unsupervised pattern mining from symbolic temporal data. *SIGKDD Explorations*, 9(1):41–55, 2007.

273. R.A. Morris and Lina AlKhatib. Languages and models for reasoning with nonconvex time intervals. In *Proceedings of the AAAI Workshop on Implementing Temporal Reasoning*, pages 93–99, San Jose, CA, 1992.

274. Robert Moskovitch, Shiva Cohen-Kashi, Uzi Dror, Iftah Levy, Amit Maimon, and Yuval Shahar. Multiple hierarchical classification of free-text clinical guidelines. *Artificial Intelligence in Medicine*, 37(3):177–190, 2006.

275. Robert Moskovitch, Susanna B. Martins, Eitan Behiri, and Yuval Shahar. A comparative evaluation of full-text, concept-based, and context-sensitive search. *Journal of American Medical Informatics Association*, 14:164–174, 2007.

276. Robert Moskovitch and Yuval Shahar. Medical Temporal-Knowledge Discovery via Temporal Abstraction. In *Proceedings of the 2009 AMIA Annual Symposium*, page in press, Philadelphia, PA, 2009. Hanley & Belfus.

277. M. A. Musen, R. W. Carlson, L. M. Fagan, and S. C. Deresinski. T-HELPER: Automated Support for Community-Based Clinical Research. In *Proceedings of the Sixteenth Annual Symposium on Computer Applications in Medical Care*, pages 719–723, Washington, D.C., 1992.

278. M. A. Musen, S. W. Tu, A. K. Das, and Y. Shahar. EON: A component-based approach to automation of protocol-directed therapy. *Journal of the American Medical Information Association*, 3(6):367–388, 1996.

279. S.B. Navathe and R. Ahmed. *Temporal Extensions to the Relational Model and SQL*, pages 92–109. In Tansell et al. [399], 1993.

280. NCEP. Summary of the Second Report of the National Cholesterol Education Program (NCEP) Expert Panel on Detection, Evaluation, and Treatment of High Blood Cholesterol in Adults (adult Treatment Panel II). *Journal of the American Medical Association*, (269):3015–3023, 1993.

281. J. H. Nguyen, Y. Shahar, S. W. Tu, A. K. Das, and M. A. Musen. Integration of Temporal Reasoning and Temporal-Data Maintenance Into A Reusable Database Mediator to Answer Abstract, Time-Oriented Queries: The Tzolkin System. *Journal of Intelligent Information Systems*, 13(1-2):121–145, 1999.

282. Peng Ning, Xiaoyang Sean Wang, and Sushil Jajodia. An Algebraic Representation of Calendars. *Ann. Math. Artif. Intell.*, 36(1-2):5–38, 2002.

283. N. F. Noy, R. Fergerson, and M. A. Musen. The knowledge model of Protege-2000: Combining interoperability and flexibility. In *Proceedings of the 2nd International Conference on Knowledge Engineering and Knowledge Management (EKAW'2000)*, 2000.

284. M. O'Connor, S. W. Tu, and M. A. Musen. The Chronus II Temporal Database Mediator. In *Proceedings of the 2002 American Medical Informatics Fall Symposium (AMIA-2002)*, pages 567–571, San Antonio, TX, 2002.

285. O. Ogunyemi, Q. Zeng, and A. Bozwala. Object-oriented guideline expression language (GELLO) specification. Technical report, Brigham and Women's Hospital, Harvard Medical School, 2002. Decision Systems Group Technical Report DSG-TR-2002-001.

286. L. Ohno-Machado, J. H. Gennari, and S. N. Murphy. The guideline interchange format: a model for representing guidelines. *Journal of the American Medical Informatics Association*, 5:357–72, 1998.

287. Barbara Oliboni, Elisa Quintarelli, and Letizia Tanca. Temporal aspects of semistructured data. In *TIME*, pages 119–127, 2001.

288. Eitetsu Oomoto and Katsumi Tanaka. Ovid: Design and implementation of a video-object database system. *IEEE Trans. Knowl. Data Eng.*, 5(4):629–643, 1993.

289. G. Ozsoyoglu and R. T. Snodgrass. Temporal and real-time databases: a Survey. *IEEE Transactions on Knowledge and Data Engineering*, 7(4):513–532, 1995.

290. Gultekin Özsoyoglu and Richard T. Snodgrass. Guest editors' introduction to special section on temporal and real-time databases. *IEEE Trans. Knowl. Data Eng.*, 7(4):511–512, 1995.

291. Gultekin Özsoyoglu and Richard T. Snodgrass. Temporal and real-time databases: A survey. *IEEE Trans. Knowl. Data Eng.*, 7(4):513–532, 1995.

292. M.T. Özsu, R. Peters, D. Szafron, B. Irani, A. Lipka, and A. Munoz. TIGUKAT: A Uniform Behavioral Objectbase Management System. *VLDB Journal*, 4:445–492, 1995.

293. José Palma, José M. Juárez, Manuel Campos, and Roque Marín. Fuzzy theory approach for temporal model-based diagnosis: An application to medical domains. *Artificial Intelligence in Medicine*, 38(2):197–218, 2006.

294. J.F. Pane and B.A. Myers. Tabular and Textual Methods for Selecting objects from a Group. In *Proceedings of VL 2000: IEEE International Symposium on Visual Languages*, pages 157–164, 2000.

295. G. Panti. *Multi-valued Logics*, pages 25–74. Kluwer, 1998.

296. Y. Papakonstantinou, H. Garcia-Molina, and J. Widom. Object Exchange Across Heterogeneous Information Sources. In *Proceedings of the Eleventh International Conference on Data Engineering*, pages 251–260. IEEE Computer Society, 1995.

297. J. Paredaens, P. Peelman, and L. Tanca. G–Log: A Declarative Graphical Query Language. *IEEE Transactions on Knowledge and Data Engineering*, 7(3):436–453, 1995.

298. R.S. Patil, P. Szolovits, and W.B. Schwartz. *Modelling knowledge of the patient in acid-base and electrolyte disorders*, pages 191–226. West View Press, 1982.

299. J. Pearl. Fusion, Propagation and structuring in belief networks. *Artificial Intelligence*, 29:241–288, 1986.

300. J. Pearl. *Probabilistic Reasoning in Intelligent Systems: networks of plausible inference.* Morgan Kaufmann, 1988.

301. N. Peek. Explicit temporal models for decision-theoretic planning of clinical management. *Artificial Intelligence in Medicine,* 15(2):135–154, 1999.

302. M. Peleg, A. Boxwala, E. Bernstam, S. W. Tu, R. A. Greenes, and E. H. Shortliffe. Sharable representation of clinical guidelines in GLIF: Relationship to the Arden syntax. *Journal of Biomedical Informatics,* 34:170–181, 2001.

303. M. Peleg, A. A. Boxwala, O. Omolola, Q. Zeng, S. W. Tu, R. Lacson, E. Bernstam, N. Ash, P. Mork, L. Ohno-Machado, E. H. Shortliffe, and R. A. Greenes. GLIF3: The Evolution of a Guideline Representation Format. In M. J. Overhage, editor, *Proceedings of the 2000 AMIA Annual Symposium,* pages 645–649, Philadelphia, 2000. Hanley & Belfus.

304. M. Peleg, O. Ogunyemi, and S. Tu. Using features of Arden syntax with object-oriented medical data models for guideline modeling. In *Proceedings of the 2001 AMIA Annual Symposium,* pages 523–527, Philadelphia, 2001. Hanley & Belfus.

305. M. Peleg, S. Tu, J. Bury, P. Ciccarese, J. Fox, R. A. Greenes, R. Hall, P. D. Johnson, N. Jones, A. Kumar, S. Miksch, S. Quaglini, A. Seyfang, E. H. Shortliffe, and M. Stefanelli. Comparing Computer-Interpretable Guideline Models: A Case-Study Approach. *Journal of the American Medical Informatics Association,* 10(1):52–68, 2003.

306. Y. Peng and J.A. Reggia. *Abductive inference models for diagnostic problem solving.* Springer-Verlag, 1990.

307. F. Pinciroli, C. Combi, and G. Pozzi. Object-orientated DBMS Techniques for Time-orientated Medical Record. *Medical Informatics,* 17:231–241, 1992.

308. F. Pinciroli, C. Combi, and G. Pozzi. ARCADIA: A System fo the Integration of Angio-cardiographic Data and Images by an Object-Oriented DBMS. *Computers and Biomedical Research,* 28:5–23, 1995.

309. F. Pinciroli, C. Combi, G. Pozzi, and R. Rossi. MS2/Cardio: Towards a Multi-Service Medical Software for Cardiology. *Methods of Information in Medicine,* 31:18–27, 1992.

310. C. Plaisant, R. Mushlin, A. Snyder, J. Liand D. Heller D, and B. Shneiderman. LifeLines: Using Visualization to Enhance Navigation and Analysis of Patient Records. In *Proceedings American Medical Informatics Association Annual Fall Symposium,* pages 76–80. Hanley & Belfus, 1998.

311. M. Pollack. The Use of Plans. *Artificial Intelligence,* 57(1):43–68, 1992.

312. D. Poole. Explanation and prediction: an architecture for default and abductive reasoning. *Computational Intelligence,* 5(2):97–110, 1989.

313. D. Poole. Normality and faults in logic-based diagnosis. In *Proc. IJCAI-89,* pages 1304–1310, 1989.

314. D. Poole. A methodology for using a default and abductive reasoning system. *International Journal of Intelligent Systems,* 5(5):521–548, 1990.

315. H. E. Pople. Heuristic methods for imposing structure on ill structured problems: the structuring of medical diagnosis. *Artificial Intelligence in Medicine,* pages 119–185, 1982. P. Szolovits (ed.), AAAA Selected Symposium Series, West View Press.

316. François Portet, Ehud Reiter, Albert Gatt, Jim Hunter, Somayajulu Sripada, Yvonne Freer, and Cindy Sykes. Automatic generation of textual summaries from neonatal intensive care data. *Artif. Intell.,* 173(7-8):789–816, 2009.

317. Francesco Pinciroliand Luisa Portoni, Carlo Combi, and Francesco Violante. Www-based access to object-oriented clinical databases: the khospad project. *Computers in Biology and Medicine,* 28(5):531–552, 1998.

318. Luisa Portoni, Carlo Combi, and Francesco Pinciroli. User-oriented views in health care information systems. *IEEE Transactions on Biomedical Engineering,* 49(12):1387–1398, 2002.

319. S.M. Powsner and E.R. Tufte. Graphical Summary of Patient Status. *Lancet,* 344:386–389, 1994.

320. J. Preece, Y. Rogers, H. Sharp, D. Benion, S. Holland, and T. Carey. *Human Computer Interaction.* Addison Wesley, 1994.

321. A.N. Prior. Diodorian modalities. In *The Philosophical Quartely 5*, pages 202–213, 1955.

322. A.N. Prior. *Time and Modality*. Oxford: Clarendon Press, 1957.

323. A.N. Prior. *Past, Present and Future*. Oxford: Clarendon Press, 1967.

324. T. A. Pryor and G. Hripcsak. Sharing MLM's: an experiment between Columbia-Presbyterian and LDS Hospital. In C. Safran, editor, *Proceedings of the Seventeenth Annual Symposium on Computer Applications in Medical Care*, pages 399–403, New York, 1994. McGraw-Hill.

325. S. Quaglini, L. Dazzi, M. Stefanelli, C. Fassino, and C. Tondini. Supporting tools for guideline development and dissemination. *Artificial Intelligence in Medicine*, 14(1-2):119–137, 1998.

326. S. Quaglini, M. Stefaneli, G. Lanzola, V. Caporusso, and S. Panzarasa. Flexible guideline-based patient careflow systems. *Artificial Intelligence in Medicine*, 22:65–80, 2001.

327. Mohammed I. Rafiq, Martin J. O'Connor, and Amar K. Das. Computational method for temporal pattern discovery in biomedical genomic databases. In *CSB*, pages 362–365. IEEE Computer Society, 2005.

328. A. Rector. AIM: a personal view of where i have been and where we might be going. *Artificial Intelligence in Medicine*, 23:111–127, 2001.

329. R. Reiter. A theory of diagnosis from first principles. *Artificial Intelligence*, 32:57–95, 1987.

330. Rescher and Urquhart. *Temporal Logic*. SpringerVerlag, New York, NY, 1971.

331. A. Riva and R. Bellazzi. Learning temporal probabilistic causal models from longitutinal data. *Artificial Intelligence in Medicine*, 8:217–234, 1996.

332. Flavio Rizzolo and Alejandro A. Vaisman. Temporal xml: modeling, indexing, and query processing. *VLDB J.*, 17(5):1179–1212, 2008.

333. E. Rose and A. Segev. TOOSQL - A Temporal Object-Oriented Query Language. In *12th International Conference of the Entity-Relationship Approach*, LNCS 823, pages 122–136. Springer, 1993.

334. Nick Roussopoulos, Christos Faloutsos, and Timos K. Sellis. An efficient pictorial database system for psql. *IEEE Trans. Software Eng.*, 14(5):639–650, 1988.

335. D.W. Rucker, D.J. Maron, and E.H. Shortliffe. Temporal Representation of Clinical Algorithms Using Expert-System and Database Tools. *Computers and Biomedical Research*, 23:222–239, 1990.

336. T.A. Russ. Using hindsight in medical decision making. In *Proc. Symposium on Computer Applications in Medical Care*, pages 38–44, New York, 1989. IEEE Computer Society Press.

337. T.A. Russ. Use of data abstraction methods to simplify monitoring. *Artificial Intelligence in Medicine*, 7:497–514, 1995.

338. Lucia Sacchi, Cristiana Larizza, Carlo Combi, and Riccardo Bellazzi. Data mining with temporal abstractions: learning rules from time series. *Data Min. Knowl. Discov.*, 15(2):217–247, 2007.

339. Apkar Salatian and Jim Hunter. Deriving trends in historical and real-time continuously sampled medical data. *J. Intell. Inf. Syst.*, 13(1-2):47–71, 1999.

340. B. Salzberg and V.J. Tsotras. A Comparison of Access Methods for Time Evolving Data. *ACM Computing Surveys*, 31:158–221, 1999.

341. D. Sauquet, P. Degoulet, M. Lavril, and F. Aime'. *LIED: A Semantic and Temporal Data Management Language*, pages 267–277. North-Holland, 1991.

342. G. Schadow, D. C. Russler, C. N. Mead, and C. J. McDonnald. Integrating medical information and knowledge in the HL7 RIM. In *Proceedings of the 2000 AMIA Annual Symposium*, pages 764–768, Philadelphia, 2000. Hanley & Belfus.

343. M. Sedlmayr, T. Rose, R. Rhring, and M. Meister. A workflow approach towards glif execution. in proceedings on workshop ai techniques in healthcare: Evidence-based guidelines and protocols. In *Proceedings on Workshop AI Techniques in Healthcare: Evidence-based Guidelines and Protocols. European Conference on Artificial Intelligence (ECAI)*, Trento, Italy, 2006.

344. Rudolf Seising. From vagueness in medical thought to the foundations of fuzzy reasoning in medical diagnosis. *Artificial Intelligence in Medicine*, 38(3):237–256, 2006.

345. S. Seyfang, R. Kosara, and S. Miksch. Asbru's Reference Manual, Version 7.3. Technical report, asgaard-tr-2002-1, Vienna University of Technology, Institute of Software Technology, Vienna, 2002.

346. Robert Moskovitch Yuval Shahar. A multiple-ontology, concept-based, context-sensitive, clinical-guideline search engine. *Journal of Biomedical Informatics*, 2008.

347. Y. Shahar. A framework for knowledge-based temporal abstraction. *Artificial Intelligence*, 90:79–133, 1997.

348. Y. Shahar. Dynamic temporal interpretation contexts for temporal abstraction. *Annals of Mathematics and Artificial Intelligence*, 22:159–192, 1998.

349. Y. Shahar. Knowledge-based temporal interpolation. *Journal of Experimental and Theoretical Artificial Intelligence*, 11:123–144, 1999.

350. Y. Shahar, D. Boaz, G. Tahan, M. Galperin, D. Goren-Bar, H. Kaizer, L.V. Basso, S.B. Martins, and M.K. Goldstein. Interactive visualization and exploration of time-oriented clinical data using a distributed temporal-abstraction architecture. In *Proceedings of the 2003 AMIA Annual Fall Symposium*, Washington, DC, 2003.

351. Y. Shahar and H. Cheng. Intelligent visualization and exploration of time-oriented clinical data. *Topics in Health Information Management*, 20:15–31, 1999.

352. Y. Shahar and H. Cheng. Model-Based Visualization of Temporal Abstractions. *Computational Intelligence*, 16:279–306, 2000.

353. Y. Shahar, H. Cheng, D.P. Stites, L.D. Basso, H. Kaizer, D.M. Wilson, and M.A. Musen. Semiautomated acquisition of clinical temporal-abstraction knowledge. *Journal of the American Medical Informatics Association*, 6:494–511, 1999.

354. Y. Shahar, D. Goren-Bar, M. Galperin, D. Boaz, and G. Tahan. Knave ii: A distributed architecture for interactive visualization and intelligent exploration of time-oriented clinical data. In *The 7th International Workshop on Intelligent Data Analysis in Medicine and Pharmacology (IDAMAP-2003)*, pages 103–110, Protaras, Cyprus, 2003.

355. Y. Shahar, S. Miksch, and P. Johnson. The Asgaard project: a task-specific framework for the application and critiquing of time-oriented clinical guidelines. *Artificial Intelligence in Medicine*, 14:29–51, 1998.

356. Y. Shahar and M. Molina. Knowledge-based spatiotemporal linear abstraction. *Pattern Analysis and Applications*, 1(2):91–104, 1998.

357. Y. Shahar and M. A. Musen. RSUM: A Temporal-Abstraction System for Patient Monitoring. *Computers and Biomedical Research*, 26:255–273, 1993. Reprinted in J.H. van Bemmel and A.T. McRay, A.T. (eds) (1994), Yearbook of Medical Informatics 1994: 443-461. Stuttgart: F.K. Schattauer and The International Medical Informatics Association.

358. Y. Shahar and M. A. Musen. Plan Recognition and Revision in Support of Guideline. Notes of the AAAI Spring Symposium on Representation Mental States and Mechanisms, 1995. Stanford, CA, 118-126.

359. Y. Shahar and M.A. Musen. Knowledge-based temporal abstraction in clinical domains. *Artificial Intelligence in Medicine*, 8:267–298, 1996.

360. Y. Shahar, E. Shalom, A. Mayaffit, O. Young, M. Galperin, S. B. Martins, and M. K. Goldstein. A distributed, collaborative, structuring model for a clinical-guideline digital-library. In *Proceedings of the 2003 AMIA Annual Fall Symposium*, Washington, DC, 2003. (Available on CD).

361. Y. Shahar, O. Young, E. Shalom, A. Mayaffit, R. Moskovitch, A. Hessing, and M. Galperin. DEGEL: A Hybrid, multiple-ontology framework for specification and retrieval of clinical guidelines. In *Proceedings the Ninth Conference on Artificial Intelligence in Medicine Europe (AIME-03)*, pages 122–130, Protaras, Cyprus, 2003.

362. Yuval Shahar. Timing is everything: Temporal reasoning and temporal data maintenance in medicine. In Werner Horn, Yuval Shahar, Gregor Lindberg, Steen Andreassen, and Jeremy C. Wyatt, editors, *AIMDM*, volume 1620 of *Lecture Notes in Computer Science*, pages 30–46. Springer, 1999.

363. Yuval Shahar. Dimension of time in illness: An objective view. *Annals of Internal Medicine*, 132(1):45–53, 2000.

364. Yuval Shahar and Carlo Combi. Temporal Reasoning and Temporal Data Maintenance in Medicine: Issues and Challenges. *Computer in Biology and Medicine, Special Issue: Time-Oriented Systems in Medicine*, 27(5):353–368, 1997.

365. Yuval Shahar and Carlo Combi. Intelligent temporal information systems in medicine. *J. Intell. Inf. Syst.*, 13(1-2):5–8, 1999.

366. Yuval Shahar, Dina Goren-Bar, David Boaz, and Gil Tahan. Distributed, intelligent, interactive visualization and exploration of time-oriented clinical data and their abstractions. *Artificial Intelligence in Medicine*, 38(2):115–135, 2006.

367. Yuval Shahar, Ohad Young, Erez Shalom, Maya Galperin, Alon Mayaffit, Robert Moskovitch, and Alon Hessing. A framework for a distributed, hybrid, multiple-ontology clinical-guideline library, and automated guideline-support tools. *Journal of Biomedical Informatics*, 37(5):325–344, 2004.

368. Yuval Shahar, Ohad Young, Erez Shalom, Maya Galperin, Alon Mayaffit, Robert Moskovitch, and Alon Hessing. A framework for a distributed, hybrid, multiple-ontology clinical-guideline library, and automated guideline-support tools. *Journal of Biomedical Informatics*, 37(5):325–344, 2004.

369. Erez Shalom, Yuval Shahar, Meirav Taieb-Maimon, Yair Liel, Eitan Lunenfeld, Guy Bar, Avi Yarkoni, Ohad Young, Susana B. Martins, Laszlo T. Vaszar, Tal Marom, Mary K. Goldstein, Yair Liel, Akiva Leibowitz, Tal Marom, and Eitan Lunenfeld. A quantitative evaluation of a methodology for collaborative specification of clinical guidelines at multiple representation levels. *Journal of Biomedical Informatics*, 2008.

370. E. H. Sherman, G. Hripcsak, J. Starren, R. A. Jender, and P. Clayton. Using intermediate states to improve the ability of the Arden Syntax to implement care plans and reuse knowledge. In Gardner [151], pages 238–242.

371. R. N. Shiffman. Representation of clinical practice guidelines in conventional and augmented decision tables. *Journal of the American Medical Informatics Association*, 4(5):382–393, 1997.

372. R. N. Shiffman, B. T. Karras, A. Agrawal, R. Chen, L. Marenco, and S. Nath. GEM: a proposal for a more comprehensive guideline document model using XML. *Journal of the American Medical Informatics Association*, 7(5):488–498, 2000.

373. B. Shneiderman. Dynamic Queries for Visual Information Seeking. *IEEE Software*, 11:70–77, 1994.

374. B. Shneiderman. The Eyes Have It: A Task by Data Type Taxonomy for Information Visualizations. In *Proceedings IEEE Symposium on Visual Languages '96*, pages 336–343. IEEE Computer Society Press, 1996.

375. B. Shneiderman, D. Feldman, A. Rose, and X. Ferre' Grau. Visualizing digital library search results with categorical and hierarchical axes. In *Proceedings ACM Digital Libraries*, pages 57–66. CM Press, 2000.

376. Y. Shoham. Temporal logics in AI: Semantical and ontological considerations. *Artificial Intelligence*, 33:89–104, 1987.

377. Y. Shoham and N. Goyal. Temporal reasoning in artificial intelligence. In E.H. Shrobe, editor, *Exploring Artificial Intelligence*, San Mateo, CA, 1988. Morgan Kaufmann.

378. E.H. Shortliffe, editor. *Computer-based medical consultations: MYCIN*. American Elsevier Publishing, 1976.

379. E.H. Shortliffe, L.E. Perreault, G. Wiederhold, and L.M. Fagan, editors. *Medical Informatics: Computer Application in Health Care*. Addison Wesley, 1990.

380. M. A. Shwe, W. Sujansky, and B. Middleton. Reuse of knowledge represented in the Arden Syntax. In M. E. Frisse, editor, *Proceedings of the Sixteenth Annual Symposium on Computer Applications in Medical Care*, pages 47–51, New York, 1993. McGraw-Hill.

381. R. Snodgrass. A Taxonomy of Time in Databases. *ACM/SIGMOD*, pages 236–246, 1985.

382. R. Snodgrass and I. Ahn. Temporal databases. *IEEE Computer*, 19(9):35–42, 1986.

383. Richard T. Snodgrass. Temporal object-oriented databases: A critical comparison. In *Modern Database Systems*, pages 386–408. Addison Wesley, 1995.

384. Richard T. Snodgrass. *Developing Time-Oriented Database Applications in SQL*. Morgan Kaufmann, 1999.

385. Richard T. Snodgrass and Ilsoo Ahn. A taxonomy of time in databases. In Shamkant B. Navathe, editor, *SIGMOD Conference*, pages 236–246. ACM Press, 1985.

386. Richard T. Snodgrass, Michael H. Böhlen, Christian S. Jensen, and Andreas Steiner. Transitioning temporal support in TSQL2 to SQL3. In *Temporal Databases, Dagstuhl*, pages 150–194, 1997.

387. R.T. Snodgrass, editor. *The TSQL2 Temporal Query Language*. Kluwer Academic Publishers, 1995.

388. Richard Mark Soley and William Kent. The omg object model. In *Modern Database Systems*, pages 18–41. Addison Wesley, 1995.

389. Michael Souillard, Carine Souveyet, Costas Vassilakis, and Anya Sotiropoulou. A flexible framework for managing temporal clinical trial data. *International Journal of Electronic Healthcare*, 1(4):453–463, 2005.

390. D.J. Spiegelhalter and R.P. Knill-Jones. Statistical and knowledge-based approaches to clinical decision-support systems with an application in gastroenterology. *Journal of the Royal Statistical Society*, 147:35–77, 1984.

391. Alex Spokoiny and Yuval Shahar. An active database architecture for knowledge-based incremental abstraction of complex concepts from continuously arriving time-oriented raw data. *J. Intell. Inf. Syst.*, 28(3):199–231, 2007.

392. Alex Spokoiny and Yuval Shahar. Incremental application of knowledge to continuously arriving time-oriented raw data. *J. Intell. Inf. Syst.*, 31(1):1–33, 2008.

393. M. Stacey and C. McGregor. Temporal abstraction in intelligent clinical data analysis: A survey. *Artificial Intelligence in Medicine*, 39:1–24, 2007.

394. J. Starren, G. Hripcsak, D. Jordan, B. Allen, C. Weissman, and P. D. Clayton. Encoding a post-operativive coronary artery bypass surgery care plan in the Arden syntax. *Computers in Biology and Medicine*, 24(5):411–417, 1994.

395. P. Struss and O. Dressler. Physical negation – integrating fault models into the general diagnostic engine. In *Proc. IJCAI-89*, pages 1318–1323, 1989.

396. S.Y.W, Su and H.M. Chen. Temporal Rule Specification and Management in Object-oriented Knowledge Bases. In *1st International Workshop on Rules in Database Systems*, Workshops in Computing, pages 73–91. Springer, 1994.

397. P. C. Tang and C. Y. Young. ActiveGuidelines: Integrating Web-Based Guidelines with Computer-Based Patient Records. In M. J. Overhage, editor, *Proceedings of the 2000 AMIA Annual Symposium*, Philadelphia, 2000. Hanley & Belfus.

398. Abdullah Uz Tansel, James Clifford, Shashi K. Gadia, Sushil Jajodia, Arie Segev, and Richard T. Snodgrass, editors. *Temporal Databases: Theory, Design, and Implementation*. Benjamin/Cummings, 1993.

399. A.U. Tansell, J. Clifford, S. Gadia, S.K. Jajodia, A. Segev, and R.T. Snodgrass, editors. *Temporal Databases: Theory, Design and Implementation*. Benjamin/Cummings Pub. Co., 1993.

400. A. Ten Teije, S. Miksch, and P. Lucas, editors. *Computer-based Medical Guidelines and Protocols: A Primer and Current Trends*. Ios Press, The Netherlands, 2008.

401. P. Terenziani. Towards a causal ontology coping with the temporal constraints between causes and effects. *Int. J. Human-Computer Studies*, 45:847–863, 1995.

402. Paolo Terenziani, Carlo Carlini, and Stefania Montani. Towards a Comprehensive Treatment of Temporal Constraints in Clinical Guidelines. In *TIME 2002*, pages 20–27, 2002.

403. D.A. Tong and L.E. Widman. Model-Based Interpretation of the ECG: A Methodology for Temporal and Spatial Reasoning. In C. Safran, editor, *17. annual symposium on computer applications in medical care*, pages 133–139, New York, 1993. McGraw-Hill.

404. P. Torasso and L. Console. *Diagnostic Problem Solving: Combining Heuristic, Approximate and Causal Reasoning*. Van Nostrand Reinhold, 1989.

405. Shuichi Toyoda, Noboru Niki, and Hiromu Nishitani. Sakura-viewer: Intelligent order history viewer based on two-viewpoint architecture. *IEEE Transactions on Information Technology in Biomedicine*, 11(2):141–152, 2007.

406. S. W. Tu, M. G. Kahn, M. A. Musen, J. C. Ferguson, E. H. Shortliffe, and L. M. Fagan. Episodic Skeletal-plan refinement on temporal data. *Communications of ACM*, 32(12):1439–1455, 1989.

407. S. W. Tu and M. A. Musen. A flexible approach to guideline modeling. In *Proceedings of the 1999 AMIA Annual Symposium*, pages 420–424, Philadelphia, 1999. Hanley & Belfus.

408. S. W. Tu and M. A. Musen. Modeling data and knowledge in the EON guideline architecture. In *Proceedings of MEDINFO 2001*, pages 280–284, London, the UK, 2001.

409. S.W. Tu, H. Eriksson, J.H. Gennari, Y. Shahar, and M.A. Musen. Ontology-based configuration of problem-solving methods and generation of knowledge-acquisition tools: Application of PROTG-II to protocol-based decision support. *Artificial Intelligence in Medicine*, 7(3):257–289, 1995.

410. E.R. Tufte. *The Visual Display of Quantitative Information*. Graphics Press, 1983.

411. E.R. Tufte. *Envisioning Information*. Graphics Press, 1990.

412. E.R. Tufte. *Visual Explanations*. Graphics Press, 1997.

413. P. van Beek. Temporal query processing with indefinite information. In *Artificial Intelligence in Medicine* [203], pages 325–339.

414. J.F.A.K. van Benthem. *The Logic of Time*. D. Reidel, 1991. 2nd ed.

415. C. Vassilakis, P. Georgiadis, and A. Sotiropoulou. A Comparative Study of Temporal DBMS Architectures. In R. R. Wagner and H. Thomas, editors, *Proc. 7th Int. Workshop on Database and Expert Systems Application*, pages 153–164. IEEE Computer Press, 1996. Los Alamitos, CA.

416. M. Vilain and H. Kautz. Constraint propagation algorithms for temporal reasoning. In *Proceedings of the Sixth National Conference on Artificial Intelligence*, pages 377–382, Los Angeles, CA, 1986. Morgan Kaufmann.

417. M. Vilain, H. Kautz, and P.G. van beek. *Constraint propagation algorithms for temporal reasoning: A revised report*, pages 373–381. Morgan Kaufmann, 1989.

418. J. Wainer and A. de Melo Rezende. A temporal extension to the parsimonious covering theory. *Artificial Intelligence in Medicine*, 10:235–255, 1997.

419. J. Wainer and S. Sandri. Fuzzy temporal/categorical information in diagnosis. *Journal of Intelligent Information Systems*, 13:9–26, 1999.

420. D. Wang and E. H. Shortliffe. GLEE - A Model-Driven Execution System for Computer-Based Implementation of Clinical Practice Guidelines. In *Proceedings of the 2002 AMIA Annual Symposium*, Philadelphia, 2002. Hanley & Belfus.

421. F. Wang and C. Zaniolo. Preserving and Querying histories of XML-published relational databases. In *Proceedings of the Second International Workshop on Evolution and Change in Data Management*, volume 1909 of *Lecture Notes in Computer Science*, pages 26–38. Springer-Verlag, Berlin, 2002.

422. X. Wang, G. Hripcsak, M. Markatou, and C. Friedman. Visual methods for analyzing time-oriented data. *J Am Med Inform Assoc*, 2009.

423. S.M. Weiss, C.A. Kulikowski, S. Amarel, and A. Safir. A model-based method for computer-aided medical decision-making. *Artificial Intelligence*, 11:145–172, 1978.

424. M. West and J. Harrison. *Bayesian Forecasting and Dynamic Models*. Springer Verlag, New York, 1989.

425. J. Widom and S. Ceri. *Active Database Systems: Triggers and Rules for Advanced Database Processing*. Morgan Kaufmann, San Mateo, CA, 1996.

426. G. Wiederhold. *Databases for Health Care*. Lecture Notes in Medical Informatics. Springer-Verlag, 1981.

427. G. Wiederhold. Mediators in the architecture of future information systems. *IEEE Computer*, 25:38–50, 1992.

428. G. Wiederhold and M. Genesereth. The Conceptual Basis of Mediation Services. *IEEE Expert*, 12(5):38–47, 1997.

429. G. Wiederhold and L.E. Perreault. *Clinical Research System*, pages 503–534. In Shortliffe et al. [379], 1990.

430. G.T. Wuu and U. Dayal. *A Uniform Model for Temporal and Versioned Object-oriented Databases*, pages 230–247. In Tansell et al. [399], 1993.

431. A. Yeh. Automatically analyzing a steadily beating ventricle's iterative behavior over time. In *Artificial Intelligence in Medicine* [203], pages 313–323.

432. O. Young and Y. Shahar. Spock: A hybrid model for runtime application of asbru clinical guidelines. In *Proceedings of the 2004 Medinfo*, San Francisco, CA, 2004.
433. Ohad Young, Yuval Shahar, Yair Liel, Eitan Lunenfeld, Guy Bar, Erez Shalom, Susana B. Martins, Laszlo T. Vaszar, Tal Marom, and Mary K. Goldstein. Runtime application of hybrid-asbru clinical guidelines. *Journal of Biomedical Informatics*, 40(5):507–526, 2007.
434. L. A. Zadeh. Fuzzy sets. *Inf Control*, 8:338–353, 1965.
435. I.M. De Zegher-Geets. IDEFIX: Intelligent summarization of a time-oriented medical database. M.S. Dissertation, Medical Information Sciences Program, Stanford University School of Medicine Knowledge Systems Laboratory Technical Report KSL-88-34, Department of Computer Science, Stanford University, Stanford, CA, June 1987.
436. I.M. De Zegher-Geets, A.G. Freeman, M.G. Walker, R.L. Blum, and G. Wiederhold. Summarization and display of on-line medical records. *M.D. Computing*, 5:38–46, 1988.
437. Li Zhou and George Hripcsak. Temporal reasoning with medical data - a review with emphasis on medical natural language processing. *Journal of Biomedical Informatics*, 40(2):183–202, 2007.
438. Li Zhou, Genevieve B. Melton, Simon Parsons, and George Hripcsak. A temporal constraint structure for extracting temporal information from clinical narrative. *Journal of Biomedical Informatics*, pages 424–439, 2006.
439. R. D. Zielstorff, J. M. Teich, and R. L. Fox. P-CAPE: A high-level tool for entering and processing clinical practice guidelines. In *Proceedings of the 1998 AMIA Annual Symposium*, pages 478–482. Hanley & Belfus, 1998.

Index

About the authors

Carlo Combi is Professor of Computer Science at the University of Verona. He has a PhD in Biomedical Engineering (1993) and a Laurea Degree in Electronic Engineering (1987) both from the Politecnico of Milan. Previously, he was Assistant Professor at the University of Udine. His research interests are in the area of information systems with a special emphasis on the management of clinical data. Specific research topics are related to temporalities in workflow systems, temporal databases, information visualization of temporal data, temporal data warehouses and temporal data mining. He is author of more than 30 papers on international journal and of more than 80 papers in the proceedings of peer reviewed international conferences. He has been member of the editorial board of the international journal Artificial Intelligence in Medicine since 1999 and has been involved in the scientific organization of several international conferences. He was the Scientific Co-chair and served as Local Chair of the 12th Conference on Artificial Intelligence in MEdicine (AIME - Italy, Verona 2009). During the period 2009-2011 he served as Chairperson of the Artificial Intelligence in MEdicine (AIME) Board.

Elpida Keravnou-Papailiou is Professor of Computer Science at the University of Cyprus and the Chairperson of the Governing Board of the Cyprus University of Technology. She served as Vice-Rector for Academic Affairs at the University of Cyprus during the period 2002-2006. Previously she was the Dean of the Faculty of Pure and Applied Sciences and the first Chairperson of the Department of Computer Science. She received a B.Tech in Computer Science from Brunel University, UK (1982) and a Ph.D. in Cybernetics from the same University (1985). She taught in the Department of Computer Science of University College London (1985-1992). She is Associate Editor of the scientific journal Artificial Intelligence in Medicine (Elsevier). She has carried out research in the areas of knowledge engineering, expert systems, deep knowledge models, diagnostic reasoning, temporal reasoning, artificial intelligence in medicine, and hybrid decision support systems. In addition to her publications in scientific journals, conference proceedings and book chapters she has co-authored (with Professor L. Johnson) three books (*Expert Systems Technology: a guide*, Abacus Press, 1985, *Competent Expert Systems: a case study* *in fault diagnosis*, Kogan, 1986, and *Expert Systems Architectures*, Kogan, 1988), she has edited a book volume on *Deep Models for Medical Knowledge Engineering* (Elsevier, 1992), and has co-edited (with Drs N. Lavrac and B. Zupan) a book volume on *Intelligent Data Analysis in Medicine and Pharmacology* (Kluwer Academic Publishers, 1997). She has also authored a book on *Artificial Intelligence and Expert Systems* for the Greek Open University. She was the Program Committee Chair of the 6th Conference on Artificial Intelligence in Medicine Europe (France, Grenoble 1997) and served as the Organizing Committee Chair for the 9th Conference on Artificial Intelligence in Medicine Europe (AIME 2003), which was hosted in Cyprus in October 2003. During the period 2003-2005 she served as Chairperson of the Artificial Intelligence in MEdicine (AIME) Board.

Yuval Shahar is a Professor of Information Systems Engineering at Ben Gurion University (BGU), Beer Sheva, Israel, and the head of BGU's Medical Informatics Research Center, which focuses on development of Artificial Intelligence methods and on their application to medicine. Prof. Shahar received his B.Sc. and M.D. degrees from the Hebrew University in Jerusalem, Israel (1981), his M.Sc. in computer science (artificial intelligence area) at Yale University, New Haven, CT, USA (1990), and his Ph.D. in Medical Information Sciences from Stanford University, Stanford, CA, USA (1994). After spending 2 years at Yale University and a decade (1990-2000) at Stanford University as a researcher and as a full-time faculty member in Medicine and Computer Science, Prof. Shahar has joined BGU while continuing to serve as a Consulting Professor (Medical Informatics) at Stanford University's School of Medicine. His research focuses on the areas of temporal reasoning, planning, decision analysis, information visualization, knowledge acquisition, knowledge representation, and knowledge-based systems (with particular emphasis on biomedical applications). He specializes in the intelligent interpretation and exploration of large amounts of time-oriented data, in particular longitudinal clinical data, and in the representation and application of procedural knowledge, such as clinical guidelines. Prof. Shahar is on the editorial board of *Artificial Intelligence in Medicine*, *The Journal of Biomedical Informatics*, *Methods of Information in Medicine*, and *Applied Ontology*. He was the Scientific Co-chair of the 12th Conference on Artificial Intelligence in MEdicine (AIME - Italy, Verona 2009). In 2005, Prof. Shahar was elected as an International Fellow of the American College of Medical Informatics (ACMI).